AF323748

energy
harvesting
materials

edited by

david l andrews

University of East Anglia, UK

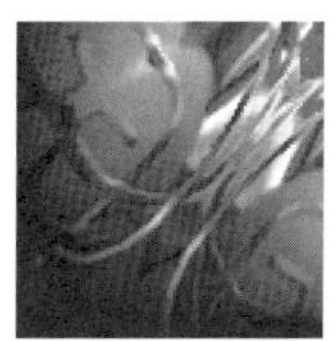
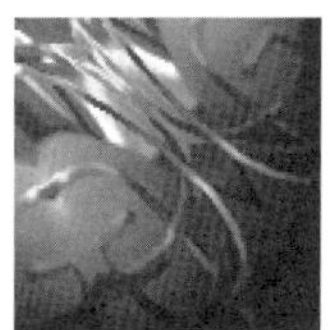
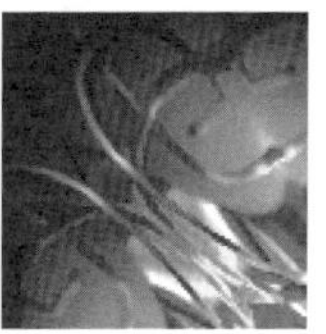

energy

harvesting

materials

World Scientific

NEW JERSEY • LONDON • SINGAPORE • SHANGHAI • HONG KONG • TAIPEI • CHENNAI

Published by

World Scientific Publishing Co. Pte. Ltd.

5 Toh Tuck Link, Singapore 596224

USA office: 27 Warren Street, Suite 401-402, Hackensack, NJ 07601

UK office: 57 Shelton Street, Covent Garden, London WC2H 9HE

British Library Cataloguing-in-Publication Data
A catalogue record for this book is available from the British Library.

ENERGY HARVESTING MATERIALS

ISBN 981-256-412-8

Editor: Tjan Kwang Wei

Printed in Singapore by B & JO Enterprise

PREFACE

The concept of energy harvesting signifies an integrated approach to the gathering of energy, usually from natural resources, with centralised collection providing for subsequent distribution according to requirements. On Earth, by far the most significant source of available energy is solar in origin. For the abundance of the environmental resource it represents, and also the extent of its geographic availability, solar energy easily outstrips any competition. Each year, in excess of 10^{24} J of solar energy impinges on the Earth, surpassing by a factor of more than a hundred the next largest source, wind energy. The scale of these resources considerably exceeds a global demand that is not expected to reach 10^{21} J until the 2020's.[1] Thus, one of key drivers in the science of energy harvesting is the pursuit of better control and collection efficiency in the global utilization of solar energy. However, the research objectives in this increasingly diverse field extend well beyond issues of the energy economy.

In nature, solar energy harvesting is the primary stage of photosynthesis, arguably the most important process on the planet. Photosynthesis represents the foundation for every food chain, the basis for life itself. Life owes its origin to the presence of sunlight and nature has been exploiting this source for biosynthetic purposes since life began. The unfolding discovery of exquisitely intricate and organized structures and functionalities operating within natural photosynthetic systems has hugely advanced our understanding of the molecular principles which nature deploys to expedite energy harvesting. Key structural and mechanistic principles have been identified in the operation of biological photosystems, and the emulation of these principles is an obvious aim in the devising of new *biomimetic* energy harvesting materials. For example each complex contains an array of chromophores with broad, intense absorption bands, held in place by a molecular superstructure. Excitation energy is conveyed through the

---

[1] S.F. Baldwin, Renewable energy: Progress and prospects, *Phys Today* **55** (4) (2002), 62-67.

system with high efficiency by a series of ultrafast steps usually involving resonance energy transfer.[2] Each such step tends to progress towards a unit that absorbs at a longer wavelength, conferring significant directionality as the energy advances through a series of different chromophores. Finally, at a trap or reaction centre the energy of arriving photons serves to trigger an ensuing process such as electron transfer.

In the pursuit of biomimetic energy harvesting using these principles, the science of capturing light energy has undergone a transformation, branching out in many previously unforeseen directions, especially through the utilization of synthetic routes to polymeric dendrimers and other multichromophore arrays.[3] Thus, most recently engineered light harvesting materials operate in ways that bear little relation to traditional solar cell technology. The purpose of such man-made systems is not usually biosynthesis, nor necessarily high efficiency, low cost energy acquisition; these are materials that present other, more immediate opportunities for implementation in devices for signal processing, optical computing and information technology applications.

Thus, at one end of the subject spectrum the characterization of structures and mechanisms in natural photosynthetic systems continues to advance at a tremendous pace. At the other, synthetic and physical chemists are devising new and increasingly sophisticated materials for specific device applications. The two ends of this spectrum are linked by the elucidation of common principles and theory. In the volume before you, an international team of experts brings together all these themes, providing the first fully comprehensive treatment of light harvesting. The coverage ranges from natural plant and bacterial photosystems, through their biomimetic analogues, to other photoactive materials. Individually, each chapter captures the state of the art of its energy harvesting theme. Together, the chapters showcase the inter-relatedness of these topics, establishing the common ground and underlying principles across the full range of light harvesting systems.

---

[2] D. L. Andrews and A. A. Demidov, *Resonance Energy Transfer* (Wiley, New York, 1999).

[3] P. Ball, Natural strategies for the molecular engineer, *Nanotechnology* **13** (2002), R15-R28: P. Ball and D.L. Andrews, Light harvesting, *Chem. World* **1** (3), 34-39 (2004).

The first chapter sets out principles for the efficient deployment of excitation transfer in energy harvesting. Detailed theoretical models are described and exhibited with reference to photosynthetic systems. Major implications for the design of artificial light harvesting systems are also discussed. In Chapter 2, an account is given of how protein-mediated, electron and energy transfer processes operate both in natural and designed structures, particularly with reference to the associated kinetic factors. Chapter 3 focuses on one of the best-known and most widely studied systems, light harvesting purple bacteria, whose high efficiency and unique combination of photophysical properties already suggest a host of device applications. In Chapter 4, highly important but often neglected issues of regulation in natural light harvesting are addressed, focusing on dynamic issues, photoprotection and photostasis. Linking principles of natural and artificial light harvesting are further highlighted in Chapter 5, the latter with reference to systems as diverse as conjugated polymers, transition metal supramolecular complexes and semiconductor films. Further electron and energy transfer principles are elaborated in the next chapter, addressing the means and effect of tailoring the electronic coupling between chromophores in a variety of bridged structures. This topic is also taken up in Chapter 7, which describes the results of kinetic studies on a series of macromolecules engineered for optimum energy transfer and trapping efficiency. Chapter 8 focuses on dendritic polymers, whose multi-branched structures represent one of the most prominent motifs in modern light harvesting materials. The theme is further developed in Chapter 9, which emphasizes the unique intramolecular energy transfer properties of light harvesting dendrimers. In the final chapter, it is shown how the incorporation of fullerenes into larger supramolecular systems offers new light harvesting opportunities.

It has been a delight to work with such enthusiastic authors on this project; thanks are also due to the staff at World Scientific for their assistance and support. It is our hope that this book will help a wide audience to appreciate both the detail and the general principles of energy harvesting. If, additionally, it represents a resource that can expedite future developments, it will have done its job.

David L. Andrews
Norwich, November 2004

Contents

Chapter 1

PHYSICAL PRINCIPLES OF EFFICIENT EXCITATION TRANSFER IN LIGHT HARVESTING

Melih K. Şener and Klaus Schulten

After light absorption the primary process in light harvesting is the transfer of excitation to a reaction center which facilitates a separation of charge across a cell membrane. The physical principles underlying excitation transfer are explained. Theoretical methods for the description of the excitation migration process, including an expansion for excitation lifetime in terms of repeated trapping and subsequent detrapping events, and the construction of representative pathways for excitation transfer based on mean first passage times, are presented. Measures for robustness and optimality of excitation transfer in terms of quantum yield are introduced. Photosystem I (PSI) is used as an example to illustrate the methods discussed. Some conclusions for the design of artificial light harvesting systems are also discussed.

Keywords: Photosynthesis, photosystem I, excitation transfer, quantum yield, mean first passage times, robustness, optimality.

1. INTRODUCTION

As the primary source of energy in the biosphere, photosynthesis is a process by which the energy of a photon is converted into increasingly more stable energy forms, first in the form of an electronic excitation of a pigment, followed by a charge separation across the cell membrane, and finally in the form of stable chemical bonds. In order to facilitate this process in an efficient manner, photosynthetic systems share some common features despite the wide variety of their actual structures (van Amerongen *et al.*, 2000, Blankenship, 2002).

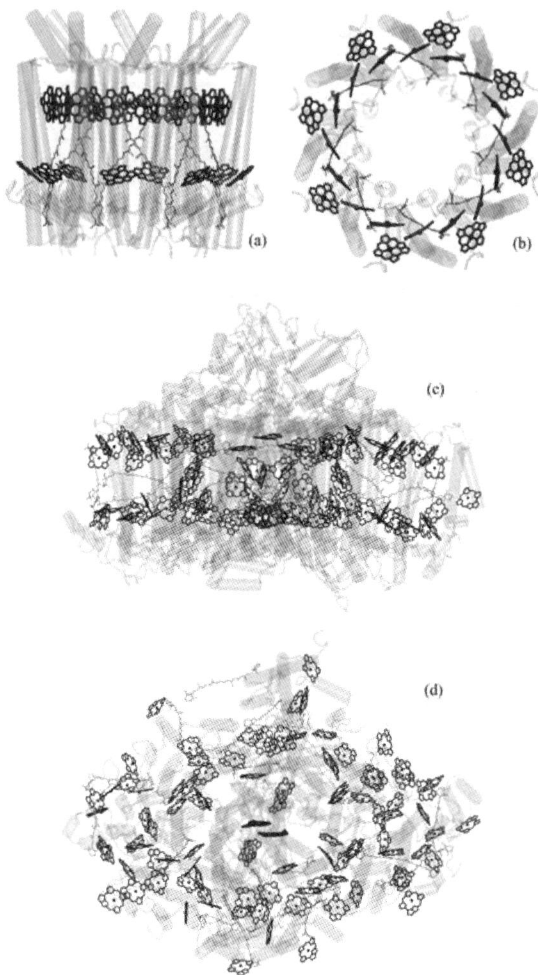

Fig. 1 Comparison of two protein-pigment complexes used in light harvesting. Peripheral light harvesting complex LH2 from the purple bacterium *Rhodospirillum molischianum*: (a) side view (from the plane of the membrane), (b) top view (normal to the plane of the membrane). Photosystem I from the cyanobacterium *Synechococcus elongatus*: (c) side view, (d) top view. Unlike LH2, PSI contains also an electron transfer chain (not shown). For both systems the protein is rendered in transparent cartoon representation, the carotenoids are shown in gray and the (bacterio-)chlorophylls are shown in black represented by their porphyrin rings. Other cofactors are not shown for simplicity. Figure made using Protein Data Bank files 1LGH and 1JB0 with the program VMD (Humphrey *et al.*, 1996).

Pigments are the primary components of a light harvesting system, responsible for converting the energy of an absorbed photon into an electronic excitation. This excitation energy is then used for the transport of an electron across the cell membrane, resulting in a voltage gradient. Most pigments are not directly involved in the charge separation process; instead their excitation energy is transferred eventually to a reaction center which facilitates the charge transfer. Thus, a light harvesting system typically comprises an array of peripheral antenna pigments surrounding a reaction center. These pigments might be located in separate antenna complexes excitonically coupled to the reaction center core, or they may constitute parts of a fused photosystem containing both a network of peripheral pigments and a reaction center.

In contrast to reaction center cores, antenna complexes display an amazing diversity (Blankenship, 2002). A typical example of an antenna complex is the peripheral light harvesting complex LH2 of the (anoxygenic) purple bacterial photosynthetic unit (Koepke *et al.*, 1996; Hu *et al.*, 2002), whereas an example of a fused photosystem is given by photosystem I (PSI), one of the two such major reaction center complexes used in oxygenic photosynthesis. The structure of PSI has recently been determined crystallographically in cyanobacteria (Jordan *et al.*, 2001) and in higher plants (Ben-Shem *et al.*, 2003). It is of interest to contrast LH2 with PSI (see Figure 1). LH2 contains 24 bacterio-chlorophylls and 8 carotenoids arranged in a cylindrically symmetric fashion vs. 96 chlorophylls and 22 carotenoids found in cyanobacterial PSI with no obvious symmetry. PSI also contains additional cofactors forming an electron transfer chain. The uniform distribution of carotenoids in both structures is indicative of their photoprotective role. The anoxygenic purple bacterial light harvesting apparatus is known to have evolved earlier than the oxygenic light harvesting systems employed by cyanobacteria and plants (Xiong *et al.*, 2000; Blankenship, 2001) indicating a trend for increased complexity.

A comparison of cyanobacterial and plant PSI structures (Ben-Shem *et al.*, 2003) reveals a high degree of conservation in the geometry of the two chlorophyll networks (see Figure 2). Except for an additional ten chlorophylls providing connections to the LHCI belt in plants (which is absent in cyanobacteria), the position and orientation of most of the chlorophylls are conserved between the two structures. This is especially interesting, as chloroplasts in plants are believed to have diverged from cyanobacteria at least one billion years ago. The degree of conservation of the chlorophyll network after such a long period of independent evolution raises a question as to whether the geometry of the chlorophyll network of PSI had reached a point of optimality in terms of facilitating efficient excitation transfer prior to the divergence of the two structures.

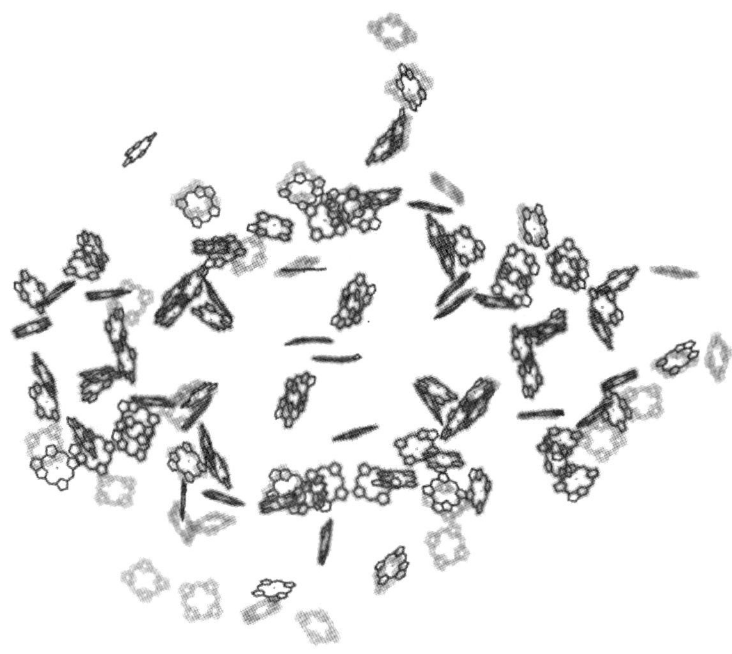

Fig. 2 Comparison of the chlorophyll networks in PSI from cyanobacteria (*Synechococcus elongatus*) represented as black lines and the higher plants (*Pisum sativum* var. alaska) represented as transparent gray bonds. The chlorophylls corresponding to the LHCI belt are not shown for plant PSI (see Figure 5). Figure made using Protein Data Bank files 1JB0 and 1QZV with the program VMD.

The excitation energy absorbed by peripheral pigments migrates to a reaction center in a sequence of resonant energy transfers via intermediate pigments. Sufficiently strong couplings and significant spectral overlap between pairs of pigments are essential for the efficient transfer of energy before the excitation is lost to dissipative processes. The excitation travels in a funnel-like fashion generally proceeding from higher energy pigments to lower energy ones. This is not always true, however, as PSI is known to contain chlorophylls that absorb light at longer wavelengths than the reaction center chlorophylls. These so-called 'red chlorophylls' are likely responsible for extending the spectral absorption profile of the complex. Spectral broadening of the pigment lineshapes due to thermal disorder makes it possible to have spectral overlap between pigments of varying energies. In fact, at lower temperatures the overall efficiency of the excitation migration process can drop significantly due to loss of resonance between neighboring pigments. Thus, thermal disorder constitutes an important ingredient for efficient excitation transfer in a light harvesting system of broad spectral profile.

The presence and significance of thermal disorder in what is essentially a quantum mechanical process provide unique challenges for the study of excitation migration. Of the two of many possible approaches for describing thermal effects in light harvesting, one is a description of 'dynamic' disorder given in the context of the energy fluctuations of the pigments along a molecular dynamics trajectory (Damjanović *et al.*, 2002a). Another possible description is that of 'static' disorder as given by a thermodynamic average over many realizations of a light harvesting system formulated in terms of random matrix theory (Şener *et al.*, 2002a).

A further challenge is provided by light harvesting complexes that are formed by the aggregation of multiple subunits. The cyanobacterial PSI, for example, is sometimes found in a trimeric form (see Figure 3) containing a total of 288 chlorophylls. Although the function of trimer formation is not yet fully understood, excitation sharing between individual monomers is found to be feasible in trimeric PSI (Şener *et al.*, 2004). Furthermore, conditions of iron deficiency are known to induce certain cyanobacteria to form even larger light harvesting assemblies comprised of a trimeric PSI core surrounded by a ring of satellite complexes containing a total of nearly five hundred chlorophylls, increasing the number chlorophylls per reaction center by nearly 60% (Bibby *et al.*, 2001).

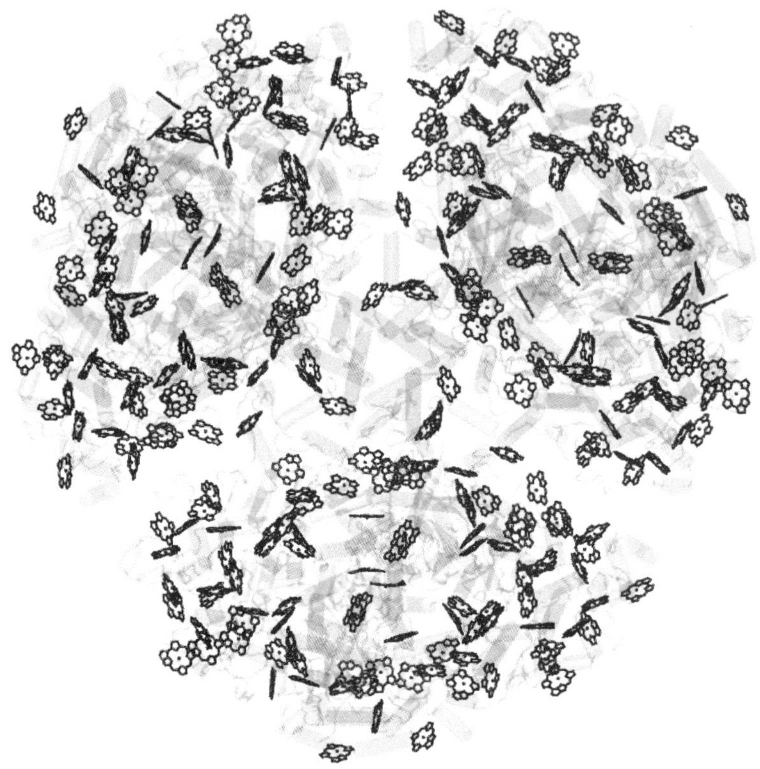

Fig. 3 Trimeric form of PSI from the cyanobacterium *Synechococcus elongatus*. Chlorophylls are shown in black, other cofactors not shown for simplicity. The trimeric complex contains a total of 288 chlorophylls. Figure made using Protein Data Bank file 1JB0 with the program VMD.

The organization of this article is as follows: in the next section we first introduce physical principles of excitation transfer based on Förster theory. In section 3 the average excitation lifetime and quantum yield are defined in terms of excitation transfer rates. In section 4, representative pathways of excitation migration are described in terms of mean first passage times to a reaction center. In section 5, an expansion method for excitation migration in terms repeated trapping and detrapping events is introduced. In section 6, some measures of robustness and optimality of a pigment network are defined. Finally, section 7 contains a discussion on principles for the design of artificial light harvesting systems.

2. PRINCIPLES OF EXCITATION TRANSFER

In this section we introduce Förster theory as a basis of excitation transfer using an effective Hamiltonian formulation as a starting point as described by Şener *et al.* (2002b). Under normal light conditions the flux of photons (~10 photons / chlorophyll / s) is sufficiently low that excitation migration in a pigment network can be satisfactorily modeled by single chlorophyll excitations. A basis set for an effective Hamiltonian can therefore be given in terms of exciton states where one pigment is at its lowest excited electronic state while all other pigments are in their ground state;

$$| i > = | \phi_1 \phi_2 \cdots \phi_i^* \cdots \phi_N >, \quad i = 1,2,\ldots,N. \tag{1}$$

Here N denotes the number of pigments; ϕ_i and ϕ_i^* denote the ground and first excited states of the ith pigment, respectively. For a chlorophyll molecule the lowest excited state is the so-called Q_y state (Scheer, 1991). In this basis set an effective Hamiltonian can be expressed as;

$$H = \begin{pmatrix} \varepsilon_1 & H_{12} & \cdots & H_{1N} \\ H_{21} & \varepsilon_2 & \cdots & H_{12} \\ \vdots & \vdots & \ddots & \vdots \\ H_{N1} & H_{N2} & \cdots & \varepsilon_N \end{pmatrix}, \tag{2}$$

where ε_i denotes the excitation energy for pigment i and H_{ij} is the electronic coupling between pigments i and j. The coupling between two pigments has two contributions corresponding to a direct Coulomb term (Förster, 1948) and an electron exchange term (Dexter, 1953). For a typical network of chlorophylls as illustrated in the previous section the distance between a pair of chlorophylls is generally large enough that the exchange term is negligible and the coupling is dominated by the Coulomb term (Damjanović *et al.*, 1999). In the lowest order approximation this coupling is;

$$H_{ij} = C\left(\frac{\mathbf{d}_i \cdot \mathbf{d}_j}{r_{ij}^{\,3}} - \frac{3(\mathbf{r}_{ij} \cdot \mathbf{d}_i)(\mathbf{r}_{ij} \cdot \mathbf{d}_j)}{r_{ij}^{\,5}} \right), \tag{3}$$

where $\mathbf{d}_i$ is the unit vector along the transition dipole moment of pigment i, $\mathbf{r}_{ij}$ is the vector connecting the pigments i and j, and C is a constant. For the coupling of chlorophyll a molecules in PSI the prefactor in (3) is $C = 116\,000$ Å^{3}cm^{-1}, where (as a customary abuse of notation) the energy is measured in terms of wavenumbers (1 cm^{-1} = 8066^{-1} eV).

The dipolar approximation given in (3) has the advantage of enabling the computation of the coupling between two chlorophylls simply from the knowledge of their relative spatial orientations. The direction of the transition dipole moment vector of the lowest excited (Q_y) state of a chlorophyll is approximately directed along a vector connecting the N$_B$ and N$_D$ atoms in the porphyrin ring of the chlorophyll and positioned at the central Mg atom (see Figure 4). The dipolar approximation becomes increasingly less reliable as the inter-chlorophyll distance becomes smaller than 10 Å, in which case higher multipole contributions need to be taken into account (Damjanović *et al.*, 1999; Şener *et al.*, 2002b).

According to the Förster theory, the rate T_{ij} of transfer of excitation energy between two pigments, i and j, depends on their respective coupling H_{ij} as well as the spectral overlap J_{ij} between the emission spectrum S_i^D of the donor and the absorption spectrum S_j^A of the acceptor;

$$T_{ij} = \frac{2\pi}{\hbar}\left|H_{ij}\right|^2 J_{ij}, \quad J_{ij} = \int S_i^D(E)S_j^A(E)dE \ . \tag{4}$$

Förster excitation transfer describes a nonradiative process and is applicable to weakly coupled pigments. Combining Eqs (3) and (4) it is seen that the transfer rate drops as R^{-6} over large distances. The Förster radius, defined to be the distance over which excitation transfer is 50% efficient, is about 80-90 Å for a pair of chlorophyll a molecules (Blankenship, 2002). Excitation transfer has typically a longer range than electron transfer which requires a direct overlap of electronic wave-functions of the two pigments. This has an important implication on the evolutionary design of reaction center cores: antenna pigments are situated away from the immediate vicinity of the electron transfer chain to prevent loss of transported electrons, while still enabling efficient excitation transfer to the reaction center.

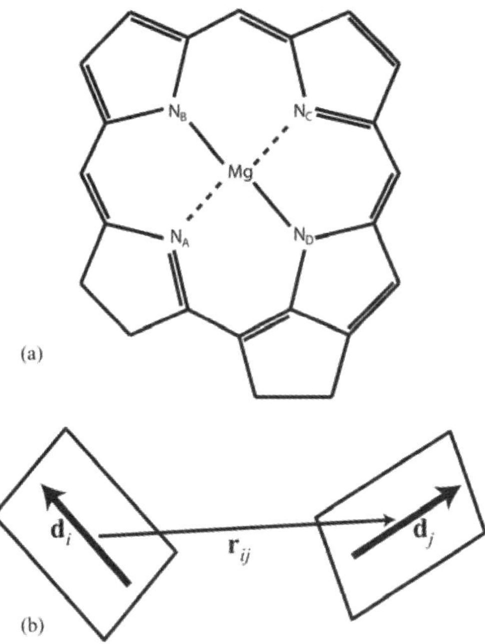

Fig. 4 Geometry of chlorophyll interactions for dipolar approximation. (a) Porphyrin ring of a chlorophyll molecule. The transition dipole moment for the Q_y state is approximately along a vector connecting the N_B and N_D atoms. (b) Inter-chlorophyll coupling between two chlorophylls in the dipolar approximation is determined by their transition dipole moments $\mathbf{d}_i$ and the vector $\mathbf{r}_{ij}$ connecting their central Mg atoms, see Eq. (3).

The transition rate matrix T_{ij} as given in Eq. (4) can be used to create a map of excitation transfer pathways by illustrating the strongest connections between chlorophylls by increasingly thicker bonds (see Figure 5). An excitation can then be viewed to follow a stochastic path along the connections until it is finally used up in a reaction center for charge separation, or dissipated. Not surprisingly, a comparison of the excitation transfer pathways of cyanobacterial and plant PSI display remarkable similarities (except for the LHCI belt in the plant system) due to the conserved geometry of chlorophylls.

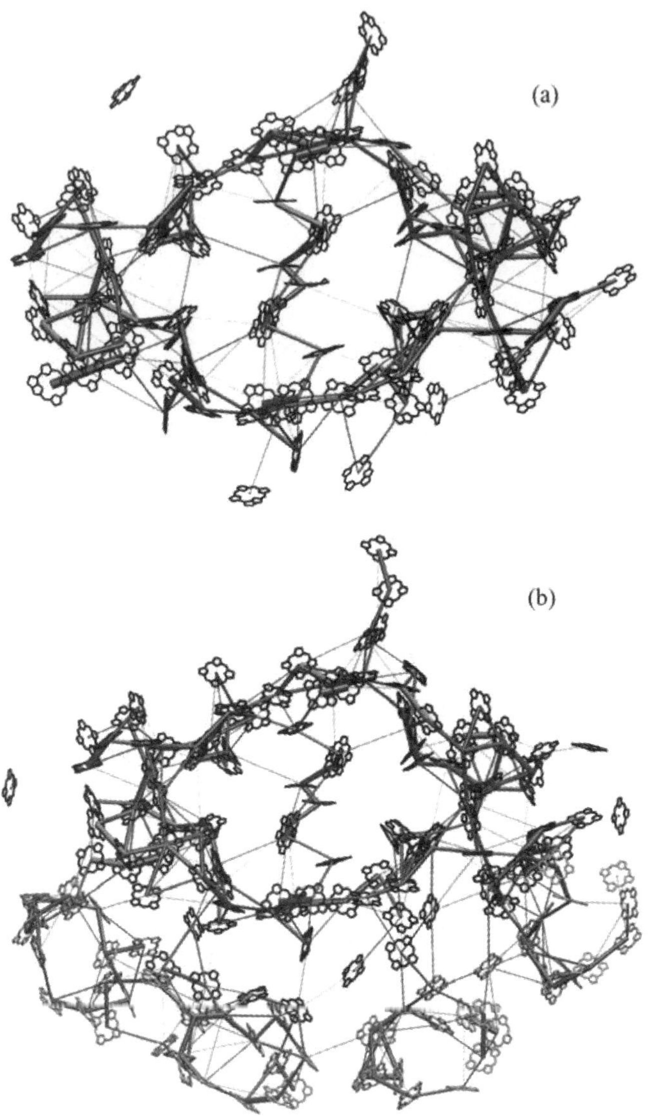

Fig. 5 Connectivity of the chlorophyll networks in PSI from (a) cyanobacteria and (b) higher plants. The thickness of a line between two chlorophylls is proportional to the logarithm of the excitation transfer rate between them. Only the strongest connections are shown, for simplicity. Chlorophylls of the LHCI belt in plant PSI are rendered in gray in (b). Figure made using Protein Data Bank files 1JB0 and 1QZV with the program VMD.

3. EXCITATION LIFETIME AND QUANTUM YIELD

The transfer rates T_{ij} between pigments can be used to describe the excitation migration process. Of particular interest are the average lifetime of an excitation after the initial absorption of a photon and the quantum yield or efficiency of the system, which is given by the probability of an excitation to cause charge separation as opposed to being dissipated. As we shall see shortly, typical quantum yields tend to be large (near unity) due to a separation of the dissipation (ns) and the excitation transfer and trapping (ps) timescales.

In order to formulate the excitation lifetime and the quantum yield in terms of transfer rates, we first introduce a master equation for the rate of change of occupation probabilities of chlorophylls. In the discussion below, a single excitation will be assumed to be localized at one of the chlorophylls and the effects of excitonic delocalization will be ignored. As a specific example we shall consider the case of excitation migration in cyanobacterial PSI (Şener *et al.*, 2002b; 2004).

Let $p_i(t)$ denote the probability of chlorophyll i being electronically excited at time t. The change in this probability is due to excitation transfer, dissipation, or charge separation events (if i is a charge separation site). The rate of change can be expressed as a master equation for the state vector $|p(t)\rangle = \sum_i p(i)|i\rangle$;

$$\frac{d}{dt}|p(t)\rangle = K|p(t)\rangle,$$

$$K_{ij} = T_{ji} - \delta_{ij}\left(k_{CS}\delta_{i,CS} + k_{diss} + \sum_k T_{ik}\right),$$

$$(5)$$

where $k_{diss} = 1$ ns^{-1} is the dissipation rate, assumed to be uniform for all chlorophylls, $k_{CS} = 1$ ps^{-1} is the charge separation rate at the reaction center, and $\delta_{i,CS}$ is equal to one if i is a charge separation site and zero otherwise. The formal solution to Eq. (5) is

$$|p(t)\rangle = e^{Kt}|p(0)\rangle .$$

$$(6)$$

Let us denote by $n(t)$ the probability that there is still an excitation in the system at time t. One can express $n(t)$ as;

$$n(t) = \sum_i p_i(t) = \sum_i \langle i | p(t) \rangle = \langle 1 | p(t) \rangle , \tag{7}$$

where $|1\rangle \equiv \sum_i |i\rangle$. Then the probability that the excitation disappears between t and $t + dt$ is given by $-\dfrac{d}{dt} n(t) dt$ and the expectation value of the excitation lifetime in the system is

$$\tau = -\int_0^\infty dt\, t \frac{d}{dt} n(t) . \tag{8}$$

The quantum yield q of the system describing the probability of charge separation can also be expressed in a form similar to Eq. (8). The probability that the excitation is used up for charge transfer between t and $t + dt$ is given by $k_{CS} \sum_{i \in CS} \langle i | p(t) \rangle dt$. Thus, the quantum yield is;

$$q = \int_0^\infty dt\, k_{CS} \sum_{i \in CS} \langle i | p(t) \rangle . \tag{9}$$

Eqs (8) and (9) can be evaluated with the help of an identity for a matrix K with negative eigenvalues;

$$\int_0^\infty dt\, e^{Kt} = -K^{-1} . \tag{10}$$

Integrating Eq. (8) by parts and combining with Eqs (6) and (10) we arrive at a final exact expression for average excitation lifetime;

$$\tau = -\langle 1 | K^{-1} | p(0) \rangle . \tag{11}$$

Similarly, combining Eqs (9), (6), and (10) results in an expression for the quantum yield;

$$q = -k_{CS} \sum_{i \in CS} \langle i | K^{-1} | p(0) \rangle. \tag{12}$$

For the excitation transfer network of cyanobacterial PSI illustrated in Figure 5(a) and for chlorophyll site energies as computed in (Damjanović *et al.*, 2002b), average excitation lifetime and quantum yield, computed from Eqs. (11) and (12), are τ = 32 ps and q = 0.97, respectively. These values compare favorably with observations (the computed average excitation lifetime is an overestimate of the reported values of 20-25 ps) and do not change significantly between monomeric and trimeric forms of PSI (Şener *et al.*, 2004).

It is important to emphasize that this structure based approach on studying excitation transfer dynamics contains no arbitrary parameters. Geometrical information about the chlorophyll network, combined with the application of basic physical principles of excitation transfer determines the values of all dynamical quantities. Thus, a comparison with observation essentially provides a test for our understanding of the physics of the light harvesting process.

4. REPRESENTATIVE PATHWAYS OF EXCITATION TRANSFER BASED ON MEAN FIRST PASSAGE TIMES

A typical excitation migration event taking place over the network illustrated in Figure 5 can contain hundreds of individual excitation transfer steps between pigments. Therefore, the random paths along which the excitation travels are not easy to visualize. A more intuitive picture of the way excitation energy is 'funneled' toward the reaction center can be constructed from paths of steepest descent based on mean first passage times of excitation from a pigment to a reaction center. Representative pathways of excitation transfer constructed in this manner are unidirectional and always terminate at a charge separation site (Park *et al.*, 2003).

Let us denote the mean first passage time of an excitation located at chlorophyll i to a charge separation site by τ_i^{MFPT}. An expression for

τ_i^{MFPT} can be constructed (Park *et al.*, 2003) in terms of the matrix K appearing in Eq. (5):

$$\tau_i^{MFPT} = -\frac{1}{\phi_i} \sum_j \phi_j \left(K^{-1}\right)_{ji},$$

$$\phi_i \equiv -\sum_j \xi_j \left(K^{-1}\right)_{ji},$$

$$\xi_i \equiv \sum_{j \in CS} K_{ji}. \tag{13}$$

In order to construct paths of steepest descent in the excitation transfer landscape, we regard the average excitation transfer time $1/T_{ij}$ from a chlorophyll i to a chlorophyll j as a measure of distance among the set of chlorophylls. Then the product $(\tau_i^{MFPT} - \tau_j^{MFPT})T_{ij}$ can be interpreted as the rate of descent in the value of the mean first passage time from chlorophyll i to chlorophyll j. Therefore, the path of steepest descent from a chlorophyll i will go to a chlorophyll k only if the value of $(\tau_i^{MFPT} - \tau_j^{MFPT})T_{ij}$ is maximized for k.

Figure 6 illustrates the representative pathways thus constructed for the chlorophyll network of the trimeric cyanobacterial PSI. The pathways in Figure 6 split naturally into three disjoint sets. This is because each chlorophyll unidirectionally connects to only one other chlorophyll, and the three separate reaction centers of the individual monomers provide termination points for the pathways. Coincidentally, the three sets of chlorophylls defined by this partition do not coincide exactly with the sets of chlorophylls belonging to the same PSI monomer. This is because some chlorophylls near the boundary are more closely coupled to their neighboring monomer instead of their own. The division of the chlorophylls into disjoint sets in this manner does not imply the absence of excitation transfer between monomers. The stochastic path followed by an excitation may connect two chlorophylls on different sides of the inter-monomer boundary as long as they are coupled sufficiently strongly. In fact, a cross-monomer excitation trapping probability of about 40% was reported for the PSI trimer (Şener *et al.*, 2004).

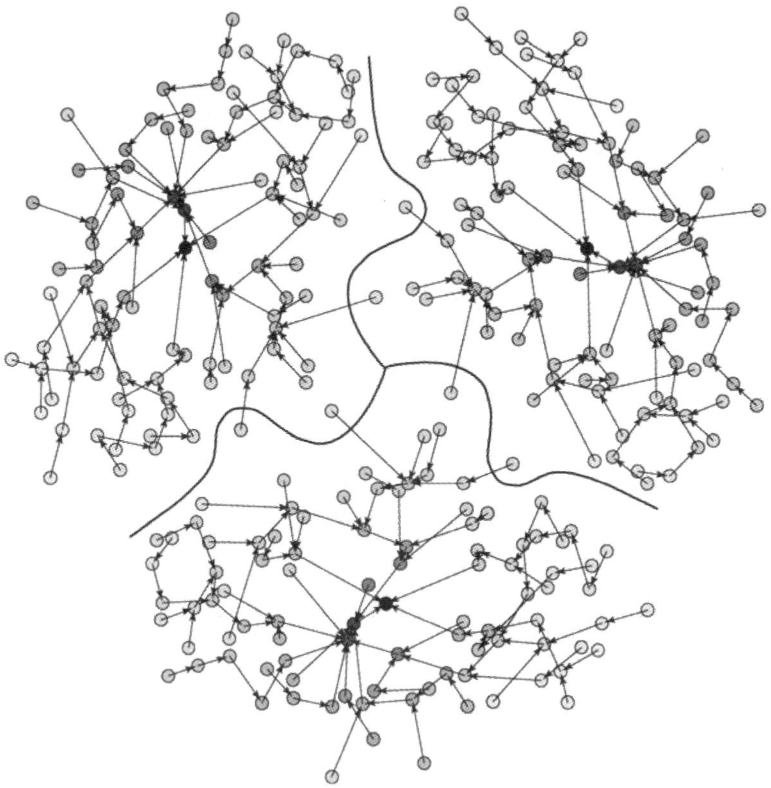

Fig. 6 Representative pathways of excitation transfer based on mean first passage times to a reaction center for the chlorophyll network of trimeric cyanobacterial PSI. Chlorophylls are represented as circles; the tone of each circle denotes the mean first passage time of that chlorophyll in decreasing order from light gray to black. The partition divides the chlorophylls according to the PSI monomer they belong to. Thus, some chlorophylls near the inter-monomer boundary are functionally part of the neighboring PSI monomer instead of their own.

5. SOJOURN EXPANSION: AN EXPANSION FOR EXCITATION MIGRATION IN TERMS OF REPEATED DETRAPPING EVENTS

The excitation migration process does not necessarily terminate with the arrival of the excitation at a charge separation site. There is a finite probability, depending on the ratios of the charge separation rate and the total detrapping rate from the charge separation site, that the excitation will escape back to the chlorophylls in the periphery, only to migrate back again to a reaction center, if not dissipated first. Thus, the excitation migration process can be expanded naturally in terms of migration, trapping, and subsequent detrapping and retrapping events. Below we present such an expansion for the average excitation lifetime, called the sojourn expansion since it describes repeated return events to the reaction center (Şener *et al.*, 2004).

Let us consider a network of N chlorophylls, M of which are charge separation sites, whose excitation transfer dynamics is described in terms of a master equation as in section 3. In order to expand Eq. (11) for the average excitation lifetime, we will separate from the matrix K in Eq. (5) the part that corresponds to detrapping processes. Let the matrix Δ denote the part of the transfer matrix describing the detrapping events from the charge separation sites. The detrapping matrix Δ can be expressed in terms of the detrapping rates as

$$\Delta \equiv \sum_{k}\sum_{j\in CS} T_{jk}|k\rangle\langle j| = \sum_{j\in CS} W_{D,j}|T_j\rangle\langle j|,$$

$$W_{D,j} \equiv \sum_{k} Tjk, \quad j \in CS, \tag{14}$$

$$|T_j\rangle \equiv \frac{1}{W_{D,j}}\sum_{k} T_{jk}|k\rangle, \quad j \in CS,$$

where we have introduced the total detrapping rate W_{Dj} from a charge separation site j, and the transient state $|T_j\rangle$ describing the distribution of occupation probabilities immediately following a detrapping event at site j. Thus, we separate K into two parts as

$$K \equiv \kappa + \Delta, \tag{15}$$

which can be inverted to yield

$$K^{-1} = \kappa^{-1} - \kappa^{-1}\Delta\kappa^{-1} + \kappa^{-1}\Delta\kappa^{-1}\Delta\kappa^{-1} - \cdots. \tag{16}$$

Eqs. (11) and (16) can be combined to yield a series for the average excitation lifetime;

$$\tau = \tau_0 + \tau_1 + \tau_2 + \cdots,$$

$$\tau_0 = -\langle \mathbf{1} | \kappa^{-1} | p(0) \rangle,$$

$$\tau_1 = \langle \mathbf{1} | \kappa^{-1} \Delta \kappa^{-1} | p(0) \rangle, \tag{17}$$

$$\tau_2 = -\langle \mathbf{1} | \kappa^{-1} \Delta \kappa^{-1} \Delta \kappa^{-1} | p(0) \rangle,$$

$$\vdots$$

The individual terms in the expansion in Eq. (17) can be evaluated explicitly using Eq. (14). For this purpose, we introduce the conditional detrapping probabilities at site j,

$$(Q)_j \equiv -W_{D,j} \langle j | \kappa^{-1} | p(0) \rangle, \quad j \in CS, \tag{18}$$

for the initial condition given by $|p(0)\rangle$ and

$$(Q_T)_{jk} \equiv -W_{D,j} \langle j | \kappa^{-1} | T_k \rangle, \quad j,k \in CS, \tag{19}$$

for the initial condition given by the transient state $|T_k\rangle$ following detrapping at site k. Additionally, we introduce the sojourn time

$$(T_{soj})_j \equiv -\langle \mathbf{1} | \kappa^{-1} | T_j \rangle, \quad j \in CS, \tag{20}$$

as the average lifetime of an excitation immediately following a detrapping event at site j, but involving no further detrapping events.

The terms Q and T_{soj} in Eqs. (18) and (20), respectively, form vectors of dimension M, whereas Q_T in Eq. (19) forms a matrix of dimension M. Using these quantities, the various terms in Eq. (17) can be evaluated in a succinct form:

$$\tau_1 = T_{soj} \cdot Q,$$

$$\tau_2 = T_{soj} \cdot Q_T \cdot Q,$$

$$\tau_3 = T_{soj} \cdot Q_T^2 \cdot Q, \tag{21}$$

$$\vdots$$

where a dot indicates an interior product between vectors and matrices of dimension M. The convergence of this expansion is proved elsewhere

(Şener *et al.*, 2004). A final expression for the average excitation lifetime is obtained by explicitly summing the terms in Eq. (21);

$$\tau = \tau_0 + T_{soj} \cdot \left(1_M - Q_T\right)^{-1} \cdot Q, \tag{22}$$

where 1_M denotes the identity matrix of size M.

As an application of the sojourn expansion to a light harvesting system with multiple reaction centers, we consider the trimeric PSI complex portrayed earlier. The trimeric symmetry results in a further simplification of Eq. (22), since the conditional probabilities $(Q)_j$ and $(Q_T)_{jk}$ and the sojourn times $(T_{soj})_j$ given in Eqs. (18), (19), and (20) are invariant under a cyclic permutation of the labels. Thus, Eq. (22) can be rewritten for the case of the PSI trimer as

$$\tau = \tau_0 + \frac{3(Q)_1 (T_{soj})_1}{1 - (Q_T)_{11} - (Q_T)_{12} - (Q_T)_{13}}. \tag{23}$$

The terms appearing in Eq. (23) are given in Table 1. It is seen that nearly 40% of the total lifetime stems from repeated detrapping events.

Table 1. Coefficients of the sojourn expansion, Eq. (23), for trimeric PSI.

τ	τ_0	$(T_{soj})_1$	$(Q)_1$	$(Q_T)_{11}$	$(Q_T)_{12}$	$(Q_T)_{13}$
32 ps	19 ps	7.5ps	0.21	0.563	0.037	0.037

6. ROBUSTNESS AND OPTIMALITY OF A LIGHT HARVESTING SYSTEM

Environmental change and competition are two major challenges that all biological systems must cope with. Adaptability to changing external conditions, or *robustness*, of a system typically manifests itself in terms of a parameter insensitivity of its dynamics and a graceful degradation of its components. Competition, on the other hand, drives a system towards *optimality*, as a less efficient system will find itself at an evolutionary disadvantage.

It is very difficult to quantify robustness and optimality in general terms for an arbitrary biological system since the fitness landscape over which adaptability needs to be judged is enormously complex. A light harvesting system, however, provides a natural, if somewhat crude, measure of its efficiency in terms of the quantum yield of the excitation migration process. It is a simple matter to model the effects of various perturbations, such as thermal disorder or loss of individual components, on the quantum yield. Similarly, questions regarding the optimality of the geometry of the chlorophyll network can be investigated by generating ensembles of alternative network configurations.

Quantum yield of excitation migration is not the best measure of robustness and optimality for a light harvesting system, merely the simplest one. Ideally, the aspects of regulation, synthesis, assembly, and repair of the light harvesting apparatus, as well as the processes of charge transfer and photoprotection, need to be taken into account before judging the adaptability of a light harvesting system. In fact, since the dissipation rates are much lower than excitation transfer rates, excitation migration is typically not a rate limiting step in the context of the overall photosynthetic function. The quantum yields for typical chlorophyll networks as investigated above are very high and disturbances on the network usually cause only small changes on the quantum yield. Nevertheless, even with these shortcomings in mind, an investigation of the excitation migration process under the influence of external perturbations provides insights into the design principles of a light harvesting complex.

Below we present results regarding the robustness and optimality of the chlorophyll network of cyanobacterial PSI (Şener *et al.*, 2002b; 2004). Similar results have also been reported (Yang *et al.*, 2003). Figure 7 illustrates that the quantum yield of PSI changes very little through fluctuations of chlorophyll site energies nor even through selective loss of individual chlorophylls from the network. The former is a case of parameter insensitivity, while the latter illustrates the aspect of graceful degradation as two major manifestations of robustness. In the case of insensitivity to site energy fluctuations (*cf.* Figure 7(a)), the consistently high quantum yields are a consequence of the broad lineshapes of pigments, which maintain significant overlap for resonant energy transfer even when chlorophyll site energies are displaced randomly. The tolerance against loss of individual chlorophylls (*cf.* Figure 7(b)) is a consequence of the Förster radius being sufficiently large compared to the typical inter-chlorophyll distances. Even with the pruning of

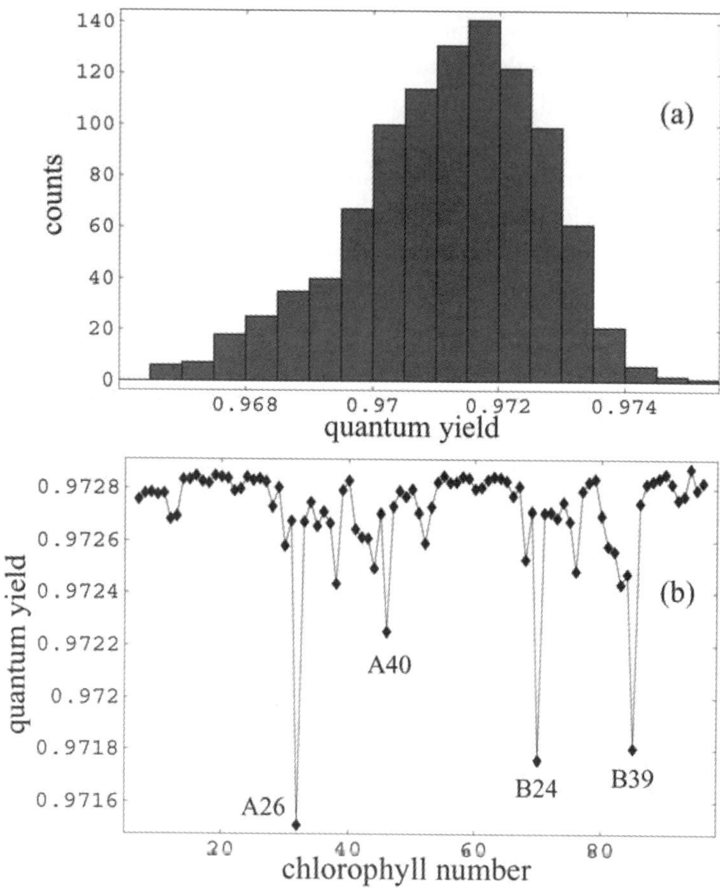

Fig. 7 Robustness of the chlorophyll network in cyanobacterial PSI. (a) Robustness against fluctuations of site energies. The histogram shows the distribution of quantum yield over an ensemble of 1000 chlorophyll configurations generated by randomly displacing the chlorophyll site energy within a width of 180 cm^{-1}. (b) Robustness against the pruning of individual chlorophylls. The average quantum yield of the remaining chlorophyll network is shown as a function of the pruned chlorophyll for all chlorophylls except for six central chlorophylls. The chlorophylls whose deletion has the highest impact on quantum yield are indicated.

individual components, the network depicted in Figure 5(a) maintains efficient excitation transfer. A similar result is seen for the case of simultaneous pruning of a large number chlorophylls (not shown); after taking into account the loss of the corresponding cross-section, the relative quantum yield of the pruned system remains high due to the slowness of dissipative processes. Expectedly, the highest impact on the quantum yield results from the pruning of the chlorophylls closest to the reaction center.

Is the geometry of the chlorophyll network in PSI depicted in Figure 5 optimized for efficient excitation transfer? Are the particular positions and orientations of individual chlorophylls critical for the light harvesting function? These questions arise naturally by contrasting the seemingly random arrangement of chlorophylls in PSI with the symmetrical arrangement of chlorophylls in LH2 illustrated in Figure 1. The distribution of quantum yields across an ensemble of alternative network geometries generated by random reorientations of chlorophylls is given in Figure 8. It is seen that the quantum yields vary only within a narrow interval in such an ensemble. Thus, individual chlorophyll orientations are not critical for maintaining a reasonable light harvesting efficiency. Yet within the narrow distribution of quantum yields the original configuration is seen to be nearly optimal. Constraining the random reorientations to peripheral chlorophylls (all except the six central chlorophylls that are part of the electron transfer chain) renders the optimality less pronounced (Yang *et al.*, 2003).

Is the apparent optimality depicted in Figure 8 a genuine result of competitive evolution or only a computational artifact? It is not obvious why a percent or less difference in the efficiency of a process would matter for the survival of an organism. Over a sufficiently large number of generations slight reproductive advantages can multiply to become discriminating. Examples of competitive advantage without the display of phenotypical differences have been reported in growth competition experiments (Ouyang *et al.*, 1998). Similarly, the remarkable conservation of geometry by the chlorophyll network in PSI in cyanobacteria and plants after a billion years of divergent evolution suggests that a degree of optimality was reached prior to the divergence of the two organisms.

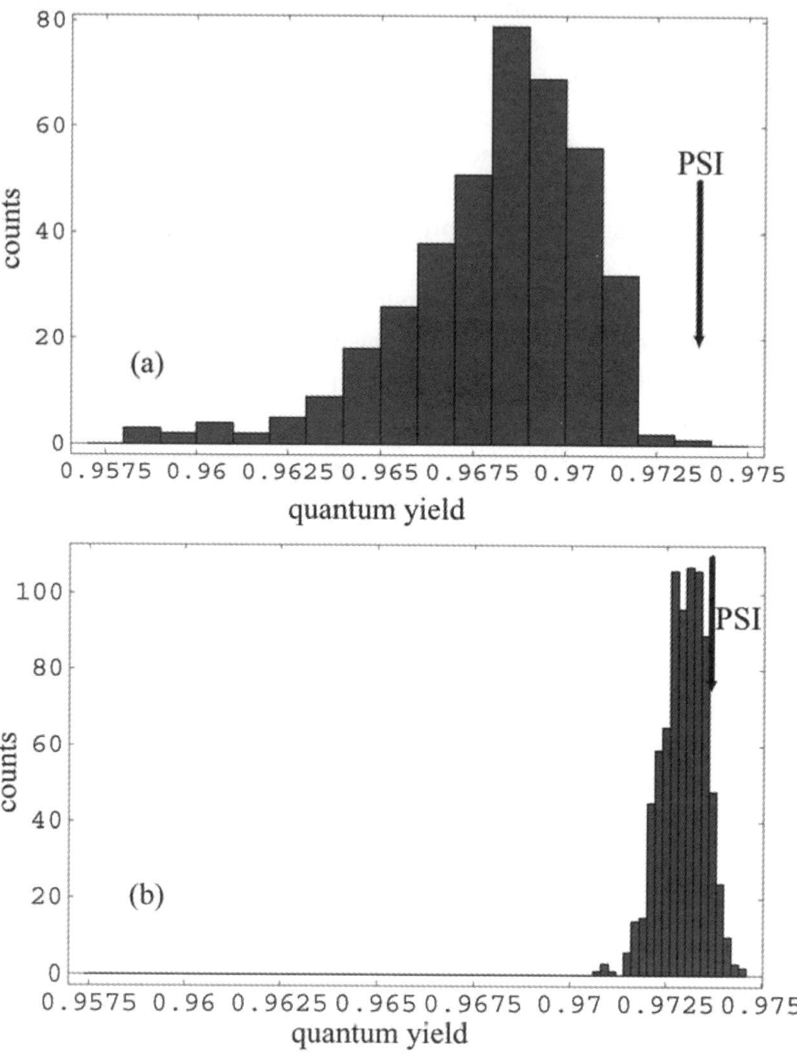

Fig. 8 Optimality of the chlorophyll network of trimeric cyanobacterial PSI. Histograms show the distribution of quantum yield over an ensemble generated by randomly rotated chlorophylls. The quantum yield of the original configuration is indicated by an arrow. (a) All chlorophylls, including the reaction center chlorophylls are reoriented (400 configurations) (b) All chlorophylls other than the six central chlorophylls are reoriented (800 configurations).

7. PRINCIPLES FOR DESIGNING ARTIFICIAL LIGHT-HARVESTING SYSTEMS

The results presented above suggest certain design principles which are relevant for the design of artificial light harvesting systems. An ideal pigment network would have to be efficient at all steps of photon absorption, excitation migration, and charge transfer. Some of the design principles of an efficient light harvesting system are:

Broad absorption profile. Natural light harvesting systems contain a variety of pigments with different absorption spectra. Efficient coupling between pigments of different energies results from thermally broadened lineshapes giving rise to resonant transfer. Pigments with higher energies tend to be located further away from the reaction centers in a funnel-like arrangement (the purple bacterial photosynthetic unit is a typical example of this (Hu *et al.*, 2002)).

Efficient excitation transfer. Most of the absorbed photons result in a charge separation event in a natural light harvesting system. This high quantum yield is a consequence of the excitation transfer rates being much larger than dissipation rates. In other words, for excitation migration processes to be efficient, the Förster radius needs to be much larger than the typical inter-pigment separation.

Efficient electron transfer. Antenna pigments are generally located far enough from the electron transfer chain to avoid a direct overlap of electronic wavefunctions. This way the loss of the transported electron is avoided.

Number of pigments per reaction center. The most efficient excitation transfer mechanism will still be wasteful if it is idle most of the time. Therefore, the number of pigments surrounding a reaction center must be chosen such that the electron transfer chain is constantly active. For example, for a charge separation time scale of 1 ms and a high light intensity of 10 photons / chlorophyll / s, on the order of 100 chlorophylls are needed to keep the reaction center supplied with electronic excitation. Monomeric cyanobacterial PSI contains 96 chlorophylls.

Robustness: Parameter insensitivity. External perturbations, such as the effects of thermal disorder, or modifications to network geometry have little effect on the overall efficiency of the light harvesting process.

Robustness: Graceful degradation. Natural light harvesting systems are tolerant to the loss of individual components. Loss of one pigment generally does not prove detrimental beyond the loss of the corresponding cross-section.

Optimality of excitation transfer network. Even though natural light harvesting systems appear to be optimized in terms of the details of their network geometry, this is probably not a high priority constraint for artificial light harvesting systems.

Protection from photodamage and repair. Natural light harvesting systems have developed mechanisms to handle excess light energy or harmful by-products of light harvesting. For example, photosystem II is known to feature a remarkable damage-repair cycle (Blankenship, 2002).

It must be noted that these principles are derived mainly from chlorophyll-based photosynthetic species and are not necessarily relevant for rhodopsin-based photosynthesis that directly couples a *cis-trans* isomerization to ion transport across the membrane.

With the availability of an increasing number of atomic resolution structures and ultrafast spectroscopy data for different photosynthetic systems, it is becoming possible to compare the details of various light harvesting mechanisms. Further modeling challenges are provided by multi-subunit light harvesting systems, where multiple protein-pigment complexes interact with one another. It is a fascinating challenge to piece together the evolutionary history of photosynthesis from comparative studies of different light harvesting systems.

ACKNOWLEDGMENTS

The authors would like to thank Sanghyun Park for his assistance with Figure 6. This work was supported by the NIH grant PHS 2 P41 RR05969 and the NSF grant MCB02-34938.

References

1. van Amerongen H, Valkunas L, and van Grondelle R. *Photosynthetic Excitons*. World Scientific, Singapore, 2000.
2. Ben-Shem A, Frolow F, and Nelson N. Crystal structure of plant photosystem I. *Nature* 2003; **426**: 630-635.
3. Bibby TS, Nield J, and Barber J. Three-dimensional model and characterization of the iron stress-induced CP43'-photosystem I supercomplex isolated from the cyanobacterium *Synechocystis* PCC 6803. *J. Biol. Chem.* 2001; **276**: 43246-43252.
4. Blankenship, R. Molecular evidence for the evolution of photosynthesis. *Trends Plant Sci.* 2001; **6**: 4-6.
5. Blankenship R. *Molecular Mechanisms of Photosynthesis*. Blackwell Science, Malden, MA, 2002.
6. Damjanović A, Ritz T, and Schulten K. Energy transfer between carotenoids and bacteriochlorophylls in a light harvesting protein. *Phys. Rev. E* 1999; **59**: 3293-3311.
7. Damjanović A, Kosztin I, and Schulten K. Excitons in a photosynthetic light-harvesting system: a combined molecular dynamics, quantum chemistry and polaron model study. *Phys. Rev. E* 2002; **65**: 031919 - 24 pages.
8. Damjanović A, Vaswani HM, Fromme P, and Fleming GR. Chlorophyll excitations in photosystem I of *Synechococcus elongatus*. *J. Phys. Chem. B* 2002; **106**: 10251-10262.
9. Dexter DL. A theory of sensitized luminescence in solids. *J. Chem. Phys.* 1953; **21**: 836-850.
10. Förster T. Zwischenmolekulare Energiewanderung und Fluoreszenz. *Ann. Phys.* (Leipzig) 1948; **2**: 55-75.
11. Hu X, Ritz T, Damjanovic A, Autenrieth F, and Schulten K. Photosynthetic apparatus of purple bacteria. *Quart. Rev. Biophys.* 2002; **35**: 1-62.
12. Humphrey W, Dalke A, and Schulten K. VMD – Visual Molecular Dynamics. *J. Mol. Graphics* 1996; **14**: 33-38.
13. Jordan P, Fromme P, Witt HT, Klukas O, Saenger W, and Krauß N. Three-dimensional structure of cyanobacterial photosystem I at 2.5 Å resolution. *Nature* 2001; **411**: 909-917.
14. Koepke J, Hu X, Muenke C, Schulten K, and Michel H. The crystal structure of the light harvesting complex II (B800-850) from *Rhodospirillum molischianum*. *Structure* 1996; **4**: 581-597.

15. Ouyang Y, Andersson CR, Kondo T, Golden SS, and Johnson CH. Resonating circadian clocks enhance fitness in cyanobacteria. *Proc. Nat. Acad. Sci. USA* 1998; **95**: 8660-8664.
16. Park S, Şener MK, Lu D, and Schulten K. Reaction paths based on mean first passage times. *J. Chem. Phys.* 2003; **119**: 1313-1319.
17. Scheer H. (ed.) *Chlorophylls.* CRC Press, Boca Raton, Florida, 1991.
18. Şener MK, and Schulten K. A general random matrix approach to account for the effect of static disorder on the spectral properties of light harvesting systems. *Phys. Rev. E* 2002; **65**: 031916 - 12 pages.
19. Şener MK, Lu D, Ritz T, Park S, Fromme P, and Schulten K. Robustness and optimality of light harvesting in cyanobacterial photosystem I. *J. Phys. Chem. B* 2002; **106**: 7948-7960.
20. Şener MK, Park S, Lu D, Damjanovic A, Ritz T, Fromme P, and Schulten K. Excitation migration in trimeric cyanobacterial photosystem I. *J. Chem. Phys.* 2004; *in press*.
21. Xiong J, Fischer WM, Inoue K, Nakahara M, and Bauer CE. Molecular evidence for the early evolution of photosynthesis. *Science* 2000; **289**: 1724-1730.
22. Yang M, Damjanović A,, Vaswani HM, Fleming GR. Energy transfer in photosystem I of cyanobacteria *Synechococcus elongatus*: model study with structure-based semi-empirical Hamiltonian and experimental spectral Density. *Biophys. J.* 2003; **85**: 140-158.
23. Zazubovich V, Matsuzaki S, Johnson TW, Hayes JM, Chitnis PR, and Small GJ. Red antenna states of photosystem I from cyanobacterium *Synechococcus elongatus*: a spectral hole burning study. *Chem. Phys.* 2002; **275**: 47-59.

Chapter 2

DESIGN AND SYNTHESIS OF LIGHT ENERGY HARVESTING PROTEINS

Dror Noy, Bohdana M. Discher, and P. Leslie Dutton

In this chapter, we look into the ways by which nature has worked within the fundamental limits of excitation energy transfer and electron tunneling in engineering robust and efficient photovoltaic conversion units. The variety of currently available high-resolution structures of photosynthetic light-harvesting complexes and reaction center proteins has enabled a detailed survey of engineering and construction guidelines for handling protein-mediated energy and electron transfer processes. We show that the basic physics of the transfer processes, namely, the time constraints imposed by various decay processes, allows for a large degree of tolerance and that energy and electron transfer rates are easily managed, primarily by controlling the distance between chromophores and redox cofactors. Next, we consider applying the lessons learned from nature to artificial protein-based photovoltaic devices. We describe two strategies for making artificial membrane-embedded proteins and the progress we made in making transmembranal BChl-binding maquettes.

Keywords: Length scales, maquettes, top-down and bottom-up protein design

Ensuring good optical properties is a central problem in engineering photovoltaic materials. Solar energy conversion relies on light absorption for initiating charge separation and, unfortunately, efficient charge separation systems are not necessarily efficient light absorbers.[1,2] Photosynthetic organisms handle the problem by using photosynthetic units (PSUs) comprised of two types of proteins: light-harvesting complexes (LHCs) containing a large number of pigments which efficiently absorb light and transfer the excitation energy to reaction center proteins (RCs) that carry out the initial charge separation process.

The primary steps of light energy transfer and charge separation in PSUs have been the focus of extensive theoretical and experimental research for many years. Experimentally, photosystems have a unique property that permits the initiation of single-turnover catalytic cycles with short light flashes at temperatures from above 300°K to 1°K. Hence, the study of physical processes and chemical reaction pathways in PSUs has capitalized on advances in pulsed laser technology which currently allows probing photoexcitation processes that occur within a few femtoseconds. Furthermore, photosynthetic proteins are stable and robust enough to allow protein structure alterations and cofactors exchange by genetic and other biochemical methods. Currently, an extensive database of experimental information is available which has been recently augmented by high-resolution crystal structures of many PSU components from a variety of organisms. This serves as a useful benchmark for testing and validating the theories of biological energy and electron transfer mechanisms that are discussed in detail in other chapters of this book. In this chapter, we draw on the deep and detailed understanding of these mechanisms in order to develop general engineering guidelines for the design of natural energy and electron transfer proteins. We then consider the application of such guidelines for constructing custom-built artificial energy conversion systems.

Although PSUs are large and complex assemblies of multi-component pigment-protein subunits, it appears that their engineering follows a limited number of relatively simple energetic and geometric considerations dictated predominantly by the time constraints imposed by various decay rates involved in the energy conversion process. Recently, we found that the dominant determinant in the evolution of energy and electron transfer proteins is the distance between cofactors, and contended that distance selection is critical and dominant in providing natural energy conversion systems with robust foundations that accommodate broad structural and energetic tolerances.[3] Therefore, the key to the successful construction of synthetic energy conversion systems that are simple and robust is the control of cofactor organization and inter-cofactor distances on a molecular level.

Ten years ago, encouraged by the pioneering work by DeGrado and others, this laboratory introduced the concept of protein maquettes as flexible, minimal working scaffolds in which to study a selected function abstracted from highly complex proteins. The first of these was a family of cytochrome *b* protein maquettes based on the natural heme binding motifs of four-α-helix bundles.[4] X-ray and NMR solution structures of apo forms are complete[5,6] and an NMR structure of a diheme-four-helix bundle is well under way.[7] Success in making cytochrome *b* maquettes naturally prompted us to apply the lessons learned to create maquettes that will support conversion of light-energy into redox/charge separation energy. Principal to this endeavor is the construction of protein scaffolds that will also bind and assemble photosynthetic cofactors such as chlorophylls (Chls), bacteriochlorophylls (BChls) and their various metal derivatives.

Making artificial devices out of natural amino acids promises unique advantages over small synthetic organic and inorganic molecules. Possibilities include inexpensive production through expression in bacterial systems, high yield and high purity as well as considerable versatility and adaptability to various construction requirements and external conditions. Nonetheless, the challenges to achieving successful designs are considerable and involve the most fundamental questions of protein sequence-structure-function relationships.[8,9] Moreover, unlike the water-soluble hydrophilic cytochrome *b* maquettes made so far, we require the new designs to be transmembranal proteins because we expect membranes or, alternatively, air-water interfaces, solid surfaces or nanoporous materials, to provide the templates for organizing the amphiphilic or lipophilic maquettes and to provide the dielectric barriers to support the desired macroscopic function. The challenge to designing transmembranal type proteins is heightened because our understanding of the principles that underlie the structure and folding design of membrane proteins lags significantly behind our understanding of water-soluble proteins. This in part results from the fewer structures of membrane proteins than of water-soluble proteins.[10] Nevertheless, it is remarkable and fortunate that LHCs, RCs and affiliated redox proteins from photosynthetic organisms currently represent a large fraction of what have so far been structurally determined to near-atomic resolution.

Here, we present the progress very recently made in designing transmembranal light-harvesting and electron transfer proteins by drawing on newly emerging structures, as well as on the wealth of physical, chemical and functional information derived from over 50 years of research. We begin with an outline of the engineering requirements for natural energy conversion systems. This is followed by a description of strategies for making artificial membrane-embedded proteins and the progress we made in making transmembranal BChl-binding maquettes.

1. ENGINEERING INSIGHTS FROM NATURAL PHOTOSYSTEM DESIGN

Currently available X-ray crystal structures provide a substantial database of molecular details for many PSU components including peripheral LHCs from various organisms such as LH2[11-13] and LH3[14] from purple bacteria, Fenna-Mathews-Olsen (FMO) protein from green sulfur bacteria,[15] peridinin-chlorophyll protein (PCP) from dinoflagellates,[16] and phycobiliproteins from cyanobacteria.[17,18] More recently, representative structures for three out of the four types of natural photosystems have become available: type I and type II oxygenic photosystems (PSI and PSII, respectively),[19-21] and type II anoxygenic photosystem of purple bacteria.[22] Type I anoxygenic photosystem structure remains to be resolved but, in its absence, model structures have been developed based on sequence homology with PSI from oxygenic systems.[23,24] Additional lower resolution structural information is available from electron and atomic force microscopy,[25-29] which provides the broader view of supramolecular organization of PSU components within the photosynthetic membrane. A few of the currently available crystal structures of PSU components are presented in Figure 1.

This wealth of structural information provides a remarkable view of the range of evolved natural PSUs that comply with the requirements for efficient photoconversion. However, it is important to realize that these protein structures were not rationally designed, but rather evolved through repeated mutation and natural selection over billions of years. The necessary features for maximal reaction yields are therefore difficult

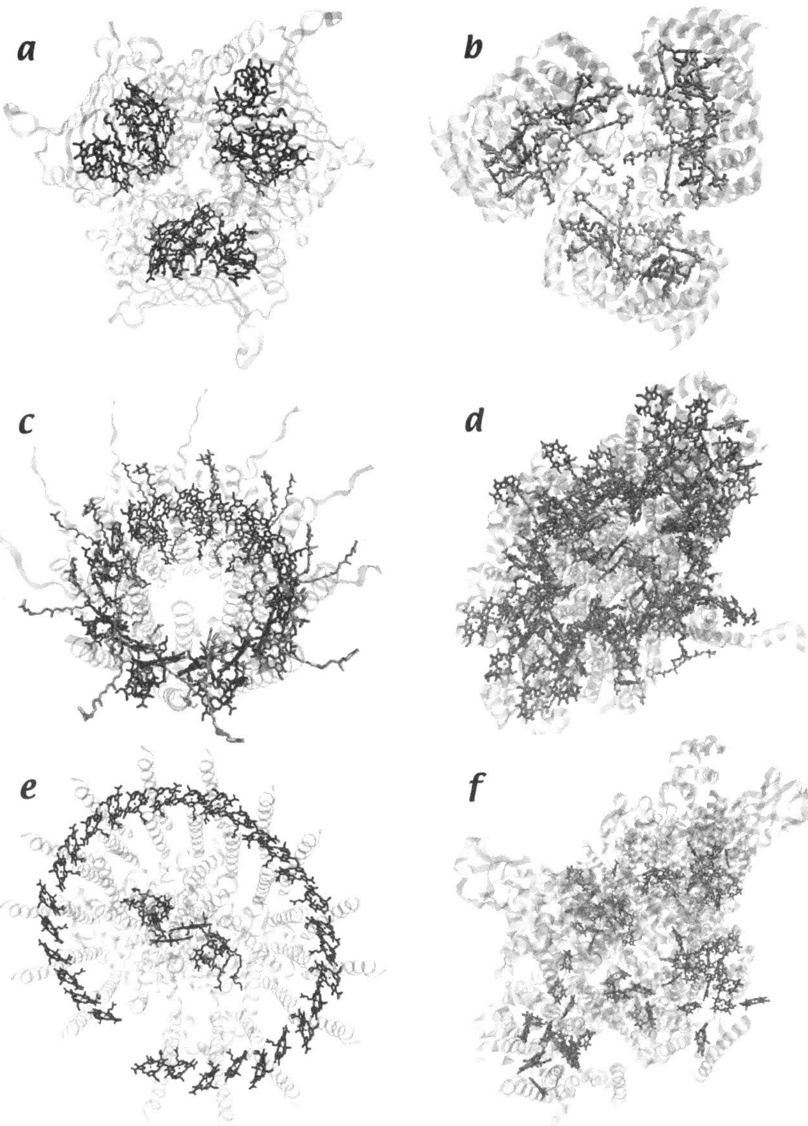

Fig. 1 Structures of some LHC proteins: (a) FMO (PDB ref. 1B50), (b) PCP (PDB ref. 1PPR), (c) LH2 (PDB ref. 1NKZ) (d) PSI (PDB ref, 1JB0), (e) LH1-RC complex (PDB ref. 1PYH), (f) PSII (PDB ref. 1IZL). Images were created with VMD [30].

to separate from a multitude of other structural features, some of which may be simply historical accidents. In fact, if the underlying physical principles of the process allow the protein to conveniently operate well away from failure boundaries, its structure may vary significantly, without compromising energy-conversion yield, either randomly or to accommodate other biological requirements.

As is clear from Figure 1, there is a broad variability in the light harvesting structures and cofactors of natural PSUs and it is hard to identify any common structural prototype from which all LHCs have been derived. Conversely, basic structural features of the RC components that carry out the initial steps of light energy conversion to electrochemical potential are remarkably similar. While this is in sharp contrast with the LHCs, again it does not necessarily imply that the conserved RC structure represents the optimal architecture for a given function of charge separation and generating oxidants and reductants. Species variation in the RC becomes evident in the energetics; light absorption properties of the RC pigments and associated redox cofactors appear to have evolved to adopt the potentials values appropriate to readily oxidize or reduce any diffusing redox products generated on each side of the supporting membrane.

Therefore, before considering protein designs that will fold to assemble pigments and redox cofactors into LHC-RC maquettes, we have endeavored to delineate the underlying chemical and physical principles of energy and electron transfer processes in PSUs and their translation into useful engineering and construction guidelines for our protein designs. In this regard we follow our recent survey of electron transfer principles in natural proteins of photosynthesis and respiration.[3] This work demonstrated how natural proteins have exploited the distinctive length scales associated with each stage of energy and charge transfer process to guarantee that the rates of desired productive processes will be faster than the unproductive ones. We concluded that the basic physics of the transfer processes, namely, the time constraints imposed by various decay processes, allow for a large degree of tolerance.

In the following, the same simple arguments of relaxation times, energy and, predominantly, distance dependence that are manifested in the length scales, are used as engineering guidelines for designing natural and artificial light-harvesting and electron transfer systems. More specifically we ask; what are the engineering principles that are essential for PSU operation at near unit quantum yield; what governs the selected engineering efficiency; what is the energetic tolerance against the onset of failure, and; how do the engineering principles challenge the construction and assembly of PSU maquettes?

1.1. Energy Guidelines

The most straightforward challenge of photovoltaic engineering is enhancing the optical properties of the charge separation unit. Natural PSUs handle this challenge by matching the absorption cross-section profile of the LHCs to the incoming photon energies in order to have sufficient absorption of the prevailing light wavelengths at the organism's habitat. The choice of antenna pigments predominantly determines the absorption profiles of PSUs. Chls, which strongly absorb blue and red visible light, are the most abundant pigments in higher plants and algae. They are complemented by strongly green light absorbing carotenes to match the absorbance profile of higher plants and algae with the solar energy spectrum. Photosynthetic bacteria growing in lakes under layers of algae have evolved to absorb light at wavelengths that are not absorbed by plants and algae. The major antenna pigments in these organisms are therefore BChls which intrinsically absorb longer wavelengths further into the NIR.

Traditionally, biological light harvesting has been discussed in terms of intermolecular resonant energy transfer, as described by Förster's exciton theory, whereby the transition dipole moments of a donor and acceptor pair of molecules are coupled by Coulombic interactions, and energy is transferred by migration of excitons in a random-walk fashion from the place of excitation until it is "trapped" by the RC.[31] In this classical view, strict control over energy terms is important for achieving high energy-transfer efficiency. Exciton formation depends on a resonance between the excited state of the donor and the ground state of the acceptor (that is the overlap integral between the donor fluorescence spectrum and the acceptor absorbance spectrum in Förster's formalism);

exciton migration relies on "funneling" from higher to lower energy sites of excitonically coupled pairs.

The energy funnel mechanism seems appealing for controlling the direction of energy flow and ensuring that excitation energy ends up on the RC to drive redox charge separation events. Design of energy funnels in LHCs is challenging because it relies on fine tuning of individual pigment energies, typically within 1000 cm^{-1}, which requires controlling specific protein-pigment interactions such as hydrogen bonding, axial ligand coordination, and bending of the Chl/BChl's macrocycle, as well as pigment-pigment interactions through excitonic coupling and π-stacking. However, recent theoretical and spectroscopic investigations, prompted by the availability of extensive, detailed structural information, suggest that the importance of energy funnels in LHC design is overemphasized (see section 3). The underlying assumptions of Förster's theory are insufficient for many LHC architectures.[32] Primarily, the assumption that initial excitation is localized on a single donor-acceptor pair of molecules is invalid in most LHCs where chromophore are packed close enough to have their electronic states mixed and thereby delocalize the excitation energy over many molecules forming large excitation domains. The chlorosomes of anoxygenic green sulfur and heliobacteria are an extreme example for this type of close packing, which may include up to 200,000 Chls.[33]

"Funneling" is not required for efficient energy transfer in LHCs under conditions of strong coupling and long-range delocalization. Recent calculations of site energies for the 90 antenna Chls in PSI could not discern a significant energy funnel from the LHC towards the RC and only minor variations in quantum yield and transfer rates upon either random changes of site energies or pruning individual Chls from the LHC ensemble.[34,35] In purple bacterial PSUs, only minor differences were found between calculated energy transfer rates to and from the core antenna complex, LH1, and the RC – suggesting that LH1 operates as an excitation energy reservoir rather than a funnel.[36] In fact, recent measurements at ambient temperature indicate that transfer rate from LH1 to the RC is almost twofold slower than the back transfer rate.[37]

Nevertheless, in extreme cases of light "starvation", organisms such as *Rps. acidophila* are capable of adjusting their PSUs to develop discernable energy funnels.[38] This appears to be accomplished by altering the hydrogen-bonding pattern around the BChls in LH2, shifting its major absorbance band from 850 nm to 820 nm.[14] However, this 2.6 fold increase in the energy gap between LH2 to LH1 (absorbing at 870 nm) only increases the overall yield of energy transfer by 5%.[36]

In conclusion, energetic considerations are most important in determining the choice of chromophores to be used in LHCs yet there are no strict requirements for the fine tuning of pigment energy levels as long as they are closely packed and their electronic states are strongly coupled.

1.2. Lifetime Guidelines

The efficiency of excitation energy transfer is determined by the rate constant ratio between the productive (energy transfer) and unproductive (excited state relaxation) processes. The excited state lifetimes of Chls and BChls are typically about 3 and 6 ns, respectively,[39] which implies that even when the rates of energy transfer between Chls or BChls are about $(5 \text{ ns})^{-1}$, PSUs may still operate at more than 90% efficiency. If chromophores such as the carotenoids with much shorter excited state lifetime, typically a few hundred of fs are to be used as an energy donor, energy transfer rate to the acceptor must be accelerated into the order of $(10 \text{ fs})^{-1}$ in order to maintain 90% yield. Most LHCs exploit carotenoids as accessory light harvesting chromophores in order to increase the absorption cross-section of visible light, and their very short excited state lifetimes require specific arrangement for strong coupling to Chls or BChls. Even when carotenoids are positioned at van der Waals contact and are strongly coupled to acceptor Chls or BChls, the energy transfer rate is at best on the same timescale of excited state relaxation and the yield is 30-70%.[40] However, some LHCs such as PCP improve the carotenoid–Chl energy transfer yield to almost 100% by making the optically forbidden but longer lived (picosecond) S_1 excited state of the carotenoid [41] accessible for energy transfer to the Chls.

The requirement for picosecond excited state lifetimes limits the choice of central metals in porphyrin type complexes suitable for LHC maquette design. Open d–shell transition metals provide efficient relaxation pathways for porphyrins excited states thereby resulting in extremely short femtosecond lifetimes.[42,43] For example, Ni substituted BChl (Ni-BChl) has an excited state lifetime of 40-100 fs[44] and therefore, replacing even a single BChl out of the 32 BChls of native LH1 complex by Ni-BChl is enough to completely quench fluorescence from this complex.[45] Choice of the central metal is therefore limited to closed shell atoms such as Mg(II) or Zn(II). This limitation makes the design of synthetic LHCs more challenging because using porphyrin-type cofactors with closed shell central metals greatly diminishes the options for axial ligation to amino-acid residues, and the most convenient means of controlling non-covalent binding of cofactors to proteins.

1.3. Distance and Orientation Guidelines

Efficient energy transfer requires electronic coupling between donor and acceptor molecules, which depends on the distance and relative orientation. The functional dependence is known only to the level of approximation applied in solving the coulomb and exchange integrals between the donor and acceptor wavefunctions.[1,32] A useful rough guideline describes the distance (r) dependence of energy transfer rate as r^{-6} for $r > 20$ Å and e^{-r} relationship for $r < 10$ Å. At shorter distances, angular dependence of energy transfer rate is obviously a complicated function of the donor and acceptor molecular shape, while at longer distances it takes on the form $\cos\alpha - 3\cos\beta_D\cos\beta_A$; here α is the angle between the transition moments of a donor and an acceptor, and β_D and β_A are the angles between the respective transition moments of donor and acceptor and a vector connecting their centers. However, for practical engineering purposes, distance is clearly the strongest determinant of energy transfer rate. This has nicely been demonstrated in two recent calculations of energy transfer rates in PSI showing that randomly changing the orientations of antenna Chls, while keeping their position fixed at the crystal structure coordinates, resulted only in minor changes in the overall transfer rate to the RC.[34,35]

The coupling of LHCs to the RC, the final destination of excitation energy, requires special attention because of electrochemically reactive intermediates that are generated along the RC electron transfer chain and in sites of substrate redox catalysis. Oxidation or reduction of LHC components by one of these intermediates will create a quenching center that will rapidly dissipate further incoming light-energy before it can reach the RC.[46] Although it seems necessary to consider the specific chemical nature, redox potentials, and lifetimes of a multitude of redox cofactors which vary significantly in different types of RCs, this problem seems to be swept away in the construction: the dramatically different distance dependence of electron and energy transfer rates provides simple and general engineering guidelines to avoid these unproductive processes.

The fundamentally different mechanisms of electron and energy transfer actually enable using identical cofactors either in electron or energy transfer chains simply by controlling inter-cofactor distances. This is illustrated in Figure 2 where we consider the transfer of excitation energy and electrons between an antenna chromophore (C), and two redox cofactors, (D and A), in a typical protein medium. Choosing tolerance levels of 5% loss due to fluorescence and 1% loss due to antenna oxidation requires the photoinduced charge separation rate, k_{CD^+}, and reduction rate of D^+, k_{CD}, (straight solid arrows) to be at least 100-fold faster than the unproductive antenna oxidation rate, k_{C^+D} (dashed straight arrows). At the same time, the rate of excitation energy transfer from photoexcited C (C^*) to D, k_{CD^*}, (zigzag line) as well as k_{CD^+} should be at least 20-fold faster than the fluorescence rates of C^* and D^*, k_{C_0} and k_{D_0}, respectively (dashed wavy line). Under these conditions, the overall photoconversion yield is more than 89%. The shaded areas in Figure 2 represent the range of rates and distances that guarantees such yield, assuming all the cofactors are BChl derivatives with a typical excited state lifetime (τ) of 3 ns, and using the empirical equation for electron tunneling rate within a protein medium,[47] $\log(k) = 15 - 0.6r - 3.1(\Delta G + \lambda)^2/\lambda$ with a typical protein reorganization energy (λ) of 0.7 eV, and reaction free energies (ΔG) of -0.7 eV. The distance dependence of electron transfer rates according to this equation is demonstrated by the solid contours for ΔG = -0.7, -0.2, 0, and +0.2 eV. Similarly, typical energy transfer rate contours (dotted lines) are plotted using the Förster equation, $k = (r_0/r)^6/\tau$, with τ = 3 ns, and r_0 = 90, 70, 50, and 30 Å.

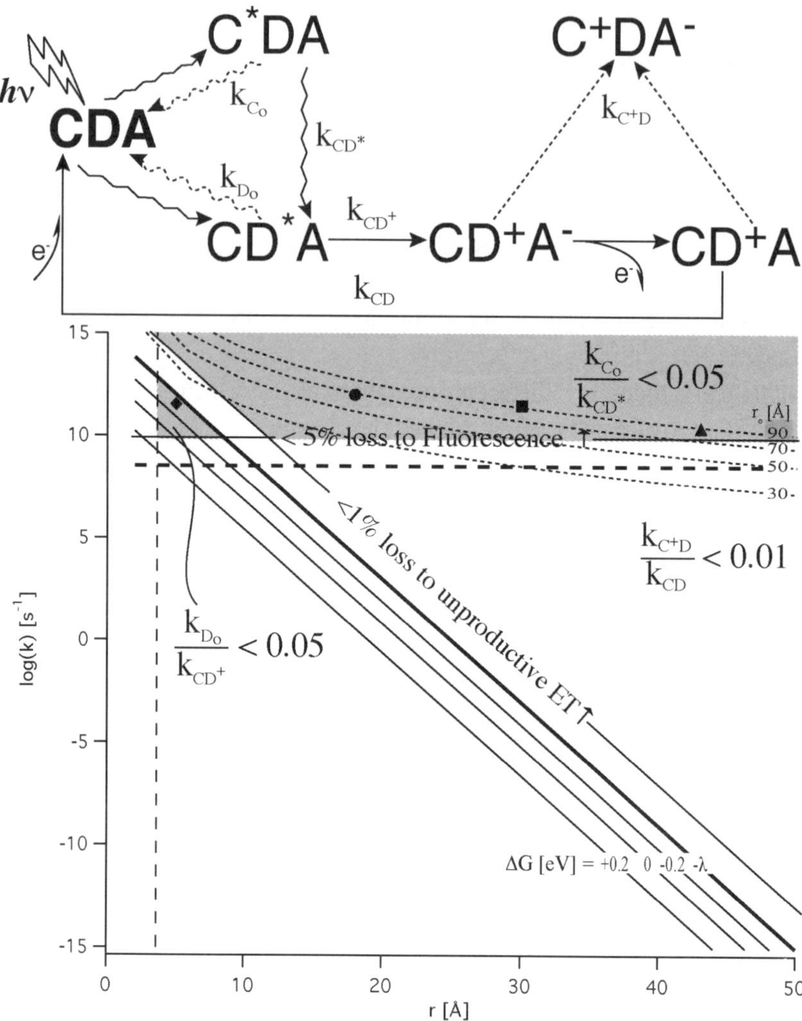

Fig. 2 Characteristic distance dependence of electron and energy transfer rate constants considering two redox cofactors, D and A, and an antenna chromophore (C). Electron transfer rate constants (solid contours) are given by the empirical formula $\log(k) = 15 - 0.6\,r - 3.1(\Delta G + \lambda)^2/\lambda$ choosing a typical reorganization energy (λ) of 0.7 eV. Excitation energy transfer rate constants (dotted contours) are given by Förster's equation $k = (r_0/r)^6/\tau$ choosing a typical fluorescence relaxation time (τ) of 3 ns. The vertical dashed line represents van der Waals contact distance (3.6 Å). The experimental rates of primary charge separation in purple bacterial RC (diamond) and energy transfer from B800 to B850* (circle), B800 to B875 (square), and B875 to RC (triangle) are plotted for comparison.

Clearly, long-range energy transfer is highly efficient even at distances of more than 40 Å whereas tunneling-mediated electron transfer requires distances shorter than 10 Å for efficient photoinduced charge separation. Thus, electron and energy transfer chains can be conveniently designed and safely coupled by controlling their relative distances. As an example, we plot the actual primary charge separation and energy transfer rates in purple bacterial PSUs all of which lie well above the acceptable lower limit of 89% high-yield photoconversion. The energy transfer rates are about five times faster than the predicted rates according to classic Förster theory using the theoretically calculated coupling between individual BChls.[48] The exact mechanism for this enhanced coupling between donors and/or acceptors comprised of pigment aggregates is a subject of extensive theoretical studies and considerable debate.[32,36,49] Nevertheless, since light-energy donor-acceptor interactions are predominantly Coulombic, the r^{-6} dependence of rate constants should be maintained as shown by the thick dashed line. As also shown in Figure 2, reducing r_0 from 90 to 50 Å results in about five-fold decrease in energy transfer rate, yet it is still possible to comfortably arrange pigments well away from the failure boundaries of the system.

1.4. Engineering Guideline Implementation by Natural LHC-RC Complexes

Currently, typical structures for three out of the four naturally occurring PSU types are known at high (2.5 Å for PSI) to moderate (3.7 and 4.8 Å for PSII and purple bacterial LH1-RC complex, respectively) resolution. Structures of the fourth type have been estimated according to sequence homology with the PSI structure. It is therefore very appealing to investigate how the engineering guidelines explained above have been deployed in natural PSU designs. Figure 3 compares the BChl/Chl arrangement in natural core PSUs. The structures are aligned by minimizing the mean of squared distances between the atoms of the two-fold symmetric pair of Chls/BChls (excluding phytyl chains) at the center of each RC. In Figure 3A, LHC pigments are represented by their central Mg atoms whereas RC pigments are presented by a stick representation

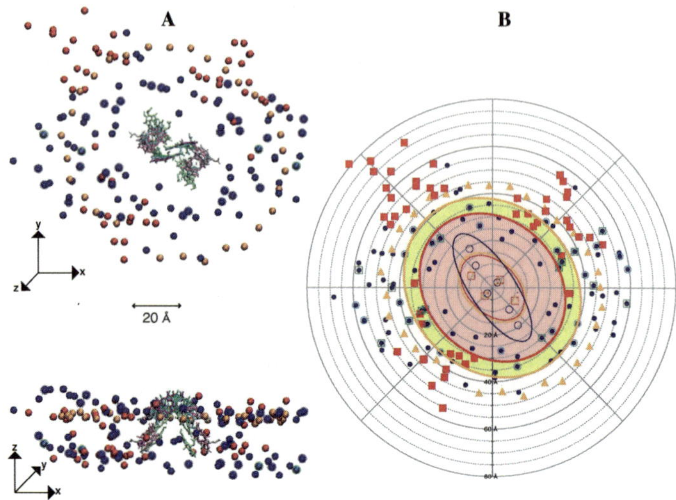

Fig. 3. A: Comparison of Chl/BChl arrangements in LH1-RC complex (orange), PSI (blue) and PSII (red). Antenna systems are represented by the central Mg atoms of their Chls/BChls, aligned at the RC primary donor; RC pigments have stick representation. Bacterial, PSI and PSII RCs are colored cyan, green, and pink, respectively. Images were created with VMD.[30] B: Polar plot representing distance and radial distributions from the middle of the central pair of RC (B)Chls. LHC and RC components are marked by full and open symbols, respectively. The pigments of PSI, PSII, and bacterial RCs are encircled by blue, red, and orange ellipses, respectively; the calculated "cordons sanitaire" for PSII and bacterial LH1-RC are represented by red and orange areas, respectively.

excluding the phytyl chain; Figure 3B presents a polar plot of the distance distributions of Mg atoms relative to the RC center which is set in the middle of the central pair of (B)Chls.

The energetics, redox potentials, and end-products, of the RCs in each PSU are very different yet their redox Chls or BChl are arranged in similar geometry. They are separated by less than 6 Å, which ensures electron tunneling rates of 10 ps or less (see Figure 2) and makes the photoinduced oxidant or reductant state capable of residing briefly on any pigment. In this setting, the energy levels of each pigment, and the energetic penalty of any uphill reverse electron transfer, determine the localization of the initial charge separated state. Other redox centers placed in the vicinity of this state would then continue charge separation in relatively longer, slower, and larger driving force electron transfer steps. Eventually, an electron is transferred 35 Å across the photosynthetic membrane leaving a delocalized oxidizing "hole" over the RC (B)Chls. This hole is typically mostly reduced within 0.1 to 10 μs by c-type cytochromes, plastocyanin, and Tyr_Z in bacterial RC, PSI and PSII, respectively. In many cases however, the kinetics of this reduction is multi-phase leaving a significant population of holes that remain unreduced for milliseconds.

Placing light-harvesting pigments too close to the oxidized (B)Chls of the RC can be expected to lead to their oxidation and possibly reduction by the redox intermediates of the electron transfer chain, thereby causing short-circuit with the energy transfer unit. While LHC reduction is not likely a problem because of rapid reoxidation by the nearby hole on the RC, oxidation is damaging because it immediately results in a neutral RC, and a cation radical on the LHC which is an effective quenching center for light excitation energy [46]. Table 1 lists the midpoint potentials of primary donors in purple bacterial, PSI, and PSII RCs, their cation radical lifetimes (when the kinetics of RC primary donor reduction involves more than one phase, we consider the slowest and take into account its amplitude), and estimated midpoint potentials of the respective LHC components. Clearly, LHC oxidation is not an issue in PSI because oxidation by $P700^+$ is energetically unfavorable, but it certainly needs to be considered in PSII where LHC oxidation by $P680^+$ is favorable ($\Delta G° = -0.27$ eV). In purple bacterial PSUs, LH1 oxidation is about 0.07 eV uphill yet it cannot be disregarded because electron transfer within the closely packed LH1 pigments is rapid and, considering site energy inhomogeniety, it is possible to trap the BChl cation radical in a lower potential site.[50,51] Using the guidelines depicted

Table 1: LHC-RC minimal coupling distances

		RC			LHC		
		E_m [V]	τ_{max}	A_{max}	E_m [V]	$\Delta G°$ [eV]	r_{min} [Å]
LH1-RC[52,53]	**P865**$^{+\cdot}$	0.50	25 μs	9%	0.57	+0.07	24
PSI [54]	**P700**$^{+\cdot}$	0.39	12 μs	100%	0.85*	+0.45	0
PSII[55,56]	**P680**$^{+\cdot}$	1.12	12 μs	18%	0.85*	-0.27	18

* Based on redox potentials of Chl a in organic solvents[57]

in Figure 2, we calculated a minimal distance, $r_{\min}$, from each of the RCs that guarantees less than 1% loss due to unproductive LHC oxidation process.

Are the aforementioned length-, energy-, and time-constraints manifested in the natural PSU architectures? Figure 3 compares the Chl/BChl organization in purple bacterial LH1-RC complex, PSI, and PSII. The pigment distances from the middle of P865, P700, and P680 are plotted in Figure 3B, with the bacterial, PSI and PSII RC pigments enclosed by the orange, blue, and red ellipses, respectively. Adding the $r_{\min}$ values in Table 1 to the axes of these ellipses defines the predicted "cordon sanitaire"[3] represented by the shaded areas. This pigment-free zone around the RC pigments ensures that LHC oxidation should occur only in 1% or less of the photocycles. Evidently, the pigment arrangement in all PSUs corresponds well with the values of $r_{\min}$. As expected, PSI antenna pigments are packed closest to the RC, some within 10 Å or less, whereas in PSII and the LH1-RC complex there are no antenna pigments within 20 and 35 Å of the RC, respectively. Although the increased distance significantly slows the energy transfer rate from ~1 ps in PSI to ~100 ps and ~30 ps in PSII and purple bacterial PSU, respectively, it is conveniently fast enough for a near unity energy transfer yield.

Beyond the "cordon sanitaire", pigments are arranged in clusters around the RC in a circular arrangement. A circle is the best geometry to pack the maximum amount of pigments at minimal average distance from to the RC. Another consequence of closely packing LHC pigments is aggregation into clusters. The strong electronic coupling between the pigments results in increased oscillator strength and radiative rates, although the exact physical mechanisms and quantum mechanical terms of these effects are not completely resolved.[58] Nevertheless, the variability in cluster geometry among species as demonstrated in Figs. 1 and 3, as well as the recently shown insensitivity of the total quantum yield to relative pigment orientation,[34,35] imply that there are no strict requirements for specific forms of pigment aggregate. Maximum pigment density per PSU is of obvious importance set at a distance from the RC to safely avoid oxidative damage.

1.5. Summary: PSU Engineering Blueprint

In conclusion, the physical and chemical principle of excitation energy and electron transfer processes allows for a high degree of flexibility in PSU engineering. Our survey of natural PSUs shows that they are designed to operate well within the margins of functional systems. The emerging constraints for PSU design are not stringent and can be summarized as follows:

* Antenna pigment absorbance should be compatible with the available light spectrum.
* Total absorbance should be maximized by high pigment density.
* Antenna pigments should have an excited state relaxation rate at least 10-100 times slower than energy transfer rates.
* The RC electron transfer unit is mediated by tunneling and therefore should consist of closely separated redox centers.
* Antenna pigments should be placed at a safe distance from the electron transfer unit to avoid oxidation but close enough for energy transfer. This range of acceptable distances is broad as shown in Figure 2.

2. DESIGNING LHC PROTEIN MAQUETTES

The underlying physical and chemical principles of energy and electron transfer provide simple engineering guidelines and a flexible template for photosynthetic organisms to produce structurally variable, robust and adaptable PSUs. Making biologically inspired artificial photovoltaic devices should therefore be possible by following the same engineering guidelines as described above. However, since distance is the predominant factor affecting PSU efficiency, the key for successful application is controlling inter-cofactor distances with sub-nanometer accuracy. In an electron transfer chain, for example, increasing the distance between two cofactors in an average protein medium by 5 Å will elicit a 1000-fold decrease in electron transfer rate. Gaining such precise control over assemblies of nanoscale dimensions is a great challenge and the focus of intensive research by many scientific disciplines.[59-61] In the field of protein *de novo* design the challenge is addressed by understanding how three-dimensional structural information is encoded by amino-acid sequences.[8] Progress in this field has enabled us to make small and simple synthetic protein maquettes that maintain a prescribed functionality. The maquette strategy has been successfully employed for investigating biological electron transfer in redox proteins by using synthetic heme binding four-helix bundle protein maquettes.[4] Their design was based on heme binding motifs from the transmembranal cytochromes b domain of cytochromes bc1 complex that were incorporated into an artificial water-soluble four-helix bundle scaffold.

In the past ten years, the initial prototype has developed into a large class of redox protein maquettes. This family has proved to readily assemble hemes and other cofactors into four-helix bundles. They have successfully reproduced key functional elements on the same energetic-, length- and time-scales as those found in natural proteins. These include the modulation of cofactor redox potentials including electrostatic coupling between redox centers and between redox centers and proton transfer sites;[62] also allosterically regulated, charge-activated conformational switching,[63] and light-activated electron transfer both within a maquette[64] and between maquette and electrode.[65] However, without a supporting membrane and its dielectric, ion, and solute impermeability, the hydrophilic nature of these maquettes cannot be used to reproduce charge separation and proton gradients, which are the fundamental functions of respiration and photosynthesis. However, if our maquette approach has any validity they should be able to form a starting point from which to convert these water-soluble hydrophilic (HP) maquettes into amphiphilic (AP) or even lipophilic (LP) units.

Unfortunately, our understanding of the principles that underlie structure and folding design of membrane proteins lags significantly behind our understanding of water-soluble proteins, partly because the emergence of high-resolution structural information about native membrane proteins is relatively recent and confined to many fewer structural examples.[10,66] Added to this, unlike water-soluble proteins in which the hydrophobic effect dominates folding, membrane protein structure is defined by what appears to be a balance between weak interactions for which a usefully clear understanding of assembly has not yet been delineated.

The absence of useful guidelines impacts the decisions regarding how designed peptides will fold and incorporate cofactors such as Chls, BChls, and carotenoids.[9] Prior to binding, there are complicated practical issues regarding cofactor self-aggregation (as a result of their own hydrophobicity) and, once bound, other complications arise from the multiple possibilities and many degrees of freedom involved in cofactor-protein interactions.[9] Moreover, compared to redox proteins, LHCs by design incorporate many more cofactors. Table 2 presents the weight percentage of pigments and proteins for the structures depicted in Figure 1. Clearly, the issue of cofactor incorporation and binding site design is critical in LHCs for which organic cofactors comprise a significant 15-30% of the total mass; much is to be learned as we embark on designing and constructing novel LHC maquettes.

Table 2: Weight percentage of Chls, BChls and Carotenoids for the structure in Figure 1

	LH2 (1NKZ)	FMO (1M50)	PCP (1PPR)	PSI (1JB0)	PSII (1IZL)	LH1-RC (1PYH)
Protein	72%	86%	83%	73%	87%	86%
(B)Chls	19%	14%	4%	22%	13%	14%
Carotenes	10%	0%	13%	3%	1%	0%*
Total Pigment	28%	14%	17%	25%	14%	14%

* Unresolved in the crystal structure

Despite our limited understanding of membrane protein folding and assembly, which is currently insufficient for designing LHC maquettes from scratch, it is still possible to make a start by following the known principles of synthetic membrane protein assembly,[10,66] drawing inspiration from the features of membrane proteins with known 3D structure, and utilizing our knowledge and experience with water-soluble maquettes.

We have followed two complementary strategies to initiate the design of transmembranal-type protein maquettes that will fold and assemble (B)Chl and other pigments and redox cofactors. Both strategies utilize natural binding site motifs and well-recognized protein folds in the hope of driving the self-assembly of BChl-peptide complexes. The main difference between the two strategies is that the first aims at miniaturizing and simplifying the native protein fold of purple bacterial LHCs, whereas the second is a modular approach based on forming chimeras of motifs from natural trans-membrane peptides and synthetic water-soluble four-helix bundles. By analogy to common nanotechnology jargon, we refer to the first and second strategies as "top-down", and "bottom-up", respectively. Bear in mind, however, that here the terminology relates to the design of protein sequences rather than the assembly of molecular subunits. Both designs rely on supramolecular self-organization for the final assembly of the protein-BChl complex; however, at the outset at least, the top-down approach starts with LHC assembly driven by pigment-pigment interactions, whereas the bottom-up approach starts with assembly driven by protein-protein and protein-pigment interactions. Therefore, the top-down approach is more suitable for assembling pigment arrays and hence better for LHC applications, whereas the bottom-up approach is more suitable for binding a few pigments within a protein binding-site, and hence better for photo-induced charge separation such as in photosynthetic RCs.

2.1. LHC Maquette Design by a Top-Down Approach: Miniaturizing Native Purple Bacterial LHCs

A principal benefit of choosing purple bacterial LHCs as a starting point for maquette design is the wealth of structural information available, primarily, high resolution three-dimensional structures of the LH2 (Figure 1c)[11-13] and LH3[14] peripheral LHC complexes and, more recently, the LH1-RC core complex (Figure 1e).[22] All three types are oligomers of two short, hydrophobic, mostly alpha-helical apo-proteins, labeled α and β, that bind BChls at α:β:BChl molar ratio of 1:1:2 in LH1 or 1:1:3 in LH2 and LH3.

Additionally, purple bacterial LHCs have been extensively studied by site-directed mutagenesis and pigment exchange methods and the specific protein-pigment interactions determining BChl binding and complex formation have been identified and characterized.[67-72] The pioneering work of Loach, Parkes-Loach and colleagues established a protocol for reconstituting purple bacterial LH1 complexes from their isolated pigment and apo-protein components in micelles of $\tilde{\beta}$-octylglucoside (OG) whereby reconstitution is conveniently monitored by following the characteristic changes in the absorption and CD spectra of BChls as they self-assemble with apo-proteins to form the LHC complex.[73,74] The series of studies that followed assayed the assembly of various native and modified LH proteins, obtained either by enzymatic cleavage or solid-phase peptide synthesis, with BChl and its analogs.[75-80] These together with mutagenesis studies and the available crystal structures have provided a detailed account of the critical requirements and the important protein-pigment interactions for self-assembly of natural purple bacterial LHCs. Most importantly, several cleaved α- and β-apo-proteins were capable of self-assembling BChls into either $\alpha_x{:}\beta_y{:}BChl_z$, $\alpha_x{:}\alpha_y{:}BChl_z$, or $\beta_x{:}\beta_y{:}BChl_z$ type complexes with spectroscopic features that were similar to native LHCs.[75] The smallest polypeptide maintaining self-assembly capability was the truncated LH1 β-apo-protein from *R. sphaeroides* containing 30 amino acids. A synthetic polypeptides with the same sequence and an additional N-terminal Glu residue was shown to be equivalent to its modified natural analog.[79] This 31 amino acids peptide, labeled sphβ31, was our starting point for a top-down minimal LHC maquette design.

The work by Loach, Parkes-Loach and colleagues identified sphβ31 as the minimal structural unit in bacterial LHCs and explored the effect of specific pigment-protein interactions on complex formation and spectral properties. Nango and colleagues continued along the same lines exploring the binding and assembly of BChl as well as Zn-substituted BChl ([Zn]-BChl), chlorin and porphyrin derivatives with truncated LH1 apo-proteins prepared by solid phase peptide synthesis.[81-85] However, so far, focus was mainly on point mutations or random truncation of the native apo-proteins determined by the availability of cleavage sites. Designing novel LHC maquettes, however, requires a more global and rigorous approach to discriminate the essential from incidental in the natural sequences and structural features. Our survey of sequences and crystal structures of bacterial LHCs from different organisms, combined with information from the reconstitution studies by the Loach and

Parkes-Loach laboratory, has led to the design scheme illustrated in Figure 4. The smallest design, a 19-residues peptide labeled LH1β19, was insoluble in OG and showed no indication of self-assembly with BChls. Adding the strictly conserved E-20 and H-18 residues of LH1 β apo-proteins in an ELHIV motif appended to the amino terminal of LH1β19 was expected to improve the helix stability and solubility within the detergent micelles. Indeed this design, labeled LH1β24, was more successful and readily assembled BChls.[86]

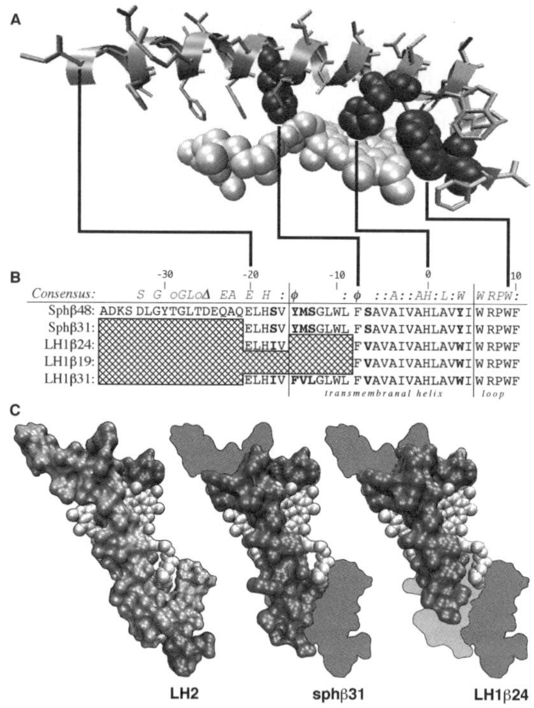

Fig. 4 Top-down design of a minimal LHC maquette. A: Binding site of BChl in natural LH2 β apo-protein of *R. molischianum* (PDB reference 1LGH). This protein shares a high sequence homology with LH1 β apo-proteins. Only 31 amino acid residues shown (medium grey). BChl and residues interacting with it are shown in space filled light and dark grey, respectively. B: Protein maquette designs based on native β apo-protein of *R. sphaeroides* (sphβ48), and the 75% consensus sequence of LH1 β apo-proteins; o-hydroxyl residues (S, T), Δ-Acidic residues (D, E), φ-Aromatic residues (F, W, Y), hydrophobic residues (I, L, V, A). C: 3D models of sphβ31 and LH1β24 subunit complexes compared to the crystal structure of native LH2 subunit. (BChls lightest grey, α and β apo-proteins medium and dark grey, respectively). Images created with VMD.[30]

The homology models of sphβ31 and LH1β24, based on the *R. molischianum* crystal structure, are shown in Fig 4C. Clearly, the LH1β24 apo-protein is just long enough to span the length of a BChl molecule and is about half the size of typical native LH1 β apo-protein. Thus, this most probably represents the smallest peptide capable of assembling BChls.

2.2. Controlling BChl Binding and Aggregation

The successful LH1β24 design as a minimal BChl-binding subunit demonstrates that as few as 24 amino acids are sufficient for designing a peptide that self-assembles together with BChl incorporation. However, reconstitution assays have shown that the spectroscopic features of the LH1β24-BChl complex are typical of B820, the common intermediate in the assembly and dissociation processes of all native bacterial LHCs. Recently, neutron scattering measurements have shown that B820 corresponds to an α:β:BChl$_2$ subunit.[87] By contrast, the B820 species formed by sphβ31-BChl complexes can further assemble into longer wavelength B850 (absorbance maximum at 850 nm) species typical of the larger oligomers formed by native LH1 and LH2 complexes.

To further explore the factors that control BChl binding and subunit aggregation we designed a longer maquette, labeled LH1β31 (Figure 4B), in which the 12 amino terminal residues of sphβ31 were added to LH1β19 but serines and methionine, suggested to be involved in hydrogen bonding between subunits, were replaced by aliphatic residues and tyrosine with phenylalanine. Reconstitution assays of LH1β24, LH1β31, and sphβ31 at different temperatures (Figure 5) indicate that although sphβ31 forms the multimeric B820 the most readily, the short LH1β24 peptide forms B820 more readily than the longer LH1β31. At lower temperatures (0 to 4°C), all three peptides form larger B850 aggregates, especially sphβ31. However, these aggregates appear to have uncomfortably large molecular weights that sediment easily after 20 minutes at 14000 rpm on a bench top centrifuge (Figures 5B, D, F). Additionally, the increased scattering from the B850 species, as well as results from sedimentation equilibrium and confocal microscopy (not

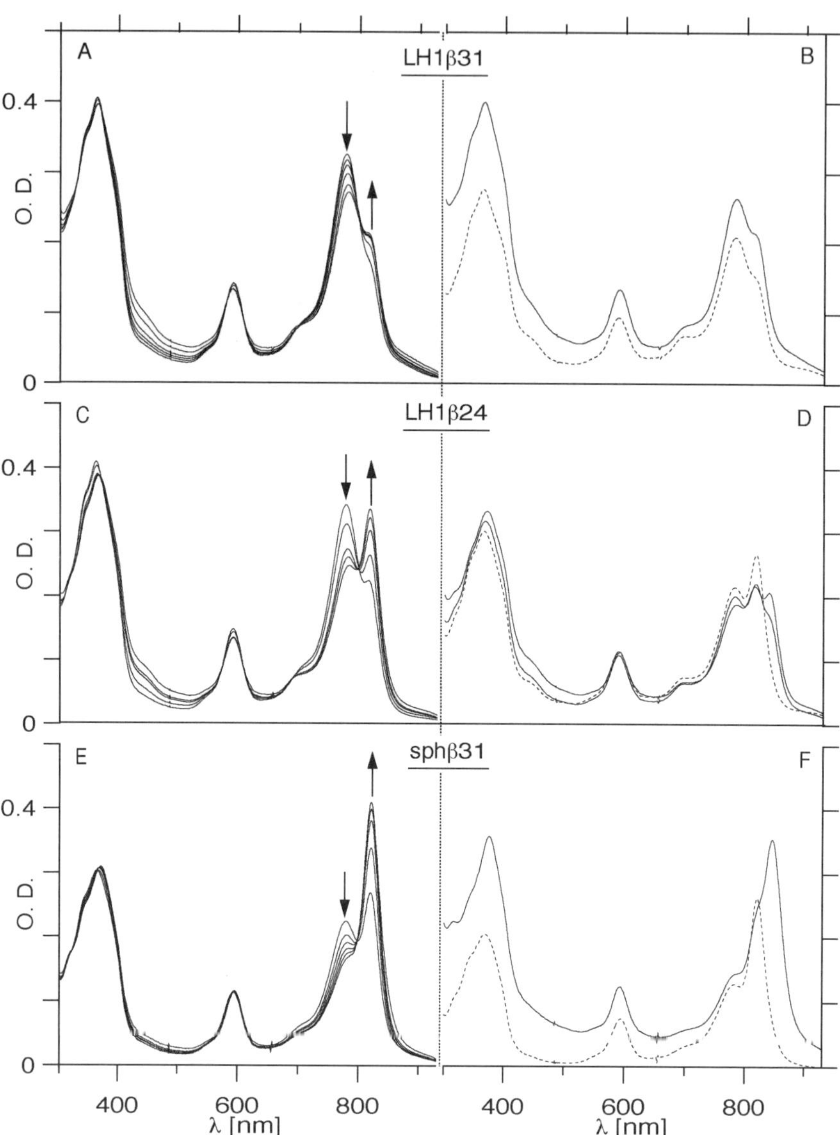

Fig. 5 Effect of temperature (left) and centrifugation (right) on the association and aggregation of LH1β31 (A,B), LH1β24 (C,D) and sphβ31 (E, F) at 0.9% OG, 50 mM phosphate buffer solution, pH 7.5. Peptide and BChl concentrations ~8 μM. A, C, E show cooling from 20°C - 6°C; B, D, F show spectra at 4°C before and after 20 minutes centrifugation at 14000 rpm (solid and dotted line respectively). LH1β24 required cooling to 0°C before significant B850 could be detected (D, additional solid line).

shown) support our observation that the B850 are very large, non-specific aggregates probably dominated by BChl self-association.

The balance between BChl self-aggregation and subunit oligomerization is a central design concern for LHC complex formation. Model studies of BChl aggregation in a 3:1 formamide/water solution have determined that the free energies for BChl dimerization and oligomerization are -4.5 and -9.25 kcal/mol, respectively,[88,89] compared to -24 kcal/mol for native LH1 subunit carotenoid free complex formation and at least -14 kcal/mol for B870 complex formation in 0.75% OG in water.[76] BChl self-aggregation is certainly a competitive process for LHC assembly on one hand, but it is conceivable that it may be organized appropriately to drive the LHC oligomerization and contribute significant stabilization energy to the LHC complex on the other hand. The most important role of the protein in such a case is not in driving the assembly but in imposing steric restrictions to limit and manage the BChl aggregation process.

2.3. LHC Maquette Design by a Bottom-Up Approach: *de novo* Designed Amphiphilic Proteins

Water-soluble cytochrome *b* maquettes are effective scaffolds for binding heme and its derivatives but despite the close chemical homology between heme, Chls, and BChls, so far we have failed to bind Chls or BChls to these maquettes, mainly because of the poor solubility of the pigments in water. Rau *et al.* have used template-assembled protein synthesis to incorporate Zn Chl derivative within a water-soluble four-helix bundle, but in this preparation the pigment was bound covalently to the protein *via* a lysine residue.[90] We have been successful, however, in binding and assembling a water-soluble Mn-substituted BChl derivative,[91] while Razeghifard and Wydrzynski have demonstrated binding Zn chlorin-e6 into very similar water soluble maquettes.[92] Using detergents with water-soluble four-helix bundle maquettes in order to increase pigment solubility was shown to induce BChl binding, but at the expense of losing most of the protein α-helical structure.[93] Similarly, Eggink and Hoober have shown that a hydrophobic

16-amino acids polypeptide based on a natural motif from plant LHCs specifically binds Chl in detergent environment only when the peptide is unfolded.[94] Nevertheless, the benefits of organizing molecular devices and creating charge gradients, expected from using membranes and interfaces, have led us to consider new transmembranal maquette designs incorporating lipophilic domains maintaining well-defined secondary structure.

Our limited understanding of transmembranal protein folding and self-assembly makes their *de novo* design a greater challenge than was faced in the initial phases of the design and construction of water-soluble proteins. In order to meet this challenge, we have developed a modular design strategy that combines the water-soluble (hydrophilic; HP) maquettes with lipophilic (LP) four-helix bundle designs to produce amphiphilic (AP) maquette scaffolds.[95] The HP domain is *de novo* designed according to the rules of water-soluble four-helix bundle engineering (exposing charged and polar residues to the aqueous environment and secluding hydrophobic residues in the maquettes' interior), whereas the LP module may be either a synthetic, *de novo* designed sequence or a motif from natural transmembranal peptides of known structure. The HP and LP blocks are connected to align the α-helical sequence according to the hydrophilicity of the residues and placement of histidine residues into the interior of the bundle.

In first members of the AP maquette family (Figure 6), the HP domain was based either on the L31M[6] or HP1[7] cytochrome *b* type maquette designs. L31M is uniquely structured in its apo-form; *i.e.* without heme, whereas the HP1 is uniquely structured in its holo-form binding two heme cofactors by ligation to histidine residues inside the bundle interior.

For the LP domain, AP maquette prototypes incorporated four-helix proton channel sequences such as the natural transmembranal segment of the M2 influenza proton channel,[96] or a synthetic channel based on *de novo* designed sequence by Lear *et al.*[97] However, for light harvesting applications, it is more desirable to engineer Chl or BChl binding sites in the LP module. This has been achieved by using a natural sequence (residues 188 to 201) from the D helix of cytochrome bc_1 which is a part of a four-helix bundle binding site of the high-potential heme (b562).[98]

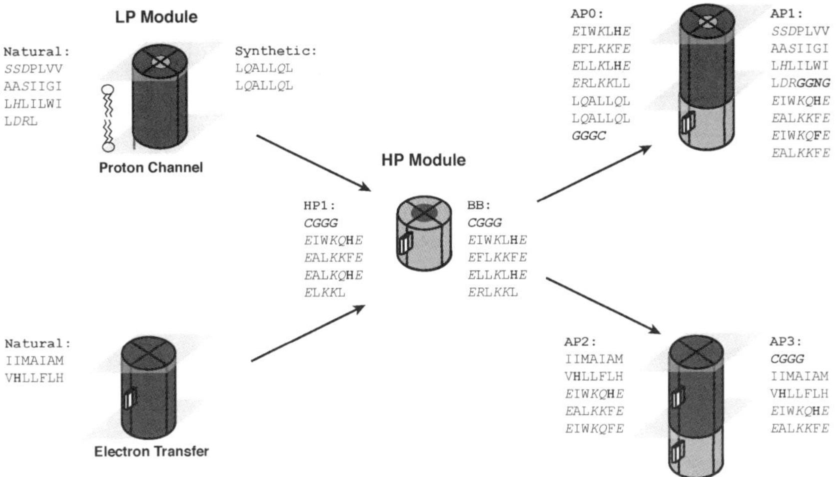

Fig. 6 A summary of the amphiphilic modular design strategy leading to the AP maquette family. Each quarter of a cylinder corresponds to an α helical peptide for which sequences are given. Hydrophilic residues are in italics, histidine ligands to heme/BChl are bold, and loop sequence italic-bold.

The novel functional characteristics of the AP maquette family make them ideal for energy transduction in light harvesting applications. So far, we were able to demonstrate their ease of purification compared to purely lipophilic transmembranal proteins, assembly into stable four-helix bundles, incorporation into phospholipid vesicles, vectorial orientation at the air-water interface, and tight bis-histidyl heme ligation to the HP and LP modules with redox potential that can be regulated by the protein environment.[99,100] Most importantly, the AP3 maquette (Figure 7) has been demonstrated capable of binding Ni and Zn substituted BChl derivatives into its LP module. [Zn]-BChl is most suitable for LHC and RC maquette applications because of its relatively long excited state lifetime.[101]

2.4. [Zn]-BChl Binding to the *de novo* Designed Amphiphilic Protein AP3

Figure 7 presents the emission and absorbance spectra of a Zn substituted BChl ([Zn]-BChl) and its AP3 maquette complex. To simplify the characterization, we disabled cofactor binding in the HP domain by substituting the histidine ligand with phenyl-alanine in the current AP3, making binding possible only at the LP module. Preliminary experiments[91] have demonstrated usefully tight binding of [Zn]-BChl; the dissociation constant (K_d) is about 200 nM and the complex stoichiometry is 1:4 [Zn]-BChl:AP3-helix (in contrast, the stoichiometrry of the respective Fe-protoporphyrin IX heme complex is 1:2 heme:AP3-helix indicating two hemes per four-helix bundle). The absorbance spectrum of [Zn]-BChl in AP3 is very similar to its spectrum in detergent except for a slight increase in the Q_y intensity and a 9 nm red shift of the Q_x transition indicative of axial ligation by a single histidine residue. The emission spectrum is significantly narrower and blue-shifted by 7 nm compared to free [Zn]-BChl. Our early success with AP3 as well as with other members of the AP family, strongly suggests that it is possible

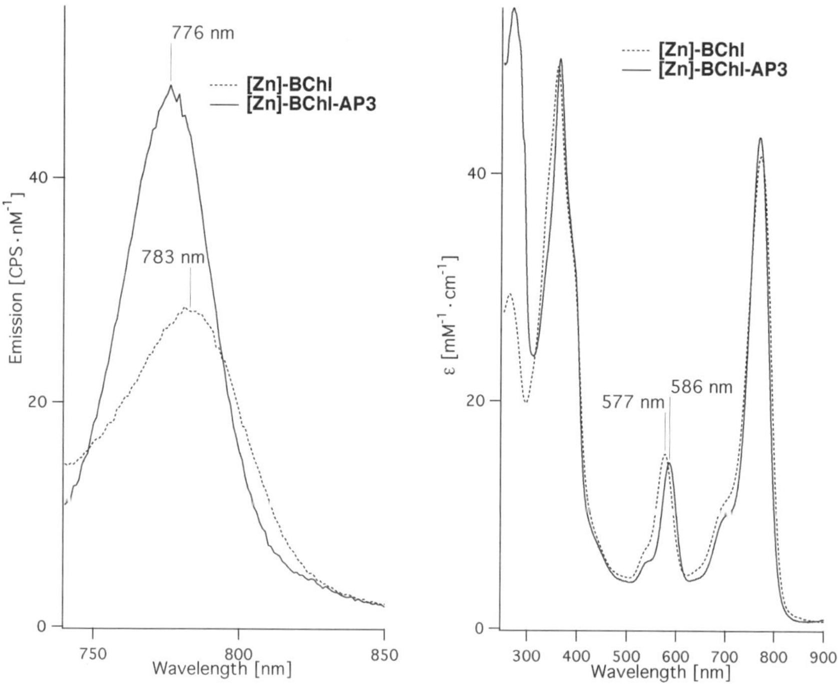

Fig. 7 Specific emission and absorption spectra of [Zn]-BChl and its AP3 complex in 0.9% OG solution containing 100 mM KCl and 20mM phosphate buffer at pH 8.0.

to design artificial transmembranal protein maquettes by using an intuitive modular approach which does not require a rigorous *de novo* design algorithm. Needless to say, further work is required in order to design a protein capable of assembling multiple cofactors in a manner similar to natural LHCs.

3. CONCLUSIONS AND PROSPECTS

The current understanding of light harvesting and photo-induced electron transfer mechanisms in natural photosynthetic organisms has become sufficient to provide the underlying engineering principles for designing robust and efficient artificial PSUs. Here, we have demonstrated how simple considerations of time, energy, and length scales of the processes involved in biological energy and electron transfer can be translated to a set of rules and guidelines for constructing protein-based molecular photoconversion devices.

We have explored two complementary strategies for designing artificial LHCs. In the first top-down approach, it was possible to create a minimal protein subunit that self-assembles with a BChl dimmer, but control over larger BChl arrays was not achieved. The second strategy is a bottom-up approach involving modular amphiphilic design of four-helix bundle maquettes comprised of *de novo* designed water-soluble protein module fused to an artificial or natural lipophilic domain. Using this strategy, it was possible to bind monomers of hydrophobic BChl derivatives, but assembling multiple BChl arrays has not yet been achieved.

With progress in understanding of the balance of interactions required for protein folding and assembly within membranal environments, as well as for packing of organic cofactors within a protein core, design and engineering of PSUs based on artificial proteins should become a feasible task.

ACKNOWLEDGMENTS

We are grateful to Christopher Moser for his useful suggestions and discussion of electron transfer mechanisms, Richard Cogdell for the coordinates of the LH1-RC complex, and Peter Heathcote for the coordinates of his anoxygenic type-I RC model. This work is supported by NIH grants GM48130 and GM41048. Dror Noy is grateful for a post-doctoral fellowship from the Human Frontiers Science Organization Program. Bohdana Discher gratefully acknowledges her support by NRSA grant GM63388.

References

1. Markvart T. Light harvesting for quantum solar energy conversion. *Prog. Quant. Electron.* 2000; **24**: 107-186.
2. Goetzberger A, Hebling C, and Schock HW. Photovoltaic materials, history, status and outlook. *Mater. Sci. Eng. R.* 2003; **40**: 1-46.
3. Moser CC, Page CC, Cogdell RJ, Barber J, Wraight CA, and Dutton PL. Length, time, and energy scales of photosystems. In *Membrane Proteins*, edited by DC Rees. *Advances in Protein Chemistry*, Academic Press, New York, 2003. **63**:71-109.
4. Robertson DE, Farid RS, Moser CC, Urbauer JL, Mulholland SE, Pidikiti R, Lear JD, Wand AJ, Degrado WF, and Dutton PL. Design and Synthesis of Multi-Heme Proteins. *Nature* 1994; **368**: 425-431.
5. Skalicky JJ, Gibney BR, Rabanal F, Urbauer RJB, Dutton PL, and Wand AJ. Solution structure of a designed four-alpha-helix bundle maquette scaffold. *J. Am. Chem. Soc.* 1999; **121**: 4941-4951.
6. Huang SS, Gibney BR, Stayrook SE, Dutton PL, and Lewis M. X-ray structure of a maquette scaffold. *J. Molec. Biol.* 2003; **326**: 1219-1225.
7. Huang SS, Koder RL, Lewis M, Wand AJ, and Dutton PL. The HP-1 maquette: from an apoprotein to a structured hemoprotein designed to promote redox-coupled proton exchange. *Proc. Natl. Acad. Sci. USA* 2004; *in press*.
8. DeGrado WF, Summa CM, Pavone V, Nastri F, and Lombardi A. *De novo* design and structural characterization of proteins and metalloproteins. *Annu. Rev. Biochem.* 1999; **68**: 779-819.
9. Barker PD. Designing redox metalloproteins from bottom-up and top-down perspectives. *Curr. Op. Struct. Biol.* 2003; **13**: 490-499.
10. Chamberlain AK, Faham S, Yohannan S, and Bowie JU. Construction of helix-bundle membrane proteins. In *Membrane Proteins*, edited by DC Rees. *Advances in Protein Chemistry*, Academic Press, New York, 2003. **63**:19-46.

11. Koepke J, Hu XC, Muenke C, Schulten K, and Michel H. The crystal structure of the light-harvesting complex II (B800-850) from Rhodospirillum molischianum. *Structure* 1996; **4**: 581-597.

12. McDermott G, Prince SM, Freer AA, Hawthornthwaite-Lawless AM, Papiz MZ, Cogdell RJ, and Isaacs NW. Crystal-Structure of an Integral Membrane Light-Harvesting Complex from Photosynthetic Bacteria. *Nature* 1995; **374**: 517-521.

13. Papiz MZ, Prince SM, Howard T, Cogdell RJ, and Isaacs NW. The structure and thermal motion of the B800-850 LH2 complex from *Rps. acidophila* at 2.0 Å resolution and 100 K: New structural features and functionally relevant motions. *J. Molec. Biol.* 2003; **326**: 1523-1538.

14. McLuskey K, Prince SM, Cogdell RJ, and Isaacs NW. The crystallographic structure of the B800-820 LH3 light-harvesting complex from the purple bacteria *Rhodopseudomonas acidophila* strain 7050. *Biochemistry* 2001; **40**: 8783-8789.

15. Camara-Artigas A, Blankenship RE, and Allen JP. The structure of the FMO protein from Chlorobium tepidum at 2.2 angstrom resolution. *Photosynth. Res.* 2003; **75**: 49-55.

16. Hofmann E, Wrench PM, Sharples FP, Hiller RG, Welte W, and Diederichs K. Structural basis of light harvesting by carotenoids: Peridinin-chlorophyll-protein from Amphidinium carterae. *Science* 1996; **272**: 1788-1791.

17. Wilk KE, Harrop SJ, Jankova L, Edler D, Keenan G, Sharples F, Hiller RG, and Curmi PMG. Evolution of a light-harvesting protein by addition of new subunits and rearrangement of conserved elements: Crystal structure of a cryptophyte phycoerythrin at 1.63-angstrom resolution. *Proc. Natl. Acad. Sci. USA* 1999; **96**: 8901-8906.

18. Nield J, Rizkallah PJ, Barber J, and Chayen NE. The 1.45 angstrom three-dimensional structure of C-phycocyanin from the thermophilic cyanobacterium *Synechococcus elongatus*. *J. Struct. Biol.* 2003; **141**: 149-155.

19. Kamiya N, and Shen JR. Crystal structure of oxygen-evolving photosystem II from *Thermosynechococcus vulcanus* at 3.7-angstrom resolution. *Proc. Natl. Acad. Sci. USA* 2003; **100**: 98-103.

20. Jordan P, Fromme P, Witt HT, Klukas O, Saenger W, and Krauss N. Three-dimensional structure of cyanobacterial photosystem I at 2.5 angstrom resolution. *Nature* 2001; **411**: 909-917.

21. Vasil'ev S, Orth P, Zouni A, Owens TG, and Bruce D. Excited-state dynamics in photosystem II: Insights from the x-ray crystal structure. *Proc. Natl. Acad. Sci. USA* 2001; **98**: 8602-8607.

22. Roszak AW, Howard TD, Southall J, Gardiner AT, Law CJ, Isaacs NW, and Cogdell RJ. Crystal Structure of the RC-LH1 Core Complex from *Rhodopseudomonas palustris*. *Science* 2003; **302**: 1969-1972.

23. Heathcote P, Jones MR, and Fyfe PK. Type I photosynthetic reaction centers: structure and function. *Phil. Trans. R. Soc. B* 2003; **358**: 231-243.

24. Fyfe PK, Jones MR, and Heathcote P. Insights into the evolution of the antenna domains of Type-I and Type-II photosynthetic reaction centers through homology modeling. *FEBS Lett.* 2002; **530**: 117-123.

25. Nield J, Morris EP, Bibby TS, and Barber J. Structural analysis of the photosystem I supercomplex of cyanobacteria induced by iron deficiency. *Biochemistry* 2003; **42**: 3180-3188.

26. Remigy HW, Stahlberg H, Fotiadis D, Muller SA, Wolpensinger B, Engel A, Hauska G, and Tsiotis G. The reaction center complex from the green sulfur bacterium Chlorobium tepidum: A structural analysis by scanning transmission electron microscopy. *J. Molec. Biol.* 1999; **290**: 851-858.

27. Remigy HW, Hauska G, Muller SA, and Tsiotis G. The reaction centre from green sulphur bacteria: progress towards structural elucidation. *Photosynth. Res.* 2002; **71**: 91-98.

28. Scheuring S, Seguin J, Marco S, Levy D, Robert B, and Rigaud JL. Nanodissection and high-resolution imaging of the *Rhodopseudomonas viridis* photosynthetic core complex in native membranes by AFM. *Proc. Natl. Acad. Sci. USA* 2003; **100**: 1690-1693.

29. Scheuring S, Seguin J, Marco S, Levy D, Breyton C, Robert B, and Rigaud JL. AFM characterization of tilt and intrinsic flexibility of *Rhodobacter sphaeroides* light harvesting complex 2 (LH2). *J. Molec. Biol.* 2003; **325**: 569-580.

30. Humphrey W, Dalke A, and Schulten K. VMD: Visual molecular dynamics. *J Mol Graphics* 1996; **14**: 33-38.

31. Pearlstein RM. Photosynthetic exciton theory in the 1960s. *Photosynth. Res.* 2002; **73**: 119-126.

32. Scholes GD. Long-range resonance energy transfer in molecular systems. *Annu. Rev. Phys. Chem.* 2003; **54**: 57-87.

33. Montano GA, Bowen BP, LaBelle JT, Woodbury NW, Pizziconi VB, and Blankenship RE. Characterization of *Chlorobium tepidum* chlorosomes: A

calculation of bacteriochlorophyll c per chlorosome and oligomer modeling. *Biophys. J.* 2003; **85**: 2560-2565.

34. Yang M, Damjanovic A, Vaswani HM, and Fleming GR. Energy transfer in photosystem I of cyanobacteria *Synechococcus elongatus*: Model study with structure-based semi-empirical Hamiltonian and experimental spectral density. *Biophys. J.* 2003; **85**: 140-158.

35. Sener MK, Lu DY, Ritz T, Park S, Fromme P, and Schulten K. Robustness and optimality of light harvesting in cyanobacterial photosystem I. *J. Phys. Chem. B* 2002; **106**: 7948-7960.

36. Ritz T, Park S, and Schulten K. Kinetics of excitation migration and trapping in the photosynthetic unit of purple bacteria. *J. Phys. Chem. B* 2001; **105**: 8259-8267.

37. Katiliene Z, Katilius E, and Woodbury NW. Energy trapping and detrapping in reaction center mutants from *Rhodobacter sphaeroides*. *Biophys. J.* 2003; **84**: 3240-3251.

38. Gardiner AT, Cogdell RJ, and Takaichi S. The Effect of Growth-Conditions on the Light-Harvesting Apparatus in *Rhodopseudomonas-Acidophila*. *Photosynth. Res.* 1993; **38**: 159-167.

39. Herek JL, Fraser NJ, Pullerits T, Martinsson P, Polivka T, Scheer H, Cogdell RJ, and Sundstrom V. B800 → B850 energy transfer mechanism in bacterial LH2 complexes investigated by B800 pigment exchange. *Biophys. J.* 2000; **78**: 2590-2596.

40. Hu XC, Ritz T, Damjanovic A, Autenrieth F, and Schulten K. Photosynthetic apparatus of purple bacteria. *Q. Rev. Biophys.* 2002; **35**: 1-62.

41. Damjanovic A, Ritz T, and Schulten K. Excitation transfer in the peridinin-chlorophyll-protein of *Amphidinium carterae. Biophys. J.* 2000; **79**: 1695-1705.

42. Ake RL, and Gouterman M. Porphyrins XIV. Theory for the luminescent state in VO, Co, Cu complexes. *Theoretica Chim. Acta* 1969; **15**: 20-42.

43. Ake RL, and Gouterman M. Porphyrins XX. Theory for states of Ni(d8) complexes. *Theoretica Chim. Acta* 1970; **17**: 408-416.

44. Musewald C, Hartwich G, Lossau H, Gilch P, Pollinger-Dammer F, Scheer H, and Michel-Beyerle ME. Ultrafast photophysics and photochemistry of [Ni]- bacteriochlorophyll a. *J. Phys. Chem. B* 1999; **103**: 7055-7060.

45. Fiedor L, Leupold D, Teuchner K, Voigt B, Hunter CN, Scherz A, and Scheer H. Excitation trap approach to analyze size and pigment-pigment

coupling: Reconstitution of LH1 antenna of *Rhodobacter sphaeroides* with Ni-substituted bacteriochlorophyll. *Biochemistry* 2001; **40**: 3737-3747.

46. Law CJ, and Cogdell RJ. The effect of chemical oxidation on the fluorescence of the LH1 (B880) complex from the purple bacterium *Rhodobium marimum. FEBS Lett.* 1998; **432**: 27-30.

47. Moser CC, Keske JM, Warncke K, Farid RS, and Dutton PL. Nature of Biological Electron-Transfer. *Nature* 1992; **355**: 796-802.

48. Ritz T, Damjanovic A, and Schulten K. The quantum physics of photosynthesis. *ChemPhysChem* 2002; **3**: 243-248.

49. Mukamel S, and Berman O. Self-consistent density matrix algorithm for electronic structure and excitations of molecules and aggregates. *J. Chem. Phys.* 2003; **119**: 12194-12204.

50. Srivatsan N, Kolbasov D, Ponomarenko N, Weber S, Ostafin AE, and Norris JR. Cryogenic charge transport in oxidized purple bacterial light-harvesting 1 complexes. *J. Phys. Chem. B* 2003; **107**: 7867-7876.

51. Srivatsan N, Weber S, Kolbasov D, and Norris JR. Exploring charge migration in light-harvesting complexes using electron paramagnetic resonance line narrowing. *J. Phys. Chem. B* 2003; **107**: 2127-2138.

52. Kropacheva TN, and Hoff AJ. Electrochemical oxidation of bacteriochlorophyll a in reaction centers and antenna complexes of photosynthetic bacteria. *J. Phys. Chem. B* 2001; **105**: 5536-5545.

53. Farchaus JW, Wachtveitl J, Mathis P, and Oesterhelt D. Tyrosine-162 of the Photosynthetic Reaction-Center L-Subunit Plays a Critical Role in the Cytochrome-C(2) Mediated Rereduction of the Photooxidized Bacteriochlorophyll Dimer in *Rhodobacter-Sphaeroides*. 1. Site-Directed Mutagenesis and Initial Characterization. *Biochemistry* 1993; **32**: 10885-10893.

54. Bottin H, and Mathis P. Interaction of Plastocyanin with the Photosystem-I Reaction Center – a Kinetic-Study by Flash Absorption-Spectroscopy. *Biochemistry* 1985; **24**: 6453-6460.

55. Jeans C, Schilstra MJ, and Klug DR. The temperature dependence of P680(+) reduction in oxygen-evolving photosystem. *Biochemistry* 2002; **41**: 5015-5023.

56. Klimov VV. Discovery of pheophytin function in the photosynthetic energy conversion as the primary electron acceptor of Photosystem II. *Photosynth. Res.* 2003; **76**: 247-253.

57. Watanabe T. Electrochemistry of chlorophylls. In *Chlorophylls*, edited by Scheer H. CRC Press, Boca Raton, 1991; 287-315.

58. Scholes GD. Designing light-harvesting antenna systems based on superradiant molecular aggregates. *Chem. Phys.* 2002; **275**: 373-386.

59. Ball P. Natural strategies for the molecular engineer. *Nanotechnology* 2002; **13**: R15-R28.

60. Lehn JM. Toward complex matter: Supramolecular chemistry and self-organization. *Proc. Natl. Acad. Sci. USA* 2002; **99**: 4763-4768.

61. Roco MC. Nanotechnology: convergence with modern biology and medicine. *Current Opinion in Biotechnology* 2003; **14**: 337-346.

62. Shifman JM, Moser CC, Kalsbeck WA, Bocian DF, and Dutton PL. Functionalized *de novo* designed proteins: Mechanism of proton coupling to oxidation/reduction in heme protein maquettes. *Biochemistry* 1998; **37**: 16815-16827.

63. Grosset AM, Gibney BR, Rabanal F, Moser CC, and Dutton PL. Proof of principle in a *de novo* designed protein maquette: An allosterically regulated, charge-activated conformational switch in a tetra-alpha-helix bundle. *Biochemistry* 2001; **40**: 5474-5487.

64. Sharp RE, Moser CC, Rabanal F, and Dutton PL. Design, synthesis, and characterization of a photoactivatable flavocytochrome molecular maquette. *Proc. Natl. Acad. Sci. USA* 1998; **95**: 10465-10470.

65. Chen XX, Discher BM, Pilloud DL, Gibney BR, Moser CC, and Dutton PL. *De novo* design of a cytochrome b maquette for electron transfer and coupled reactions on electrodes. *J. Phys. Chem. B* 2002; **106**: 617-624.

66. Bowie JU. Understanding membrane protein structure by design. *Nature Struct. Biol.* 2000; **7**: 91-94.

67. Fowler GJS, Sockalingum GD, Robert B, and Hunter CN. Blue Shifts in Bacteriochlorophyll Absorbency Correlate with Changed Hydrogen-Bonding Patterns in Light-Harvesting 2 Mutants of Rhodobacter-Sphaeroides with Alterations at Alpha-Tyr-44 and Alpha-Tyr-45. *Biochem. J.* 1994; **299**: 695-700.

68. Fowler GJS, Hess S, Pullerits T, Sundstrom V, and Hunter CN. The role of beta Arg(-10) in the B800 bacteriochlorophyll and carotenoid pigment environment within the light-harvesting LH2 complex of *Rhodobacter sphaeroides*. *Biochemistry* 1997; **36**: 11282-11291.

69. Olsen JD, Sturgis JN, Westerhuis WHJ, Fowler GJS, Hunter CN, and Robert B. Site-directed modification of the ligands to the bacteriochlorophylls of the light-harvesting LH1 and LH2 complexes of *Rhodobacter sphaeroides*. *Biochemistry* 1997; **36**: 12625-12632.

70. Sturgis JN, and Robert B. Thermodynamics of Membrane Polypeptide Oligomerization in Light-Harvesting Complexes and Associated Structural-Changes. *J. Molec. Biol.* 1994; **238**: 445-454.

71. Sturgis JN, and Robert B. The role of chromophore coupling in tuning the spectral properties of peripheral light-harvesting protein of purple bacteria. *Photosynth. Res.* 1996; **50**: 5-10.

72. Braun P, Olsen JD, Strohmann B, Hunter CN, and Scheer H. Assembly of light-harvesting bacteriochlorophyll in a model transmembrane helix in its natural environment. *J. Molec. Biol.* 2002; **318**: 1085-1095.

73. Parkes-Loach PS, Sprinkle JR, and Loach PA. Reconstitution of the B873 Light-Harvesting Complex of *Rhodospirillum-Rubrum* from the Separately Isolated Alpha- Polypeptide and Beta-Polypeptide and Bacteriochlorophyll-A. *Biochemistry* 1988; **27**: 2718-2727.

74. Parkes-Loach PS, Michalski TJ, Bass WJ, Smith U, and Loach PA. Probing the Bacteriochlorophyll Binding-Site by Reconstitution of the Light-Harvesting Complex of *Rhodospirillum-Rubrum* with Bacteriochlorophyll-a Analogs. *Biochemistry* 1990; **29**: 2951-2960.

75. Meadows KA, Iida K, Tsuda K, Recchia PA, Heller BA, Antonio B, Nango M, and Loach PA. Enzymatic and Chemical Cleavage of the Core Light-Harvesting Polypeptides of Photosynthetic Bacteria – Determination of the Minimal Polypeptide Size and Structure Required for Subunit and Light-Harvesting Complex-Formation. *Biochemistry* 1995; **34**: 1559-1574.

76. Loach PA, and Parkes-Loach PS. Structure-function relationships in core light-harvesting complexes (LHI) as determined by characterization of the structural subunit and by reconstitution experiments. In *Anoxygenic photosynthetic bacteria*, edited by Blankenship RE, Madigan MT, and Bauer CE. Kluwer, New York, 1995; 437-471.

77. Davis CM, Parkes-Loach PS, Cook CK, Meadows KA, Bandilla M, Scheer H, and Loach PA. Comparison of the structural requirements for bacteriochlorophyll binding in the core light-harvesting complexes of *Rhodospirillum rubrum* and Rhodobacter sphaeroides using reconstitution methodology with bacteriochlorophyll analogs. *Biochemistry* 1996; **35**: 3072-3084.

78. Kehoe JW, Meadows KA, Parkes-Loach PS, and Loach PA. Reconstitution of core light-harvesting complexes of photosynthetic bacteria using chemically synthesized polypeptides. 2. Determination of structural features that stabilize complex formation and their implications for the structure of the subunit complex. *Biochemistry* 1998; **37**: 3418-3428.

79. Meadows KA, Parkes-Loach PS, Kehoe JW, and Loach PA. Reconstitution of core light-harvesting complexes of photosynthetic bacteria using chemically synthesized polypeptides. 1. Minimal requirements for subunit formation. *Biochemistry* 1998; **37**: 3411-3417.

80. Todd JB, Recchia PA, Parkes-Loach PS, Olsen JD, Fowler GJS, McGlynn P, Hunter CN, and Loach PA. Minimal requirements for in vitro reconstitution of the structural subunit of light-harvesting complexes of photosynthetic bacteria. *Photosynth. Res.* 1999; **62**: 85-98.

81. Nango M, Kashiwada A, Watanabe H, Yamada S, Yamada T, Ogawa M, Tanaka T, and Iida K. Molecular assembly of bachteriochlorophyll a using light-harvesting model 1 alpha-helix polypeptides and 2 alpha-helix polypeptide with disulfide-linkage. *Chem. Lett.* 2002: 312-313.

82. Kashiwada A, Takeuchi Y, Watanabe H, *et al.* Molecular assembly of covalently-linked mesoporphyrin dimers with light-harvesting polypeptides. *Tetrahedron Lett.* 2000; **41**: 2115-2119.

83. Kashiwada A, Watanabe H, Mizuno T, Iida K, Miyatake T, Tamiaki H, Kobayashi M, and Nango M. Structural requirements of zinc porphyrin derivatives on the complex-forming with light-harvesting polypeptides. *Chem. Lett.* 2000: 158-159.

84. Kashiwada A, Watanabe H, Tanaka T, and Nango M. Molecular assembly of zinc bachteriochlorophyll a by synthetic hydrophobic 1 alpha-helix polypeptides. *Chem. Lett.* 2000: 24-25.

85. Nagata M, Nango M, Kashiwada A, Yamada S, Ito S, Sawa N, Ogawa M, Iida K, Kurono Y, and Ohtsuka T. Construction of photosynthetic antenna complex using light-harvesting polypeptide-alpha from photosynthetic bacteria, *R. rubrum* with zinc substituted bacteriochlorophyll alpha. *Chem. Lett.* 2003; **32**: 216-217.

86. Noy D, Moser CC, and Dutton PL. Bacteriochlorophyll protein maquettes. In *Biochemistry and Biophysics of Chlorophylls*, edited by Grimm B, Porra R, Rüdiger W, and Scheer H. Kluwer Academic, New York, 2004.

87. Wang ZY, Muraoka Y, Nagao M, Shibayama M, Kobayashi M, and Nozawa T. Determination of the B820 subunit size of a bacterial core light-harvesting complex by small-angle neutron scattering. *Biochemistry* 2003; **42**: 11555-11560.

88. Fisher JRE, Rosenbachbelkin V, and Scherz A. Cooperative Polymerization of Photosynthetic Pigments in Formamide-Water Solution. *Biophys. J.* 1990; **58**: 461-470.

89. Scherz A, Rosenbachbelkin V, and Fisher JRE. Distribution and Self-Organization of Photosynthetic Pigments in Micelles - Implication for the Assembly of Light-Harvesting Complexes and Reaction Centers in the Photosynthetic Membrane. *Proc. Natl. Acad. Sci. USA* 1990; **87**: 5430-5434.

90. Rau HK, Snigula H, Struck A, Robert B, Scheer H, and Haehnel W. Design, synthesis and properties of synthetic chlorophyll proteins. *Eur. J. Biochem.* 2001; **268**: 3284-3295.

91. Noy D. *Unpublished results.* 2004.

92. Razeghifard AR, and Wydrzynski T. Binding of Zn-chlorin to a synthetic four-helix bundle peptide through histidine ligation. *Biochemistry* 2003; **42**: 1024-1030.

93. Kashiwada A, Nishino N, Wang ZY, Nozawa T, Kobayashi M, and Nango M. Molecular assembly of bacteriochlorophyll a and its analogues by synthetic 4 alpha-helix polypeptides. *Chem. Lett.* 1999; 1301-1302.

94. Eggink LL, and Hoober JK. Chlorophyll binding to peptide maquettes containing a retention motif. *J. Biol. Chem.* 2000; **275**: 9087-9090.

95. Discher BM, Koder RL, Moser CC, and Dutton PL. Hydrophilic to amphiphilic design in redox protein maquettes. *Curr. Op. Chem. Biol.* 2003; **7**: 741–748.

96. Nishimura K, Kim SG, Zhang L, and Cross TA. The closed state of a H^+ channel helical bundle combining precise orientational and distance restraints from solid state NMR-1. *Biochemistry* 2002; **41**: 13170-13177.

97. Lear JD, Wasserman ZR, and Degrado WF. Synthetic Amphiphilic Peptide Models for Protein Ion Channels. *Science* 1988; **240**: 1177-1181.

98. Iwata S, Lee JW, Okada K, Lee JK, Iwata M, Rasmussen B, Link TA, Ramaswamy S, and Jap BK. Complete structure of the 11-subunit bovine mitochondrial cytochrome bc(1) complex. *Science* 1998; **281**: 64-71.

99. Ye S, Discher BM, Strzalka JW, Noy D, Zheng S, Dutton PL, and Blasie JK. Amphiphilic 4-helix bundles designed for biomolecular materials applications. *Langmuir* 2004: *submitted.*

100. Discher BM. *Unpublished results.* 2004.

101. Teuchner K, Stiel H, Leupold D, Scherz A, Noy D, Simonin I, Hartwich G, and Scheer H. Fluorescence and excited state absorption in modified pigments of bacterial photosynthesis - A comparative study of metal-substituted bacteriochlorophylls a. *J. Luminesc.* 1997; **72-4**: 612-614.

Chapter 3

HOW PURPLE BACTERIA HARVEST LIGHT ENERGY

Christopher J. Law, Alastair T. Gardiner, June Southall, Aleksander W. Roszak, Tina D. Howard, Neil W. Isaacs and Richard J. Cogdell

Purple bacteria have circular light-harvesting complexes that absorb incident solar radiation and rapidly and efficiently transfer that excitation energy to reaction centres where it is trapped. This chapter describes our current understanding of the structure and function of these antenna complexes and places particular emphasis on the general design principles that the purple bacteria have adopted to solve the light- harvesting problem.

Keywords: Membrane proteins, photosynthesis, light-harvesting, photochemical reaction centre, LH2, LH1, core complex, bacteriochlorophyll.

1. INTRODUCTION

1.1. The Initial Light Reactions

Purple photosynthetic bacteria are ideal model organisms in which to study photosynthetic light-harvesting because, in contrast to plant and algal systems, the different groups of pigments involved in their light reactions have well separated absorption spectra making them amenable to study using spectroscopic techniques. There is also a relative wealth of information about the 3-D structures of the major proteins involved in the purple bacterial light reactions (Deisenhofer *et al.*, 1984; Allen *et al.*, 1987; McDermott *et al.*, 1995; Koepke *et al.*, 1996; McLuskey *et al.*, 2001). These two factors, combined with recent technological advances in laser flash photolysis, have allowed their light reactions to be resolved from a few femtoseconds out to longer timescales (Fleming & van Grondelle, 1997; Sundström *et al.*, 1999). In this review we focus on the structure and function of the light-harvesting system from purple bacteria, with particular emphasis on those features which are of general significance for photosynthetic light-harvesting.

In the initial light reactions, the energy of solar radiation is harvested as an electronic excitation of the light-harvesting pigment molecules. Subsequently, this excitation energy is rapidly and efficiently transferred to the photochemical reaction centre (RC) where it induces charge separation across the membrane. Before we discuss in detail the purple bacterial light-harvesting complexes we will make a few general remarks about the physics of light-harvesting which set the photophysical framework within which biology has to work. When the ranges of light-harvesting complexes for which we have structures are compared, the major conclusion is that biology has produced a wide range of solutions to the problem of producing an efficient light-harvesting complex. This is in stark contrast to the photosynthetic RCs, all of which appear to be built using a very similar blueprint. The variability in design of light-harvesting complexes is clearly illustrated by comparing the structures of cyanobacterial photosystem I (PSI) (Jordan *et al.*, 2001), a purple bacterial light-harvesting complex 2 (LH2) (McDermott *et al.*, 1995; Papiz *et al.*, 2003) and the peridinin-chlorophyll-protein (PCP) from a dinoflagellate (Hofmann *et al.*, 1996) (Figure 1). Although all three complexes contain a predominance of α-helical secondary structure the protein folds of the complexes are vastly different. PSI and LH2 are both integral membrane complexes that bind the pigments chlorophyll *a* (Chl *a*) and bacteriochlorophyll *a* (Bchl *a*) respectively, (as well as carotenoids), whereas PCP is water-soluble and principally binds just carotenoids. The pigment organization in each complex also shows marked differences, the pigments in LH2 and PCP displaying a much more ordered distribution than the (apparently) random distribution of Chl *a* in PSI. What is it about the light-harvesting process that permits such a wide range of possible design solutions, all of which result in efficient photosynthetic light-harvesting?

The kinetics of the energy transfer process (and hence its efficiency) is governed by: (i) the distance between pigments; (ii) the spectral

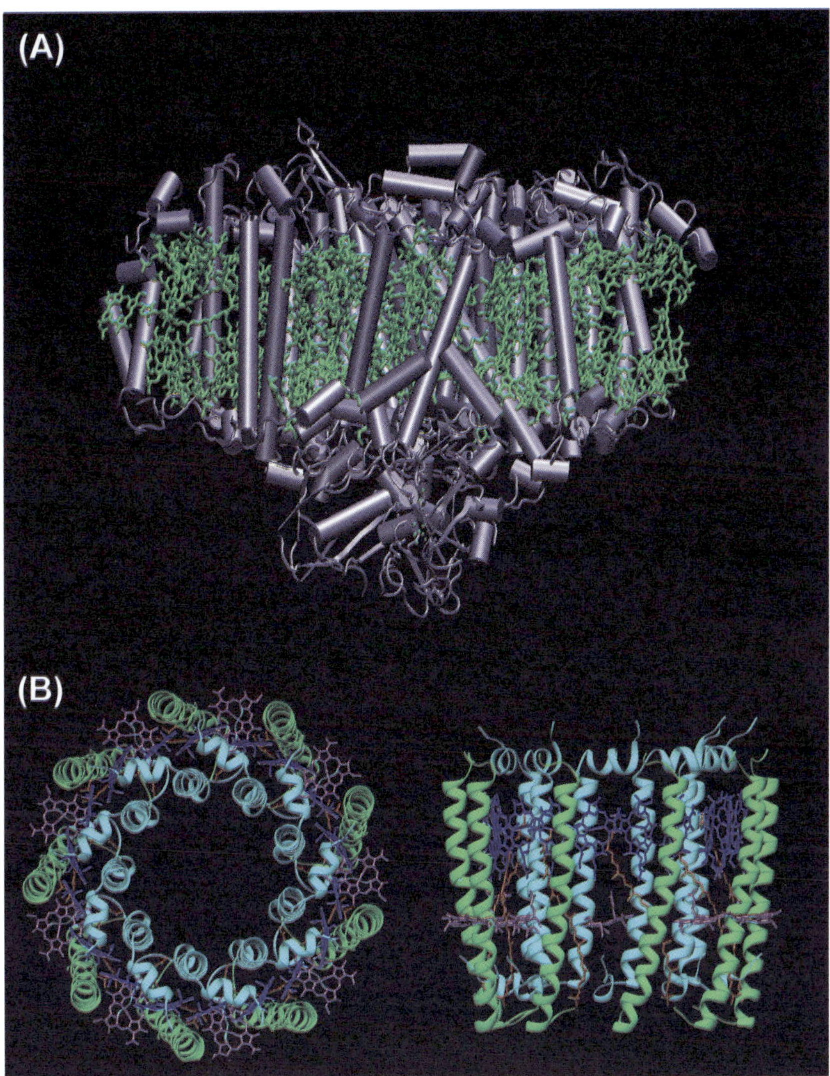

Fig. 1 Structures of (A) a monomer of PSI from *Synechococcus elongatus* at 2.5Å resolution viewed from the side with the stromal side at the bottom and the lumenal side at the top. Protein is coloured ice blue and the pigments green. Figure produced from PDB file 1JBO using VMD 1.8.1 (Humphrey *et al.*, 1996). (B) Top (left) and side (right) views of the LH2 complex from *Rps. acidophila* strain 10050. The β–polypeptides are coloured green and the α–polypeptides light blue. The monomeric B800 Bchl *a* molecules are mauve whereas the B850 Bchls and carotenoids are blue and orange, respectively. Figure courtesy of Iain Mitchell and produced from PDB file 1KNZ.

Fig. 1 (C) Structure of the complete peridinin-chlorophyll-protein trimer complex from the dinoflagellate *Amphidinium carterae*. The protein is depicted in ice blue, the carotenoids in red and the chlorophylls in green. Figure kindly produced by Robielyn Ilagan from PDB file 1PPR using VMD.

overlap between the fluorescence emission of the donor pigment and the absorption of the acceptor pigment; and (iii) the excited state lifetimes and the relative geometric arrangement of the transition dipole moments of the pigments involved (Förster, 1948). Once a molecule absorbs a photon and becomes excited, in effect a clock starts ticking and this clock stops when the energy stored in the excited state is lost by fluorescence or other competing radiationless processes which return the excited electron back to the ground state. Typically, excited state lifetimes for monomeric Bchl are in the order of a few nanoseconds (Zankel *et al.*, 1968). All that efficient energy transfer requires is that it betters this rate by a factor of about 10. Bchl molecules in the purple bacterial LH complexes are typically separated by distances ranging from a few Å up to 40-50 Å. Even at these longest distances the measured energy transfer lifetimes are less than 50 ps (Hunter *et al.*, 1989). It can readily be seen, therefore, that there is a great deal of flexibility inherent in the system if the overall efficiency can remain better than 95% when the Bchl molecules can be arranged anywhere up to about 50 Å apart. However, a striking exception to this situation involves the carotenoids. Their excited state lifetimes are much shorter with the S_2 and S_1 states, having lifetimes of 200 fs (Shreve *et al.*, 1991) and 1-25 ps (Frank *et al.*, 1997a; Frank *et al.*, 1997b), respectively; hence the distance between carotenoid molecules and Bchl molecules must always be much shorter than those found between Bchls alone if efficient energy transfer is to occur. In fact, carotenoids are always found in van der Waal's contact with Bchl molecules in the purple bacterial light-harvesting complexes. For example, the carotenoid to Bchl *a* distance in LH2 is 3.40 Å at the closest point of approach (Freer *et al.*, 1996).

Light-harvesting complexes must absorb light in the region of the spectrum available to them in the ecological niche that the organism that contains them occupies. Dinoflagellates are usually found in layers of marine environments where much of the incident solar radiation is in the blue region of the spectrum. These organisms synthesize PCP complexes where most of the light absorption is conducted by carotenoids that absorb this blue light (Haxo *et al.*, 1976; Sharples *et al.*, 1996). In contrast, purple bacteria inhabit water layers below those occupied by oxygenic (*i.e.* Chl *a* based) photosynthetic organisms and so grow photosynthetically by absorbing light in the green and far-red regions of the spectrum. Hence, they utilize Bchl *a* (to absorb the red light) and carotenoid (to absorb light in the green region) pigments (van Gemerden & Mas, 1995). Manipulating factors such as (i) the type of pigment used, (ii) pigment-pigment interactions, and (iii) pigment-protein interactions can select the wavelength(s) of light absorbed by light-harvesting complexes. Using this strategy the photosynthetic Bchl *a*-containing purple bacteria can effectively absorb light in the spectral range from 800 nm to 963 nm (Permentier *et al.*, 2001). The LH2 complex from *Rhodopseudomonas (Rps.) acidophila* (McDermott *et al.*, 1995) (Figure 1B) provides a good example of how light absorption can be tuned by a combination of pigment-pigment and pigment-protein interactions. In contrast to monomeric Bchl *a*, which has an absorption maximum at 770 nm in most organic solvents, LH2 possesses two spectrally distinct populations of Bchl *a* with near-infrared absorption peaks located at about 800 nm and 850 nm. These Bchl *a* molecules are termed B800 and B850 (Cogdell *et al.*, 1985). The large absorption red-shift and appearance of an extra absorption band are a direct result of pigment-protein and pigment-pigment interactions (Thornber *et al.*, 1978; Prince *et al.*, 1997). The 800 nm absorption arises from a ring of monomeric Bchl *a* molecules that are non-covalently bound to the outer ring of transmembrane helices of the protein. The observed red-shift of these particular pigments is due solely to pigment-protein interactions. On the other hand, the 850 nm absorption band arises from a ring of excitonically coupled Bchl *a* molecules. In this case the red-shift is due to a combination of pigment-pigment and pigment-protein interactions.

1.2. The Purple Bacterial Photosynthetic Unit

The quality and intensity of light in the environment are not constant and they can (and do) change. Photosynthetic bacteria are able to regulate both their light-harvesting capacity and the spectral range of light absorbed to suit the prevailing environmental conditions. Purple bacteria achieve this by controlling both the size of their photosynthetic unit (PSU) and the number of PSUs per cell. The concept of the PSU, which forms the basis of modern thinking of the primary processes of photosynthesis, originated from the studies of Emerson and Arnold (1932). The PSU can be defined as the combination of a RC and the light-harvesting pigments that contribute excitation energy to that RC. The PSU of most purple bacteria is relatively simple, consisting of the photochemical RC and two major types of antenna complex referred to as LH1 and LH2. The LH1 complex is intimately associated with the photochemical RC in a 1:1 stoichiometry (to form the RC-LH1 'core' complex) and all wild-type purple bacteria contain it (Aagard & Sistrom, 1972; Sistrom, 1978). LH2, which can be present in variable amounts, is arranged around the periphery of the core complexes. In some species such as *Rhodospirillum* (*Rs.*) *rubrum*, which possess LH1 as the only antenna complex, the RC-LH1 core complex represents the maximal size of the PSU. However, in species such as *Rhodobacter* (*Rb.*) *sphaeroides*, which can synthesize peripheral LH2 as well as core complexes, the core complex represents the minimal size of the PSU. Yet other species, such as *Rhodopseudomonas acidophila*, can adapt to low light conditions by synthesizing different types of LH2 complex (Cogdell *et al.*, 1983; Stadtwaldt-Demchick *et al.*, 1990; Halloren *et al.*, 1995). Depending upon light intensity the purple bacterial PSU can vary in size from about 30 to 330 Bchls per RC (Table 1).

Table 1. Ratio of total Bchl to RC Bchl in cells of *Rb. sphaeroides* grown at different light intensities (adapted from Aagard & Sistrom, 1972). *The light intensities used in this experiment were never high enough to reduce the total Bchl:RC Bchl to ~30.

Light intensity	Molar ratio (Total Bchl:RC Bchl)
HIGH	72*
	77
	100
	222
LOW	330

This is a clear demonstration of the ability of some purple bacteria to tune their capacity for photon capture, not only in terms of the wavelengths of light absorbed, but also by increasing the amount of light absorbed (and thereby the physical area occupied) by antenna complexes. The aim of such a strategy is to keep the RC supplied with sufficient photons for bacterial growth while at the same time keeping energy expenditure on protein and pigment biosynthesis to a minimum.

2. GENERAL ASPECTS OF ANTENNA COMPLEX STRUCTURE

In the ten years prior to the publication of the first 3-D structure of a purple bacterial antenna complex, a considerable accumulation of biochemical and biophysical data was used to construct models of these complexes. Included in this work was the determination of the primary structures of a large number of antenna polypeptides from both LH1 and LH2 (Zuber & Brunisholz, 1991). Analyses of these primary structures showed that all purple bacterial antenna complexes are built on the same modular principle. The minimal structural unit of antenna complexes is a heterodimer of low molecular weight (4-7 kDa) α and β-polypeptides, each consisting of 40-70 amino acid residues (Zuber & Brunisholz, 1991; Zuber & Cogdell, 1995). Typically, both polypeptides have a tripartite structure consisting of a central hydrophobic domain and polar (charged) N and C-terminal domains. Polarized infrared spectroscopy and UV circular dichroism indicated that the hydrophobic domain forms a transmembrane α–helix and surface labeling and protease digestion experiments revealed that the N and C-terminal domains are located in the polar head region or surface of the membrane (Zuber, 1985a, 1985b). The C-terminal domain lies on the periplasmic side of the membrane and the N-terminal domain on the cytoplasmic side. Using this information, models (although we now know them to be incorrect) did make a good job of describing the molecular organization of the Bchl molecules (Kramer *et al.*, 1984; Olsen & Hunter, 1994). Two such models are shown in Figure 2.

We now know that native antenna complexes are ring-like structures composed of oligomers of the $\alpha\beta$-polypeptide minimal subunit that non-covalently bind Bchl a and carotenoid pigment molecules. Generally, each pair of LH2 α and β-polypeptides bind three Bchl and one or two carotenoid molecules (Kramer *et al.*, 1984; Cogdell *et al.*, 1999; Papiz *et al.*, 2003) whereas each pair of LH1 antenna polypeptides bind two Bchl and a single carotenoid (Cogdell *et al.*, 1982). The α and β-polypeptides of each type of antenna complex contain a highly conserved histidine (His) residue located 1-2 turns down the transmembrane α-helix from the boundary between the central membrane spanning domain and the polar C-terminal domain (Zuber & Brunisholz, 1991). These His residues are liganded to the central magnesium atoms of the Bchl macrocycles that give rise to the absorption bands at ~850 nm and ~875 nm in LH2 and LH1, respectively. A second, usually conserved His residue is found near the N-terminal domain of the β-polypeptide although it is not directly involved in binding of Bchl (Robert & Lutz, 1985). This conserved His is also present in the LH1 β-polypeptide, even though the additional Bchl is absent in this complex.

(A)

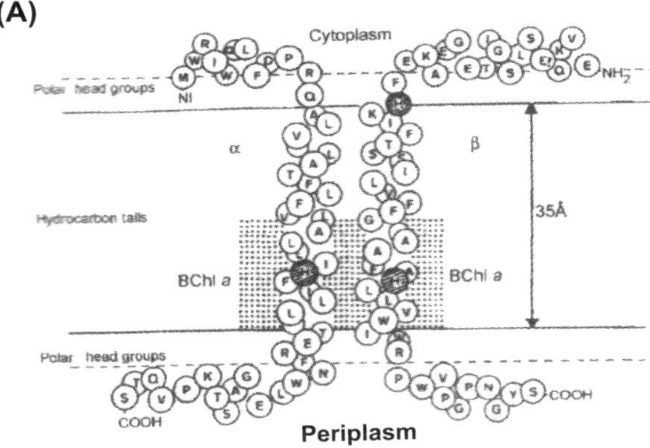

(B)

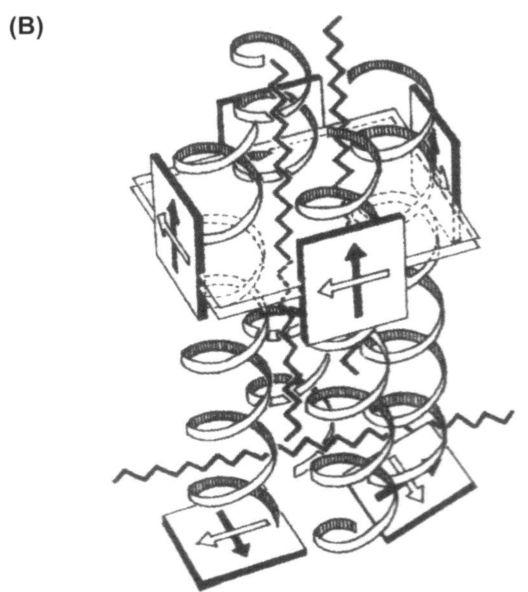

Fig. 2 (A) A model for the transmembrane arrangement of a bacterial LH1 complex αβ–subunit. Adapted from Zuber (1987). (B) The Kramer model for the bacterial LH2 complex αβ–subunit. The squares represent the bacteriochlorin rings of the Bchl *a* molecules. The Q$_x$ and Q$_y$ dipole moments are represented as solid and open arrows, respectively. From Kramer *et al*. (1984).

Several other conserved amino acid residues are found in the vicinity of the conserved central His residue in both the α and β-polypeptides. There is usually a leucine (α-polypeptide) or large aromatic residue (β-polypeptide) located at His +4 (numbered relative to the conserved His, considered as position 0) and, located about one α-helix turn away at His -4, a small hydrophobic residue (Brunisholz *et al.*, 1984; Theiler & Zuber, 1984). Conserved tyrosine or tryptophan residues are found at positions His +4, +6 and +9 in the LH1, or His +9 in the LH2 β-polypeptides and at position His +11 in the LH1 and His +9 and +14 in the LH2 α-polypeptides (Zuber & Cogdell, 1995). These aromatic amino acid residues create a particular microenvironment for the His-bound Bchl molecules, thereby influencing their spectral characteristics (Fowler *et al.*, 1992; Sturgis *et al.*, 1995). The structures of the LH2 and RC-LH1 complexes are described in detail below.

3. THE STRUCTURE OF LH2

3.1.   The Protein Scaffold

Until 1995 our understanding of light-energy capture and transfer in the purple bacteria was hampered by the lack of detailed structural information about the light-harvesting complexes. However, this situation changed dramatically with the elucidation, by X-ray crystallography, of high resolution 3-D structures of LH2 complexes from *Rps. acidophila* (McDermott *et al.*, 1995) and *Rs. molischianum* (Koepke *et al.*, 1996). As can be seen in Figure 1(B) the LH2 complex from *Rps. acidophila* consists of a nonamer of $\alpha\beta$-apoproteins arranged in a ring-like structure. The transmembrane helices of the nine α- and β-apoproteins form two concentric rings with radii of 18 Å and 34 Å, respectively. The hole in the centre of the structure is filled with lipid (Prince *et al.*, 2003). The α–apoprotein helices lie perpendicular to the membrane plane while those of the β–apoproteins are tilted by about 15° with respect to it. Although there are strong helix-helix interactions between the α-polypeptides, the structure is dominated by extensive pigment-pigment interactions. There are no $\alpha\beta$-apoprotein helix-helix interactions within the transmembrane domain of the complex and this may explain the failure of detailed models of LH2 to describe the complex prior to determination of the crystal structure. The N- and C-termini of both apoproteins, located on the cytoplasmic and periplasmic sides of the membrane respectively, fold over and interact with each other to enclose the top and bottom of the structure. Large aromatic residues located at the C-termini of both apoproteins interact, via hydrogen bonds to the Bchl *a* molecules, to interlock the whole structure (Cogdell *et al.*, 1997; Prince *et al.*, 1997). It is this protein structure that acts as a scaffold for the attachment of the photosynthetic pigments.

3.2. Binding and Organization of the Bchl *a* Molecules

The Bchl *a* molecules of the LH2 complex are arranged into two distinct populations. In *Rps. acidophila* LH2, one of these populations consists of eighteen tightly coupled Bchl *a* molecules that are sandwiched, in a very hydrophobic environment, between the concentric rings formed by the transmembrane helices of the α and β-apoproteins (McDermott *et al.*, 1995) (Figure 1B). These Bchls are liganded to the conserved α-(His 31) or β–(His 30) residues of the apoproteins and their bacteriochlorin rings lie perpendicular to the plane of the membrane. The centres of the bacteriochlorin rings are located about 10 Å from the periplasmic surface of the membrane. The individual environments and conformations of the B850 Bchls are not all equivalent. The Mg^{2+}-Mg^{2+} distance between the two B850 Bchls within each $\alpha\beta$-apoprotein dimer is 9.7 Å, whereas the Mg^{2+}-Mg^{2+} distance between the nearest B850 Bchls of adjacent $\alpha\beta$-apoprotein dimers is only 8.7 Å (Freer *et al.*, 1996; Papiz *et al.*, 2003). The orientation of the Bchls also alternates going around the ring. The face of the α-B850 Bchl bacteriochlorin ring is presented to the inside of the complex whereas that of the β-B850 Bchl is presented to the outside. The configuration of each type of Bchl also differs. The bacteriochlorin ring of the α-B850 is almost planar while that of the β-B850 Bchl shows a significant 'bowing' along the direction of the Q_y transition (Prince *et al.*, 1997).

Nine other Bchl *a*s are arranged between the β–apoprotein α–helices about 16.5 Å further into the membrane in a binding pocket that is rather hydrophilic. The planes of the bacteriochlorin rings lie more or less parallel to the presumed plane of the membrane. The bacteriochlorins of these peripheral B800 Bchls are slightly 'domed' (Prince *et al.*, 1997). In contrast to the eighteen tightly coupled Bchls, the central Mg^{2+} atoms of the nine monomeric Bchls are not liganded to a His residue, but rather to a carboxyl moiety on the N-terminal amino group of α–Met1 (Papiz *et al.*, 2003). The Mg^{2+}-Mg^{2+} distance between each of these Bchls is 21.2 Å. Several surrounding residues (N-αAsn2, N-αGln3 and NE2-β-His12) donate several H-bonds to the O2 oxygen atom of the COO-Met1 (Figure3). Based on spectroscopic studies of the complex, the group of eighteen Bchls were identified as those absorbing at about 863 nm (B850s) and the group of nine Bchls as those absorbing at about 801 nm (B800s) (Cogdell & Scheer, 1985; Robert & Lutz, 1985).

3.3. The Bacteriochlorophyll Phytyl Tails

The highly hydrophobic phytyl chains of the Bchl molecules are very important, and often overlooked, structural moieties of the LH2 antenna that play a crucial role in aligning the Bchl molecules correctly within the complex. The Bchl molecules need their optical transition dipole moments oriented in such a way as to optimize energy transfer between donor and acceptor molecules. The phytyl tail of each Bchl starts this alignment process by providing the protein scaffold with a handle for the correct orientation of the bacteriochlorin macrocycles. The Bchl molecules are then locked into place by coordination of the central Mg^{2+} and H-bonding to adjacent phytyl tails and pigments. A closer look at the pigment molecules assembled as a unit reveals an obvious hole in the arrangement. The totally conserved β-apoprotein Phe22 residue (which is present in all species of purple bacteria) protrudes into this hole and is cradled by a bed of oxygen atoms from nearby Bchl molecules and phytyl tail ester oxygens (Freer *et al.*, 1996).

In *Rps. acidophila* LH2, the B800 phytyl chain travels through the complex, wraps around the phytyl chain of the β-B850 Bchl then passes across the face of the β-B850 Bchl macrocycle. The phytyl tail of the β-B850 Bchl passes across the face of the B800 Bchl, effectively holding the latter in place within the complex. In contrast, the phytyl chain of the α-B850 Bchl is almost fully extended and does not pass

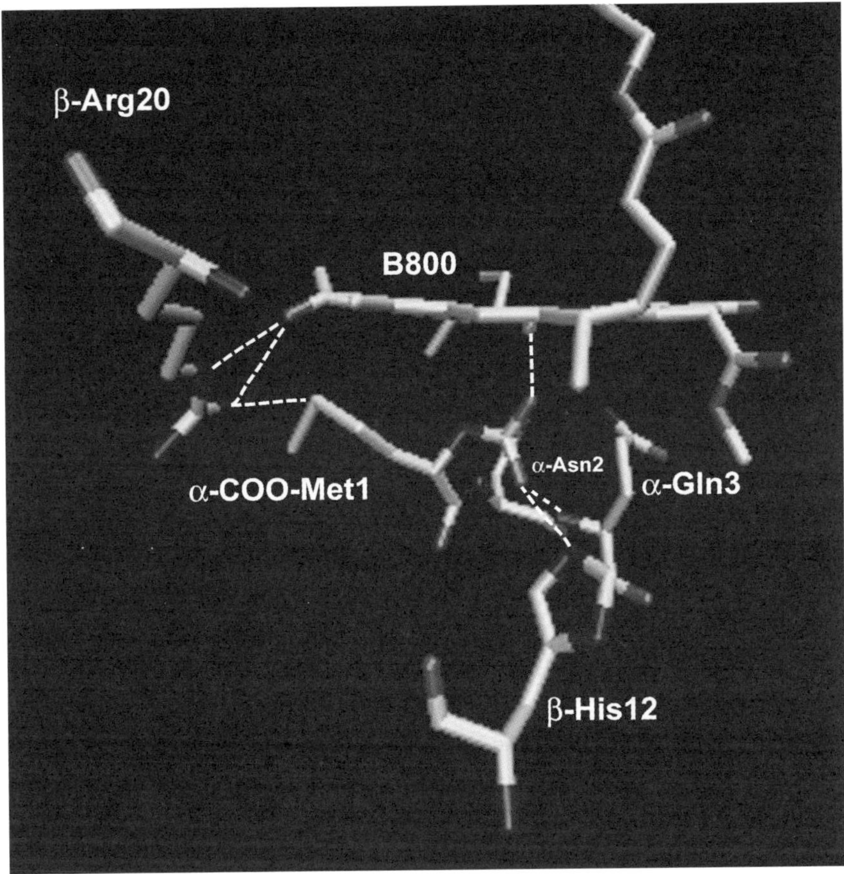

Fig. 3 The hydrogen bonding network (white dashed lines) surrounding a B800 Bchl of LH2. The O1 of the α–Met1 carboxyl group ligates the central Mg^{2+} of the B800 molecule forming a bond 2.04 Å long, while the O2 hydrogen bonds to α–Asn2-N (3.08 Å), α-Gln3-N (2.90 Å) and β–His12-NE2 (2.95 Å). β–Arg20-NE and NH2 form hydrogen bonds with the C3' acetyl OBB of B800 (2.79 Å and 2.97 Å, respectively). A possible weak H-bond of 3.34 Å could exist between the SD atom of COO-α-Met1 and NH_2 of β–Arg20 to provide extra stability to the ligating amino acid (Papiz *et al.*, 2003).

across the face of a bacteriochlorin ring. Instead, it makes three close contacts (of 3.70, 3.68 and 4.13 Å) with the isoprenoid chain of a carotenoid molecule and probably functions to correctly orient the carotenoid within the complex (Prince *et al.*, 1997).

3.4. **Structure and Arrangement of the Carotenoids**

The carotenoids of the LH2 complex function as accessory light-harvesting pigments as well as playing an important role in photoprotection and structure stabilization (Frank & Cogdell, 1995). The original electron density map of the LH2 complex only identified a single molecule of carotenoid – in this case rhodopin glucoside – per $\alpha\beta$-dimer (McDermott *et al.*, 1995). This carotenoid, which is well ordered, has a typical all-*trans* conformation and spans the depth of the complex. The glucoside ring interacts with polar residues at the N-terminus of the α–apoprotein, and the hydrocarbon chain then passes across the face of the β-B850 macrocycle in the adjacent $\alpha\beta$-dimer. In this way, the carotenoid acts as a cross-strut to lock the adjacent $\alpha\beta$-dimers in place within the structure. Viewed down its axis, the carotenoid is twisted and this gives rise to the strong circular dichroism signal observed in the visible region of the spectrum (Cogdell *et al.*, 1997).

A higher resolution structure of LH2 at 2.0 Å revealed the existence of a second rhodopin glucoside molecule per $\alpha\beta$-dimer (Papiz *et al.*, 2003). This second carotenoid lies on the periphery of the complex between the β-polypeptides and is severely bent. To accommodate this bend, two *cis* bonds are required and these have been tentatively assigned to the C12 and C15 carbons of the carotenoid. The second carotenoid is oriented in the opposite direction to the first with its glucoside head group located in the periplasmic, rather than the cytoplasmic, surface of the complex. The glucoside is located in a pocket created by α-Trp40, Ala43-Tyr44 and β-Leu40-His41 with H-bonds to the glucoside hydroxyl groups through a network of eight water molecules. In contrast, the first carotenoid H-bonds directly to α-Lys5 and β-Glu10. The isoprenoid chain of the second carotenoid travels over the outer surface of α-B850 while that of the first terminates at its inner surface. Both carotenoids are in van der Waals contact with three Bchl *a* pigments. In the first carotenoid these contacts are distributed evenly between all the types of Bchl *a* molecules. In the second, the contacts occur mostly on the outer macrocycle surface of α–B850. The presence of a second carotenoid surrounding the B850 pigments offers the complex greater protection against photo-oxidative damage. The stabilizing role played by mutual interactions between the carotenoids and Bchl molecules is underlined in studies using carotenoid deletion mutants of purple bacteria (Griffiths & Stanier, 1956; Zurdo *et al.*, 1993; Lang & Hunter, 1994). These studies have shown that LH2 either fails to assemble or is rapidly turned over in the absence of carotenoid.

4. THE STRUCTURE OF THE RC-LH1 CORE COMPLEX

4.1. Electron Microscopy Studies

RC-LH1 core complexes were first visualized in electron micrographs of the photosynthetic membranes of the Bchl *b*-containing purple bacterium *Rps. viridis* (Miller, 1979; Welte & Kreutz, 1982; Stark *et al.*, 1984). This species is useful for electron microscopy studies as its flat, lamellar photosynthetic membranes contain extensive, quasi-crystalline 2-D arrays of core complexes. 20 Å resolution images of the membranes revealed the RC-LH1 complex to be a roughly circular structure, 100-120 Å in diameter, with hexagonal symmetry. The complex consisted of a core of 45 Å diameter surrounded by a ring. The central core was postulated to represent the RC and the surrounding ring was proposed to be the LH1 antenna that consisted of twelve subunits (Stark *et al.*, 1984). Similar results were obtained from EM studies of the photosynthetic membrane of *Ectothiorhodospira halochloris*, suggesting this type of structure was a feature common to all Bchl *b*-containing photosynthetic membranes (Engelhardt *et al.*, 1983). A much later cryo-electron microscopy study of 2-D crystals of *Rps. viridis* core complex produced a higher resolution (10 Å) projection map of the RC-LH1 structure that confirmed the earlier findings of a twelve subunit LH1 ring surrounding the RC (Ikedu-Yamasaki *et al.*, 1998).

Until very recently, the best information we had about the structure of the LH1 antenna came from an 8.5 Å resolution projection map of the LH1 complex from Bchl *a*-containing *Rs. rubrum* (Karrasch *et al.*, 1995). This appeared to show LH1 as a closed ring consisting of sixteen *αβ*–subunits with a hole in the centre large enough to house a RC *in vivo*. Lower resolution processing of the data produced a pseudo-six-fold symmetry that broke down to eight-fold at higher resolution. It must be noted that in this work the LH1 complexes were reconstituted from their $\alpha\beta$–subunits prior to crystallization and, therefore, the ring-like structure may not reflect the actual *in vivo* structure of LH1. However, this problem was addressed when image analysis of 2-D crystals of the native RC-LH1 complex from a carotenoidless strain of *Rs. rubrum* supported a model of the core complex in which the LH1 ring completely surrounds the RC (Walz & Ghosh, 1997). Further confirmation of such an arrangement came from projection structures of 2-D crystals of RC-LH1 from *Rb. sphaeroides* (Walz *et al.*, 1998).

4.2. The Size of the LH1 Ring

The number of subunits that constitute the LH1 complex has been a matter of debate amongst researchers in the field for a long time. Although the projection maps of LH1 from *Rs. rubrum* and *Rb. sphaeroides* indicated it consisted of sixteen subunits, other studies of core complexes have suggested the LH1 ring is, like that of *Rps. viridis*, composed of twelve $\alpha\beta$–subunits (Boonstra *et al.*, 1994; Meckenstock *et al.*, 1994). Since there is a definite relationship between the size of LH1 and its capacity to completely surround a RC, the stoichiometry of LH1 antenna Bchls per RC provides a method to calculate the LH1 ring size (assuming each $\alpha\beta$–subunit binds two Bchl molecules). Indeed, biochemical methods have been used in studies to determine LH1 ring size in several species of purple bacteria. Some variability was observed in these measurements with values ranging from an average of $33 \pm 4{:}1$ (Gall, 1994) to $25 \pm 2{:}1$ (Francke & Amesz, 1995) being reported. The former data would be consistent with the 16 $\alpha\beta$–subunit model of the LH1 complex described by Karrasch *et al.* (1995) with the latter being consistent with a ring composed of 12 $\alpha\beta$–subunits. These differences have led some to question whether LH1 actually forms a complete ring *in vivo*. Electron microscopy analysis of tubular membranes from an LH2-null mutant of *Rb. sphaeroides* has shown that in this particular strain the LH1 rings are incomplete and form arcs around the RC (Jungas *et al.*, 1999). However, another way to justify a LH1 Bchl:RC ratio of less than 32:1 while simultaneously addressing the possibility of LH1 to completely encircle the RC would be to include an additional component in the ring. A strong body of evidence suggests that a protein, termed PufX, may be such a component in the LH1 antenna of *Rb. capsulatus* and *Rb. sphaeroides* (Farchaus & Oesterhelt, 1989; Lilburn & Beatty 1992; Lilburn *et al.*, 1992; Barz & Oesterhelt, 1994; Pugh *et al.*, 1998).

4.3.   The PufX Protein

PufX is necessary for photosynthetic growth in *Rb. sphaeroides* and *Rb. capsulatus* (Lilburn *et al.*, 1992; Barz & Oesterhelt, 1994) and its presence is a requirement for both efficient ubiquinone/ubiquinol exchange between the RC and cytochrome bc_1 as well as light driven electron-transfer and photophosphorylation (Barz *et al.*, 1995a; 1995b). PufX is only necessary for photosynthetic growth in strains that contain native LH1 antenna. Photosynthetic growth is not compromised by deletion of the *pufX* gene in RC-only mutants or those in which the LH1 antenna is reduced in size (McGlynn *et al.*, 1994). Additionally, suppressor mutants of *Rb. capsulatus* and *Rb. sphaeroides* that lacked *pufX* apparently compensated for its loss by point mutations in the α and β LH polypeptides (Barz & Oesterhelt, 1994). These observations, along with another which showed that the absence of PufX caused LH1 to increase in size by approximately two subunits per RC, support the suggestion that PufX plays an important role in the structural organization of the PSU (Lilburn *et al.*, 1992; Barz *et al.*, 1995a).

The PufX protein from *Rb. capsulatus* and *Rb. sphaeroides* has been isolated and used in reconstitution experiments to examine its effect on LH1 antenna formation *in vitro* (Recchia *et al.*, 1998). The isolated protein was shown to have a specific, high affinity for the LH1 α-polypeptide and was inhibitory to LH1 formation at low concentrations. Further studies demonstrated that it was the putative membrane-spanning region of PufX that caused this inhibition (Parkes-Loach *et al.*, 2001). To explain these results it was proffered that PufX interrupts the molecular architecture of the LH1 ring (presumably at a position adjacent to the Q_B site of the RC) by binding to the LH1 α-polypeptide in the presence of Bchl *a*, thus preventing LH1 from impeding the free passage of ubiquinol from the RC to cytochrome bc_1. Not only has PufX been implicated in organizing a specific orientation of the RC within the LH1 ring but, in LH2-minus mutants of *Rb. sphaeroides* that were grown under dark, partially aerobic conditions, it has also been shown to be responsible for the formation of long-range regular arrays of core complexes in the photosynthetic membrane (Frese *et al.*, 2000). However, the exact structural role of PufX and its interactions with other components of the PSU will probably have to await the arrival of a high resolution, 3-D structure of a PufX-containing core complex.

4.4. The 3-D Structure of RC-LH1

Very recently, our laboratory has published a 4.8 Å resolution 3-D structure of the RC-LH1 core complex from *Rps. palustris* (Roszak *et al.*, 2003). This structure shows the RC surrounded by an oval, rather than circular LH1 complex consisting of 15 $\alpha\beta$–subunits and their associated pigment molecules (Figure 4). The elliptical LH1 complex has approximate dimensions of 110 Å by 95 Å for the outer ring. The longest dimension of the inner LH1 ellipsoid is approximately 78 Å. This allows the RC (whose in-membrane longest dimension is about 70 Å) to be accommodated. The orientation of the long axis of the LH1 ellipse coincides with the long 'axis' of the RC so that LH1 appears to be wrapped tightly around the RC. The LH1 oval is prevented from completely encircling the RC by a single transmembrane helix (called protein W) that is out of register with the array of inner α-apoproteins. The presence of protein W raises the question as to whether it is a 16th α-apoprotein of the LH1 inner ring or a part of a PufX-like protein that acts to facilitate ubiquinone exchange. An equivalent gene for the PufX protein has not been found in the *Rps. palustris* genome. This is not entirely surprising because the PufX protein sequences, even for such two closely related species as *Rb. sphaeroides* and *Rb. capsulatus*, show only 23% identity and even for the α-helical membrane-spanning region of PufX the identity is only 38% (Parkes-Loach *et al.*, 2001). More work is required to unambiguously identify the protein W gene.

The structure reveals a second interesting and important feature at the location of protein W. Inspection of Figure 4 reveals the elliptical LH1-structure to have a unique orientation with respect to the RC. Both

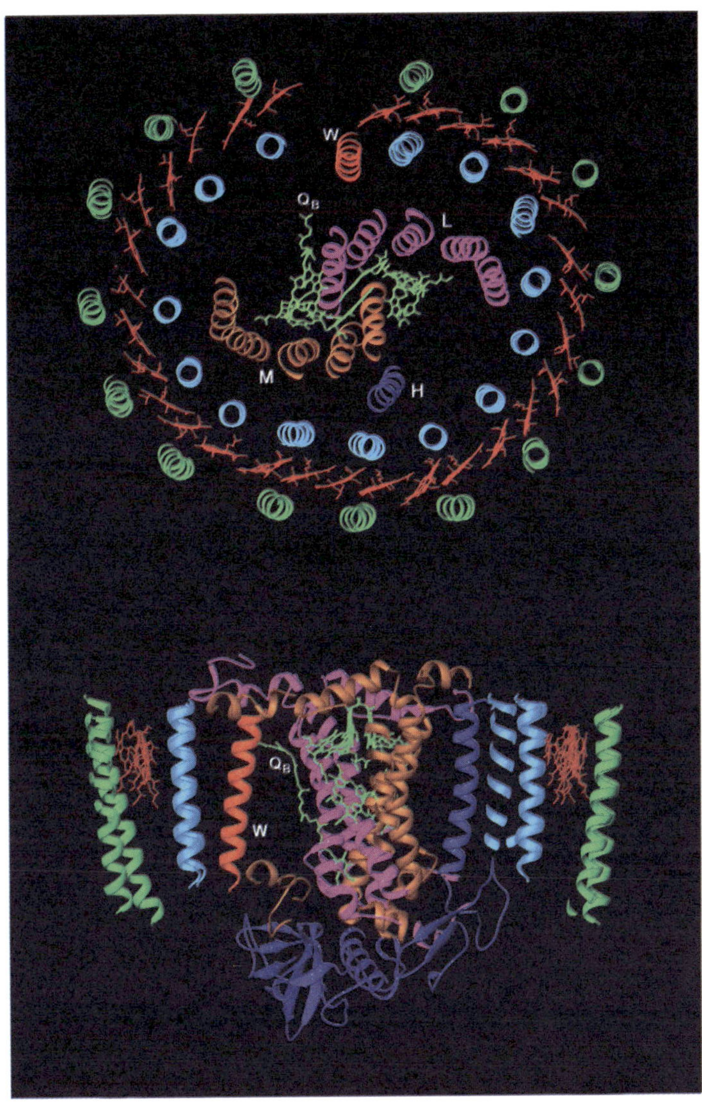

Fig. 4 Views perpendicular (top) and parallel (bottom) to the membrane plane of a schematic model of the RC-LH1 core complex from *Rps. palustris* with the transmembrane helices represented by ribbons and the bacteriochlorophylls and bacteriopheophytins represented by their respective macrocycles. The α–helices are depicted in light blue and the β–helices in green. Helix W and the B880 Bchl molecules are shown in red. The RC H, L and M subunits are shown in blue, mauve and orange, respectively (Roszak *et al.*, 2003). Drawn using the program RIBBONS (Carson, 1997).

the break of the outer ring of β-apoprotein helices and helix W of the LH1 complex are positioned on the opposite side of RC with respect to the single transmembrane helix of the RC H subunit. The latter itself breaks the overall twofold symmetry of the RC. The W helix is therefore located adjacent to the groove in the RC through which the tail of the secondary UQ_B projects. The hydrophobic tail of UQ_B points towards the gap in the LH1 complex next to the W helix, strongly suggesting that it forms a 'portal' through which fully reduced UQ_B can communicate with the UQ pool located in the membrane lipid phase outside the LH1. It further suggests that the location of helix W imparts a significant role in the unique positioning of the RC within the RC-LH1 complex. Concomitantly, the location of the single transmembrane α-helix of the H subunit of the RC indicates that it could also play an important role in the orientation of the RC within the LH1 complex.

Although this structure could explain much of the biophysical and biochemical data obtained for RC-LH1 core complexes, it must be asked if it actually reflects the *in vivo* structure. 10 Å resolution AFM images of core complexes in native membranes from *Rps. viridis* have shown that LH1 consists of a closed ellipsoid of 16 subunits (Scheuring *et al.*, 2003). However, the low resolution of the AFM images prevents visualisation of the break in the LH1 structure. Interestingly, these experiments have also shown that the LH1 subunits rearrange into a circular ring structure after removal of the RC from the core complex. These AFM studies clearly show that the oval structure for the RC-LH1 complex from *Rps. palustris* observed in the crystalline state is indicative of its *in vivo* condition.

5. ENERGY TRANSFER WITHIN THE PSU

The relative wealth of detailed structural information available for the components of the bacterial PSU in combination with sophisticated spectroscopic techniques has allowed dissection of the energy transfer events that occur upon absorption of a photon by LH2 and the subsequent separation of charge in the RC. Energy transfer within the PSU is a directed process that is 'guided' by the energy gradient going from LH2 to LH1 and then to the RC (B800$\rightarrow$B850$\rightarrow$B875) (Pullerits &

Sundström, 1996; Fleming & van Grondelle, 1997). In this way the photosynthetic light-harvesting system acts as a funnel to direct excitation energy to the RC and it is this directionality that is the key to its efficiency. The importance of this type of directionality is illustrated by large, artificial antenna systems that lack it. In these cases, once a critical size is exceeded, the absorbed light energy gets 'lost' and never performs useful work (Swallen *et al.*, 1999).

Examination of the structure of LH2 reveals an organization that is beautifully adapted to optimize the orientation of the Bchl *a*s for a rapid and efficient energy transfer. Kinetic studies using low energy pump-probe techniques have shown that energy transfer from B800 to B850 takes place with a time constant of 0.9 ps at room temperature (Kennis *et al.*, 1996). This energy transfer reaction is remarkably temperature insensitive and decreases to only 1.8 ps at 77K and 2.4 ps at 1.4 K (Reddy *et al.*, 1993; Vulto *et al.*, 1999). Measurements of the decay of the anisotropy of the excited B800 population at room temperature indicated a rapid, but limited B800 to B800 energy transfer step. The time constant for this decay was 0.3 ps, which translated into a B800 to B800 transfer time of about 0.5 ps (Sundström *et al.*, 1999).

When the excitation energy arrives at the B850 ring it can remain there for >1ns provided no other antenna complex is nearby (van Grondelle *et al.*, 1994). The B850 molecules are strongly interacting and the excited state is rapidly delocalized, with transfers between Bchls occurring on the 50-150 fs timescale (De Caro *et al.*, 1994). The extent of this delocalization is not known but a consensus view at present is that it is probably delocalized over just a part of the ring (Alden *et al.*, 1996; Koolhaas *et al.*, 1996; Sauer *et al.*, 1996). However, the exact number of Bchl *a* molecules involved is controversial and much more experimentation needs to be done to resolve this issue.

The rapid delocalization of excitation energy around the circumference of the LH2 ring has important implications for the function of the PSU. Within the excited state lifetime of LH2, every B850 molecule has the possibility of being visited by the excited state many times and so the rings of B850 molecules can be thought of as 'storage rings'. As a result of this, the probability of energy transfer out of the ring is equal from each and every B850 molecule. This means that energy is available for transfer from any part of the ring to any part of a neighbouring ring, provided they are close enough. As such, a precise arrangement of LH2 and LH1 antenna complexes is not a prerequisite for the efficient function of the PSU. LH1 complexes simply need to lie close to excited LH2 complexes for energy transfer to occur with high efficiency.

The next energy transfer step, from B850 of LH2 to B880 of LH1, has a multi-exponential rate with the major phase having a time constant of 2-4 ps (Sundström *et al.*, 1999). Once again, when the energy reaches the circular array of B880 molecules it is rapidly delocalized, with B880 to B880 transfers occurring in about 80 fs (Visser *et al.*, 1995). The final and slowest energy transfer step, from B880 to the RC, occurs in about 30-50 ps (Otte *et al.*, 1993; Kennis *et al.*, 1994; van Grondelle *et al.*, 1994). The relatively slow rate of this transfer is a consequence of the distance between the RC special pair Bchls (P870) and the antenna Bchls. Why is this transfer rate so slow, and why are the LH1 Bchl *a* molecules not positioned nearer to the RC special pair Bchls? Oxidation of just a single Bchl *a* molecule in LH1 results in a strong quenching of the fluorescence yield (which is equivalent to the singlet excited state lifetime) thereby preventing LH1 from acting as an effective antenna for the RC (Law & Cogdell, 1998). As oxidized P870 is strong enough to oxidize antenna Bchl *a* molecules, it makes sense for the LH1 antenna Bchls to be located sufficiently far away to prevent this oxidation from occurring. Therefore, the actual positioning of the LH1 antenna Bchls relative to P870 is a compromise: they are sufficiently close enough to allow efficient energy transfer but not too close as to allow the possibility of electron transfer.

The exact mechanisms of the LH2 to LH1 to RC energy transfer steps are all assumed to follow a simple Förster process. Although some experimentation has been performed to test this (Hess *et al.*, 1994; Pullerits & Sundström, 1996) more critical testing is required to determine if this is the true mechanism.

6. CONCLUSION

We hope this review has shown that purple bacteria provide a very useful model system both to study the processes of light-harvesting and to give us clues as to the design principles required for the *de novo* construction of efficient artificial light-harvesting complexes. We expect to see real progress in this endeavor over the next few years.

ACKNOWLEDGMENTS

The authors would like to thank the BBSRC, NEDO and the Wellcome Trust for funding.

References

1. Aagard J and Sistrom WR. Control of synthesis of reaction centre bacteriochlorophyll in photosynthetic bacteria. *Photochem. Photobiol.* 1972; **15**: 209-225.
2. Alden RG, Johnson E, Nagarajan V, Parson WW, Law CJ and Cogdell RJ. Calculations of the spectroscopic properties of the LH2 bacteriochlorophyll-protein antenna complex from *Rhodopseudomonas acidophila. J. Phys. Chem B.* 1996; **101**: 4667-4680.
3. Allen, JP, Feher G, Yeates TO, Komiya H and Rees DC. Structure of the reaction center from *Rhodobacter sphaeroides* R-26: the co-factors. *Proc. Natl. Acad. Sci. USA* 1987; **84**: 5730–5734.
4. Barz WP and Oesterhelt D. Photosynthetic deficiency of a pufX deletion mutant of *Rhodobacter sphaeroides* is suppressed by point mutations in the light-harvesting complex genes pufB or pufA. *Biochemistry* 1994; **33**: 9741-9752.
5. Barz WP, Vermeglio A, Francia F, Venturoli G, Melandri BA and Oesterhelt D. Role of the PufX protein in photosynthetic growth of *Rhodobacter sphaeroides*. 2. PufX is required for the efficient ubiquinone/ubiquinol exchange between the reaction centre Q_B site and the cytochrome bc_1 complex. *Biochemistry* 1995a; **34**: 15248-15258.
6. Barz WP, Francia F, Venturoli G, Melandri BA, Vermeglio A and Oesterhelt D. Role of the PufX protein in photosynthetic growth of *Rhodobacter sphaeroides*. 1. PufX is required for efficient light-driven

electron transfer and photophosphorylation under anaerobic conditions. *Biochemistry* 1995b; **34**: 15235-15247.

7. Boonstra AF, Germeroth L and Boekma EJ. Structure of the light-harvesting antenna from *Rhodospirillum molischianum* studied by electron microscopy. *Biochimica et Biophysica Acta* 1994; **1184**: 227-234.

8. Brunisholz RA, Wiemken V, Suter F, Bachofen R and Zuber H. The light-harvesting polypeptides of *Rhodospirillum rubrum*: II. Localization of amino terminal regions of the light-harvesting polypeptides B870-α and B870-β and the reaction center subunit L at the cytoplasmic side of the photosynthetic membrane of *Rs. rubrum* G-9$^+$. *Hoppe-Seyler's Z. Physiol. Chem.* 1984; **365**: 689-701.

9. Carson, M. RIBBONS. *Methods Enzymol.* 1997; **277**:493-505.

10. Cogdell RJ and Scheer H. Circular dichroism of light-harvesting complexes from purple photosynthetic bacteria. *Photochem. Photobiol.* 1985; **42**: 669-678.

11. Cogdell RJ, Lindsay JG, Valentine J and Durant I. A further characterisation of the B890 light-harvesting pigment-protein complex from *Rhodospirillum rubrum* strain S1. *FEBS Lett.* 1982; **150**: 151-154.

12. Cogdell RJ, Durant I, Valentine J, Lindsay JG and Schmidt K. The isolation and partial characterisation of the light-harvesting pigment protein complement from *Rhodopseudomonas acidophila*. *Biochimica et Biophysica Acta* 1983; **722**: 427-455.

13. Cogdell RJ, Zuber H, Thornber PJ, Drews G, Gingras G, Niederman RA, Parson WW and Feher G. Recommendations for the naming of photochemical reaction centres and light-harvesting pigment-protein complexes from purple photosynthetic bacteria. *Biochimica et Biophysica Acta* 1985; **806**: 185-186.

14. Cogdell RJ, Isaacs NW, Freer AA, Arrelano J, Howard TD, Papiz MZ, Hawthornthwaite-Lawless AM and Prince SM. The structure and function of the LH2 (B800-850) complex from the purple photosynthetic bacterium *Rhodopseudomonas acidophila* strain 10050. *Prog. Biophys. Molec. Biol.* 1997; **68**: 1-27.

15. Cogdell RJ, Fyfe PK, Howard TD, Fraser N, Isaacs NW, Freer AA, McLuskey K and Prince SM. The structure and function of the LH2 complex from *Rhodopseudomonas acidophila* strain 10050, with special reference to the bound carotenoid. In: *The Photochemistry of Carotenoids* (Frank HA, Young AJ, Britton G and Cogdell RJ, eds.). 1999; pp. 71-80. Kluwer Academic Publishers, Dordrecht.

16. De Caro C, Visschers RW, van Grondelle R and Volker S. Inter- and intraband energy transfer in LH2 antenna complexes of purple bacteria. A fluorescence line-narrowing and hole-burning study. *J. Phys. Chem.* 1994; **98**: 10584-10590.

17. Deisenhofer J, Epp O, Miki K, Huber R and Michel H. The structure of the protein subunits in the photosynthetic reaction centre of *Rhodopseudomonas viridis* at 3Å resolution. *Nature* 1984; **318**: 618–624.

18. Drews G. Structure and functional organization of light-harvesting complexes and photochemical reaction centers in membranes of phototrophic bacteria. *Microbiol. Revs.* 1985;**49**: 59-70.

19. Emerson R and Arnold WA. The photochemical reaction in photosynthesis. *J. Gen. Physiol.* 1932; **16**: 191-205.

20. Engelhardt H, Engel A and Baumeister W. Stoichiometric model of the photosynthetic unit of *Ectothiorhodospira halochloris*. *Proc. Natl. Acad. Sci. USA* 1986; **83**: 8972–8976.

21. Farchaus JW and Oesterhelt D. A *Rhodobacter sphaeroides* pufL, M and X deletion mutant and its complementation in trans with a 3.5kb puf operon shuttle fragment. *EMBO J.* 1989; **8**: 47-54.

22. Fleming GR and van Grondelle R. Femtosecond spectroscopy of photosynthetic light-harvesting systems. *Curr. Op. Struct. Biol.* 1997; **7**: 738-748.

23. Förster T. Intermolecular energy transfer and fluorescence. *Annals of Physics* 1948; **2**: 55-75.

24. Fowler GJS, Visschers RW, Grief GG, van Grondelle R and Hunter CN. Genetically modified photosynthetic antenna complexes with blue shifted absorbance bands. *Nature* 1992; **355**: 848-850.

25. Francke C and Amesz J. The size of the photosynthetic unit in purple bacteria. *Photosynth. Res.* 1995; **46**: 347-352.

26. Frank HA and Cogdell RJ. Carotenoids in photosynthesis. *Photochem. Photobiol.* 1995; **63**: 257-264.

27. Frank HA, Desamero RZB, Chynwat V, Gebhard R, van der Hoef I, Jansen FJ, Lugtenburg J, Gosztola D and Wasielewski R. Spectroscopic properties of sphaeroidene analogs having different extents of π–electron conjugation. *J. Phys. Chem A.* 1997a; **101**: 149-157.

28. Frank HA, Chynwat V, Desamero RZB, Farhoosh R, Erickson J and Bautista JA. On the photophysics and photochemical properties of carotenoids and their role as light-harvesting pigments in photosynthesis. *Pure & Applied Chemistry* 1997b; **69**: 2117-2124.

29. Freer A, Prince S, Sauer K, Papiz M, Hawthornthwaite-Lawless A McDermott G, Cogdell RJ and Isaacs NW. Pigment-pigment interactions and energy transfer in the antenna complex of the photosynthetic bacterium *Rhodopseudomonas acidophila*. *Structure* 1996; **4**: 449-462.

30. Frese RN, Olsen JD, Branwall R, Westerhuis WHJ, Hunter CN and van Grondelle R. *Proc. Natl. Acad. Sci. USA* 2000; **97**: 5197–5202.

31. Gall, A. Purification, characterisation and crystallisation from a range of *Rhodospirillineae* pigment-protein complexes. 1994; PhD Thesis, University of Glasgow.

32. Griffiths M and Stanier RY. Some mutational changes in the photosynthetic pigment system of *Rhodopseudomonas sphaeroides*. *J. Gen. Microbiol.* 1956; **14**: 698-715.

33. Halloren E, McDermott G, Lindsay JG, Miller C, Freer AA, Isaacs NW and Cogdell RJ. Studies on the light-harvesting complexes from the thermotolerant purple bacterium *Rhodopseudomonas cryptolactis*. *Photosynth. Res.* 1995; **44**: 149-155.

34. Haxo FT, Kycia JH, Somers GF, Bennett A and Siegelman HW. Peridinin-chlorophyll *a* proteins of the dinoflagellate *Amphidinium carterae* (Plymouth 450). *Plant Physiol.* 1976; **57**: 297-303.

35. Hess S, Visscher KJ, Sundström V, Fowler GJS and Hunter CN. Enhanced rates of subpicosecond energy transfer in blue-shifted light-harvesting LH2 mutants of *Rhodobacter sphaeroides*. *Biochemistry* 1994; **33**: 8300-8305.

36. Hofmann E, Wrench PM, Sharples FP, Hiller RG, Welte W and Diederichs K. Structural basis of light harvesting by carotenoids: Peridinin-chlorophyll-protein from *Amphidinium carterae*. *Science* 1996; **272**: 1788-1791.

37. Humphrey W, Dalke A and Schulten K. VMD- Visual Molecular Dynamics. *J. Molecular Graphics* 1996; **14**: 33-38.

38. Hunter CN, van Grondelle R and Olsen JD. Photosynthetic antenna proteins: 100ps before photochemistry starts. *Trends Biochem. Sci.* 1989; **14**: 72-76.

39. Ikedu-Yamasaki I, Odahara T, Mitsuoka K, Fujiyoshi Y and Murata K. Projection map of the reaction center-light harvesting 1 complex from *Rhodopseudomonas viridis* at 10Å resolution. *FEBS Lett.* 1998; **425**: 505-508.

40. Jordan P, Fromme P, Witt HT, Klukas O, Saenger W and Krauss N. Three-dimensional structure of cyanobacterial photosystem I at 2.5 Å resolution. *Nature* 2001; **411**: 909-917.

41. Jungas C, Ranck JL, Rigaud JL, Joliot P and Vermeglio A. Supramolecular organisations of the photosynthetic apparatus of *Rhodobacter sphaeroides*. *EMBO J.* 1999; **18**: 534-542.

42. Kennis JTM, Aartsma TJ and Amesz J. Energy trapping in the purple sulphur bacteria *Chromatium vinosum* and *Chromatium tepidum*. *Biochimica et Biophysica Acta* 1994; **1188**: 278-286.

43. Kennis JTM, Streltsov AM, Aartsma TJ, Nozawa T and Amesz J. Energy transfer and exciton coupling in isolated B800-850 complexes of the photosynthetic purple sulfur bacterium *Chromatium tepidum*. The effect of structural symmetry on bacteriochlorophyll excited states. *J. Phys. Chem. B.* 1996; **100**: 2438-2442.

44. Koepke J, Hu X, Muenke C, Schulten K and Michel H. The crystal structures of the light-harvesting complex II (B800-850) from *Rhodospirillum molischianum*. *Structure* 1996; **4**: 581-597.

45. Koolhaas MHC, van der Zwan G, Frese RN and van Grondelle R. Structure-based calculations of absorption and CD spectra of the LH2 antenna system of *Rhodopseudomonas acidophila*. *J. Phys. Chem. B.* 1996; **101**: 7262-7270.

46. Kramer HJM, van Grondelle R, Hunter CN, Westerhuis WHJ and Amesz J. Pigment organization of the B800-850 antenna complex of Rhodopseudomonas sphaeroides. *Biochimica et Biophysica Acta* 1984; **765**: 156-165.

47. Lang HP and Hunter CN. The relationship between carotenoid biosynthesis and the assembly of light-harvesting LH2 complex in *Rhodobacter sphaeroides*. *Biochem. J.* 1994; **298**: 197-205.

48. Law CJ and Cogdell RJ. The effect of chemical oxidation on the fluorescence of the LH1 (B880) complex from the purple bacterium *Rhodobium marinum*. *FEBS Lett.* 1998; **432**: 27-30.

49. Lilburn TG, Haith CE, Prince RC and Beatty JT. Pleiotropic effects of pufX gene deletion on the structure and function of the photosynthetic apparatus of *Rhodobacter capsulatus*. *Biochimica et Biophysica Acta* 1992; **1100**: 160-170.

50. McDermott G, Prince SM, Freer AA, Hawthornthwaite-Lawless AM, Papiz M, Cogdell RJ and Isaacs NW. Crystal structure of an integral membrane light-harvesting complex from photosynthetic bacteria. *Nature* 1995; **374**: 517-521.

51. McGlynn P, Hunter CN and Jones MR. The *Rhodobacter sphaeroides* PufX protein is not required for photosynthetic competence in the absence of a light harvesting system. *FEBS Lett.* 1994; **349**: 349-253.

52. McLuskey K, Prince SM, Cogdell RJ and Isaacs NW. The crystallographic structure of the B800-820 light-harvesting LH3 light-harvesting complex from *Rhodopseudomonas acidophila* strain 7050. *Biochemistry* 2001; **40**: 8783-8789.

53. Meckenstock RU, Krusche K, Staehlin LA, Cyrklaff M and Zuber H. The six-fold symmetry of the B880 light-harvesting complex and the structure of the photosynthetic membranes of *Rhodopseudomonas marina*. *Biol. Chem. Hoppe-Seyler.* 1994; **375**: 429-438.

54. Miller KR. Structure of a bacterial photosynthetic membrane. *Proc. Natl. Acad. Sci. USA* 1979; **76**: 6415–6419.

55. Olsen JD and Hunter CN. Protein structure modelling of the bacterial light-harvesting complex. *Photochem. Photobiol.* 1994; **60**: 521-535.

56. Otte SCM, Kleinherenbrink FAM and Amesz J. Energy transfer between the reaction centre and the antenna in purple bacteria. *Biochimica et Biophysica Acta* 1993; **1143**: 84-90.

57. Papiz M, Prince SM, Howard T, Cogdell RJ and Isaacs NW. The structure and thermal motion of the B800-850 LH2 complex from *Rps. acidophila* at 2.0Å resolution and 100K: New structural features and functionally relevant motions. *J. Mol. Biol.* 2003; **326**: 1523-1538.

58. Parkes-Loach PS, Law CJ, Recchia PA, Kehoe J, Nehrlich S, Chen J and Loach PA. Role of the core region of the PufX protein in inhibition of reconstitution of the core light-harvesting complexes of *Rhodobacter sphaeroides* and *Rhodobacter capsulatus*. *Biochemistry* 2001; **40**: 5593-5601.

59. Permentier HP, Neerkin S, Overmann J and Amesz J. A bacteriochlorophyll *a* antenna complex from purple bacteria absorbing at 963nm. *Biochemistry* 2001; **40**: 5573-5578.

60. Prince SM, Papiz MZ, Freer AA, McDermott G, Hawthornthwaite-Lawless AM, Cogdell RJ and Isaacs NW. Apoprotein structure in the LH2 complex from *Rhodopseudomonas acidophila* strain 10050: Modular assembly and protein-pigment interactions. *J. Mol. Biol.* 1997; **268**: 412-423.

61. Prince SM, Howard TD, Myles DAA, Wilkinson C, Papiz MZ, Freer AA, Cogdell RJ and Isaacs NW. Detergent structure in crystals of the integral membrane light-harvesting complex LH2 from *Rhodopseudomonas acidophila* strain 10050. *J. Mol. Biol.* 2003; **326**: 307-315.

62. Pugh RJ, McGlynn P, Jones MR and Hunter CN. The LH1-RC core complex of *Rhodobacter sphaeroides*: Interaction between components, time dependent assembly, and topology of the PufX protein. *Biochimica et Biophysica Acta* 1998; **1366**: 301-316.

63. Pullerits T and Sundström V. Photosynthetic light-harvesting pigment-protein complexes: toward understanding how and why. *Accounts of Chemical Research* 1996; **29**: 381-389.

64. Recchia PA, Davis CM, Lilburn TG, Beatty JT, Parkes-Loach PS, Hunter CN and Loach PA. Isolation of the PufX protein from *Rhodobacter capsulatus* and *Rhodobacter sphaeroides*: Evidence for its interaction with the α–polypeptide of the core light-harvesting complex. *Biochemistry* 1998; **37**: 11055-11063.

65. Reddy NRS, Cogdell RJ and Small GJ. Non photochemical hole burning of the B800-850 complex of *Rhodopseudomonas acidophila*. *Photochem. Photobiol.* 1993; **37**: 35-39.

66. Robert B and Lutz M. Structure of antenna complexes of several *Rhodospirillales* from their resonance Raman spectra. *Biochimica et Biophysica Acta* 1985; **807**: 10-23.

67. Roszak AW, Howard TD, Southall J, Gardiner AT, Law CJ, Isaacs NW and Cogdell RJ. Crystal structure of RC-LH1 core complex from *Rhodopseudomonas palustris*. *Science* 2003; **302**: 1969-1972.

68. Sauer K, Cogdell RJ, Prince SM, Freer AA, Isaacs NW and Scheer H. Structure-based calculations of the optical spectra of LH2 bacteriochlorophyll-protein complexes from *Rhodopseudomonas acidophila*. *Photochem. Photobiol.* 1996; **64**: 564-576.

69. Scheuring S, Seguin J, Marco S, Levy D, Robert B and Rigaud J-L. Nanodissection and high-resolution imaging of the *Rhodopseudomonas viridis* photosynthetic core complex in native membranes by AFM. *Proc. Natl. Acad. Sci. USA* 2003; **100**: 1690–1693.

70. Sharples FP, Wrench PM, Ou K and Hiller RG. Two distinct forms of the peridinin-chlorophyll a-protein from *Amphidinium carterae*. *Biochimica et Biophysica Acta* 1996; **1276**: 117-123.

71. Shreve AP, Trautman JK, Frank HA, Owens TG and Albrecht AC. Femtosecond energy-transfer processes in the B800-850 light-harvesting complex of *Rhodobacter sphaeroides* 2.4.1. *Biochimica et Biophysica Acta* 1991; **1058**: 280-288.

72. Sistrom, WR. Control of antenna pigment components in photosynthetic bacteria. In: *The Photosynthetic Bacteria* (Clayton RE and Sistrom WR, eds.) 1978. pp. 841-848. Plenum Press, New York.

73. Stadtwald-Demchick R, Turner FR and Gest H. *Rhodopseudomonas cryptolactis*, sp. nov., a new thermotolerant species of budding phototrophic purple bacteria. *FEMS Microbiol. Lett.* 1990; **71**: 177-182.

74. Stark W, Kühlbrandt W, Wildhaber, I, Wehrli E and Mühlethaler K. The structure of the photoreceptor unit of *Rhodopseudomonas viridis*. *EMBO J.* 1984; **3**: 777-783.

75. Sturgis JN, Jiraskova V, Reiss-Husson F, Cogdell RJ and Robert B. Structure and properties of the bacteriochlorophyll binding site in peripheral light-harvesting complexes of purple bacteria. *Biochemistry* 1995; **34**: 517-523.

76. Sundström V, Pullerits T and van Grondelle R. Photosynthetic light harvesting: Reconciling dynamics and structure of purple bacterial LH2 reveals function of photosynthetic unit. *J. Phys. Chem B.* 1999; **103**: 2327-2346.

77. Swallen SF, Kopelman R and Moore JS. Excited rate dynamics in organic dendrimer supermolecules. *Electrochemical Society Proceedings* 1999; **98-25**: 85-92.

78. Theiler R and Zuber H. The light-harvesting polypeptides of *Rhodopseudomonas sphaeroides* R-26-1. II. Conformational analyses by attenuated total reflection infrared spectroscopy and possible molecular structure of the hydrophobic domain of the B850 complex. *Hoppe-Seyler's Z. Physiol. Chem.* 1984; **365**: 721-729.

79. Thornber PJ, Trosper T and Strouse CE. Bacteriochlorophyll *in vivo*: relationships of spectral forms to specific membrane components. In: *The Photosynthetic Bacteria* (Clayton RE and Sistrom WR, eds.) 1978. pp. 133-160. Plenum Press, New York.

80. Van Gemerden H and Mas J. Ecology of phototrophic sulfur bacteria. In: *Anoxygenic Photosynthetic Bacteria* (Blankenship RE, Madigan MT and Bauer CE, eds.) 1995. pp. 49-85. Kluwer, Dordrecht.

81. Van Grondelle R, Decker JP, Gillbro T and Sundström V. Energy transfer and trapping in photosynthesis. *Biochimica et Biophysica Acta Bioenergetics* 1994; **1187**: 1-65.

82. Visser HM, Somsen OJG, van Mourik F, Lin S, van Stokkum IHM and van Grondelle R. Direct observation of sub-picosecond equilibration of

excitation energy in the light-harvesting antenna of *Rhodospirillum rubrum.*
Biophys. J. 1995; **69**: 1083-1099.

83. Vulto SIE, Kennis JTM, Streltsov AM, Amesz J and Aartsma TJ. Energy relaxation within the B850 absorption band of the isolated light-harvesting complex LH2 from *Rhodopseudomonas acidophila* at low temperature. *J. Phys. Chem B.* 1999; **103**: 787-883.

84. Walz T and Ghosh R. Two-dimensional crystallization of the light-harvesting I-reaction centre photounit from *Rhodospirillum rubrum. J. Mol. Biol.* 1997; **265**: 107-111.

85. Walz T, Jamieson SJ, Bowes CM, Bullough PA and Hunter CN. Projection structures of three photosynthetic complexes from *Rhodobacter sphaeroides*: LH2 at 6Å, LH1 and RC-LH1 at 25Å. *J. Mol. Biol.* 1998; **282**: 833-845.

86. Welte W and Kreutz W. Formation, structure and composition of a planar hexagonal lattice composed of specific protein-lipid complexes in the thylakoid membranes of *Rhodopseudomonas viridis. Biochimica et Biophysica Acta* 1982; **692**: 479-488.

87. Zankel KL, Reed DW and Clayton RK. Fluorescence and photochemical quenching in photosynthetic reaction centers. *Proc. Natl. Acad. Sci. USA* 1968; **61**: 1243.

88. Zuber H. Structure and function of light-harvesting complexes and their polypeptides. *Photochem. Photobiol.* 1985a; **42**: 821-844.

89. Zuber H. Structural organisation of tetrapyrrole pigments in light-harvesting pigment-protein complexes. In: *Optical Properties and Structure of Tetrapyrroles* (Blauer G and Sund H, eds.) 1985b; pp. 425-441. Walter de Gruyter, Berlin.

90. Zuber H and Brunisholz RA. Structure and function of antenna polypeptides and chlorophyll-protein complexes: Principles and variability. In: *The Chlorophylls* (Scheer, H ed.) 1991; pp. 627-703. CRC Press, Boca Raton, Florida.

91. Zuber H and Cogdell RJ. Structure and organisation of purple bacterial antenna complexes. In: *Anoxygenic Photosynthetic Bacteria* (Blankenship RE, Madigan MT and Bauer CE, eds.) 1995; pp315-348. Kluwer, Dordrecht.

92. Zurdo J, Cabrera-Fernandez C and Ramirez JM. A structural role of the carotenoid in the light-harvesting II protein of *Rhodobacter capsulatus. Biochem. J.* 1993; **290**: 531-537.

Chapter 4

REGULATION OF LIGHT HARVESTING IN PHOTOSYSTEM II OF PLANTS, GREEN ALGAE AND CYANOBACTERIA

Norman P.A. Huner, Kenneth E. Wilson, Ewa Miskiewicz, Denis P. Maxwell, Gordon R. Gray, Marianna Krol, Alexander G. Ivanov

Photosystem II of terrestrial plants, green algae and cyanobacteria exhibit specialized light harvesting systems associated with their thylakoid membranes. In eukaryotes, these light harvesting complexes are integral membrane pigment-protein complexes whereas in cyanobacteria these structures are extrinsic pigment-protein complexes called phycobilisomes. These structures not only act as molecular antenna to harvest light energy but also act to regulate energy transfer to photosystem II reaction centres. We summarize the dynamic nature of light harvesting in eukaryotes and prokaryotes in response to short-term or long-term changes in light intensity, temperature and nutrient availability. This is discussed in the context of photoprotection and the requirement for photostasis, that is, a balance in budget. We conclude that the photosynthetic apparatus has a dual function: not only does it function as an essential energy transformer, it also functions as a primary environmental sensor in plants, green algae and cyanobacteria.

Keywords: Photosynthesis, photosystem II, light-harvesting complexes, reaction centre, photostasis, photoprotection, acclimation.

1. INTRODUCTION

Life is an endergonic process characterized by structural and functional order, the maintenance of which reflects homeostasis. The complex, integrated metabolic pathways characteristic of all living cells not only represent the mechanism by which cells biosynthesize and degrade a multitude of cellular constituents but also represent the mechanism by which cells regulate and transform energy within the cell. The ultimate source of this energy for most of the biosphere is the sun. Photoautotrophs link all other organisms to the sun through their ability to use visible light as their sole energy source and CO_2 as their sole source of carbon. The mechanism for the reduction of CO_2 derived though photosynthesis, in turn, sustains all living organisms with respect to most of their bioenergetic and structural needs. Thus, photosynthesis is considered the single most important process on earth.

The thylakoid membrane systems of plants, green algae and cyanobacteria contain specialized photosystems to absorb and trap light energy which is subsequently transformed into usable chemical energy in the form of ATP and NADPH for the biochemical reduction of C, N and S. The major light harvesting complexes (LHC) are supramolecular, integral membrane, pigment-protein complexes associated with photosystem I (PSI) and photosystem II (PSII) to which the photosynthetic pigments are bound non-covalently. The bulk of the chlorophyll *a* (Chl *a*), chlorophyll *b* (Chl *b*) and carotenoids found in most plants and green algae are bound to the LHC *b* and LHC *a* family of nuclear encoded, LHC polypeptides associated with PSII and PSI respectively (Jansson, 1994; Green and Durnford, 1996). The LHCs associated with PSI and PSII, called LCHI and LHCII respectively, bind Chl *a*, Chl *b* and carotenoids and can be physically separated and purified by non-denaturing detergent extraction.

Upon removal of LHCII, the remaining PSII consists of the core antenna polypeptides of PsbB (CP47) and PsbC (CP43) which bind only chlorophyll *a* and caroteniods. This core antenna is associated with the PSII reaction center polypeptides PsbA (D1), PsbD (D2) which bind the redox carriers P680, Pheo, Q_A and Q_B. Thus, PSII is considered to be a Q-type or Type 2 reaction centre (Blankenship 2002) and was crystallized from the cyanobacterium, *Synechococcus elongatus*, to a 3.8 Å resolution (Zouni *et al.*, 2001). In contrast to PSII, removal of LHCI leaves the PSI reaction center polypeptides PsaA and PsaB which not only bind redox carriers but also bind the core antenna chlorophyll *a* and carotenoids associated with PSI. Thus, PSI is considered to be a Fe-S

Type or Type 1 reaction centre (Blankenship, 2002). The structure of cyanobacterial PSI has been resolved to 2.5 Å resolution by x-ray crystallography resolution (Jordan *et al.*, 2001).

The primary photochemical reactions of photosynthesis are dependent upon the photophysical processes of light absorption and energy transfer within the LHC pigment bed and the subsequent exciton migration from the light harvesting pigments, through the core antenna to reaction center pigments, P680 and P700 of PSII and PSI respectively. Excitation energy transfer occurs on a timescale of femtoseconds to picoseconds (10^{-15} to 10^{-12} s) which makes this probably the fastest process in biology. When the excitation energy reaches either a PSII or a PSI reaction center, photo-oxidation of PSII (P680 $\rightarrow$ P680$^+$)and PSI (P700 $\rightarrow$ P700$^+$) occurs on a timescale of nanoseconds to microseconds (10^{-9} to 10^{-6} s). The electrons generated by the photo-oxidation of P700 are used to reduce NADP$^+$ to NADPH via ferredoxin (Fdx) and the enzyme, FNR. Photosynthetic reducing power is consumed in a variety of biosynthetic reactions including CO_2 fixation through the reductive pentose phosphate cycle (RPPC), nitrate and sulphate reduction as well as lipid biosynthesis.

However, to process excitation energy from the light harvesting and core antenna pigments on a continuous basis, a cycle of photo-oxidation followed by reduction of the reaction centers of PSI and PSII occurs continuously by linear photosynthetic electron transport. Cyt b_6f is the major thylakoid complex which connects PSII photochemistry with PSI photochemistry while concomitantly contributing to the generation of the transthylakoid ΔpH. Recently, this complex was crystallized to a 3.0 Å resolution from both the thermophilic cyanobacterium, *Mastigocladus luminous* (Kirusi *et al.*, 2003) as well as the model green alga, *Chlamydomonas reinhardtii* (Strobel *et al.*, 2003). Electrons from Q_A^- are transferred via Q_B and convert plastoquinone (PQ) to plastoquinol (PQH_2) which is subsequently oxidized by the Cyt b_6f complex. This is considered to be the rate-limiting step in photosynthetic electron transport and occurs on the time scale of milliseconds (10^{-3} s) (Haehnel, 1984). P700$^+$ oxidizes the Cyt b_6f complex via plastocyanin (PC) which converts P700$^+$ toP700. The oxidation of PQH_2 via the Cyt b_6f complex occurs concomitantly with the vectorial transport of protons from the stroma to the thylakoid lumen. In the case of PSII, $P680^+$ exhibits a sufficiently positive reduction potential to oxidize H_2O *via* the lumenal oxygen evolving complex (OEC). This results in the reduction of $P680^+$ to P680, the evolution of O_2 and the release of protons into the thylakoid lumen. The proton gradient generated by photosynthetic electron transport is used for the chemiosmotic synthesis of ATP by the chloroplast H^+-dependent ATP synthase. Thus light energy initially captured by the photosynthetic light harvesting systems is transformed into ATP and NADPH.

Cyanobacteria are a large and diverse group of prokaryotes which perform oxygenic photosynthesis because they exhibit PSI as well as PSII with its associated OEC and an intersystem electron transport chain comparable to that of eukaryotic photoautotrophs. In contrast to the intrinsic, major Chl *a/b* light harvesting pigment-protein complex found in chloroplast thylakoid membranes of plants and green algae, the light harvesting complex of cyanobacteria is an extrinsic pigment-protein complex called a phycobilisome which is bound to the outer, cytoplasmic surface of cyanobacterial thylakoids (Sidler, 1994; Glazer, 1994). Phycobilisomes (PBSs) are rod-shaped chromoproteins called phycobiliproteins which may constitute up to 40% of the total cellular protein. The phycobiliproteins usually associated with PBS include allophycocyanin (AP), phycocyanin (PC) and phycoerythrin (PE). In addition to PBS, PSII of cyanobacteria include the Chl *a* core antenna CP47 and CP43 similar to that found in eukaryotic organisms. Cyanobacteria are distinct from chloroplasts because the redox carriers involved in respiratory as well as photosynthetic electron transport are located in the cyanobacterial thylakoid membranes where they share a common PQ pool and a common Cyt b_6f complex (Scherer, 1990; Cooley *et al.*, 2000; Cooley and Vermaas, 2001).

In this review, we focus on the regulation of the process of light harvesting associated with PSII in plants, green algae and cyanobacteria and link these regulatory processes to sensing cellular energy balance and acclimation to a changing environment. Our primary goal is to illustrate the dynamic nature of photosynthetic light harvesting in response to an ever changing environment with respect to light quality, light intensity, temperature and nutrient status. For a more detailed review of different light harvesting structures the reader is referred to the chapter by Cogdell in the present volume as well as the recent volume focussed on light harvesting systems in prokaryotic and eukaryotic photosynthetic organisms edited by Green and Parson (2003).

2. COMPOSITION, STRUCTURE AND FUNCTION OF LIGHT HARVESTING COMPLEXES

2.1 Plants and Green Algae

Table 1 lists the 31 genes associated with PSII supramolecular complexes and their corresponding gene products. However, only fifteen of these genes are encoded by chloroplast DNA whereas 16 genes are encoded in the nuclear genome. All of the genes encoding the polypeptides associated with the major light harvesting complex are nuclear encoded. These polypeptides are biosynthesized in the cytosol on 80S ribosomes and must be transported into the chloroplast and properly inserted and assembled in the thylakoid membrane.

LHCII is considered the major peripheral antenna complex of PSII which can be separated and purified from the PSII core. Crystals of purified LHCII indicate that this major light harvesting complex is organized as trimers of LHCb1, LHCb2 and LHCb3 (Kuhlbrandt *et al.*, 1994). LHCII isolated from leaves in the dark-adapted state typically exhibit the presence of 7-8 Chl *a*, 5-6 Chl *b* and the xanthophylls lutein, neoxanthin and violaxanthin in the ratio of 2:1:1 per monomer (van Amerongen and Dekker, 2003). Subpicosecond transient absorption spectroscopy of trimeric spinach LHCII indicates that lutein and violoxanthin transfer energy to Chl *a* exclusively whereas neoxanthin transfers energy to Chl *b*, which subsequently transfers to Chl *a* (Gradinaru *et al.*, 2000). CP24, CP26 and CP29 are considered to represent minor peripheral PSII antenna complexes which are present in one copy per PSII reaction centre and bind approximately 15% of the total Chl per PSII. Given their lower abundance relative to LHCII, these minor complexes probably play a less significant role in light harvesting but appear to be more important in the regulation of energy transfer through the light-dependent xanthophyll cycle which converts the light-

Table 1. PSII subunit identities, their associated genes and intracellular location in eukaryotic organisms. PSII, photosystem II; LHCII, Chl *a/b* light harvesting complex of photosystem II; C, chloroplast; N, nucleus. Adapted from Blankenship (2000).

Subunit	Gene	Location
PSII-A (D1)	*psbA*	C
PSII-B (D2)	*psbD*	C
PSII-B (CP47)	*psbB*	C
PSII-C (CP43)	*psbC*	C
PSII-E	*psbE*	C
PSII-F	*psbF*	C
PSII-I	*psbI*	C
PSII-H	*psbH*	C
PSII-J	*psbJ*	C
PSII-K	*psbK*	C
PSII-L	*psbL*	C
PSII-M	*psbM*	C
PSII-N	*psbN*	C
PSII-T	*psbT*	C
PSII-X	*psbX*	C
PSII-Z	*psbZ*	C
PSII-O	*PsbO*	N
PSII-P	*PsbP*	N
PSII-Q	*PsbQ*	N
PSII-R	*PsbR*	N
PSII-S(CP22)	*PsbS*	N
PSII-U	*PsbU*	N
PSII-V	*PsbV*	N
PSII-W	*PsbW*	N
PSII-Y	*PsbY*	N
LHCII-peripheral	*Lhcb1*	N
LHCII-peripheral	*Lhcb2*	
LHCII-peripheral	*Lhcb3*	
CP 29	*Lhcb4*	
CP 26	*Lhcb5*	

harvesting xanthophylls, violaxanthin, to the energy quenching xanthophyll, zeaxanthin (Demmig-Adamd and Adams, 1996; Demmig-Adams and Adams, 1999; Horton *et al.*, 1999). CP47 and CP43 are the core PSII antenna complexes, each binding 14 Chl *a* molecules (Barbato *et al.*, 1991) and are the only PSII antenna complexes encoded in the chloroplast DNA (Table 1). Thus, energy is transferred from the PSII peripheral Chl *a/b* light harvesting complexes to the Chl *a* core antenna complexes and finally to the PSII reaction centre where charge separation is initiated.

Energy transfer is usually described as occurring by two primary physical mechanisms, depending on the distance between the pigment molecules within the antenna. When the pigment molecules are weakly coupled, excitation energy transfer occurs via Förster resonance energy transfer (Förster, 1965). The critical distance (R_0) at which Förster energy transfer is 50% efficient for Chl *b* to Chl *a*, Chl *a* to Chl *a* and β-carotene to Chl *a* energy transfer varies from 50 to 100 Å (Blankenship, 2002). When pigments are very close, that is separated by a distance of 10 Å or less, energy transfer occurs by exciton coupling. In this case, the excitation energy is spread over and shared by several adjacent pigment molecules at any moment in time. Thus, exciton coupling is applicable over short distances and strong pigment-pigment interactions, whereas the Förster mechanism is a non-radiative energy transfer mechanism applicable at long distances and weak pigment-pigment interactions. Both mechanisms are thought to be important in overall energy transfer from the antenna complexes to the reaction centre.

The Butler model for the distribution of excitation energy within the photochemical apparatus of photosynthesis assumes that the rate constant for energy transfer to PSII reaction centres exceeds the rate constant for the back-transfer of energy from the reaction centre to the antenna (Butler, 1978). This led to the concept of the PSII unit as an energy funnel. However, more recent fluorescence lifetime measurements indicate that the equilibration of excitation energy between antennae and reaction centres is one order of magnitude faster than charge separation. Thus, PSII appears to be trap limited and the reaction centre appears to act not as a funnel but as a shallow trap (Schatz *et al.*, 1988; Blankenship, 2002).

Several models for the organization of antenna pigments relative to reaction centres have been investigated functionally. At one extreme is the "puddle" or "separate units" model where arrays of antenna pigment molecules are permanently associated with one reaction centre only. In this model, a reaction centre with its associated light harvesting antenna function completely independent of any other neighbouring photosynthetic unit. Thus, excitation energy captured within an antenna can only be transferred to its associated reaction centre and there is no possibility of energy transfer between photosynthetic units. In contrast, the "lake" model assumes that all reaction centres co-exist within a common pigment bed. In this case, if excitation energy reaches a closed reaction centre, the excitation energy may be transferred to an open reaction centre. However, in most photosynthetic organisms, the "connected units" model best illustrates the organization of the photosynthetic unit. In this case, "puddles" are interconnected such that excitation energy can be transferred between photosynthetic units but with lower probability than energy transfer within a photosynthetic unit (Blankenship, 2002).

The minor Chl *a/b* complexes, CP24, CP26 and CP29 are enriched in violaxanthin compared to LHCII, which is consistent with reports that photoprotection through non-photochemical quenching (NPQ) of excess energy through the xanthophyll cycle (Demmig-Adams and Adams, 1999; Horton *et al.*, 1996; Horton *et al.*, 1999) occurs in the minor Chl *a/b* antenna complexes (Falk *et al.*, 1994; Jahns and Krause, 1994). Sandonà *et al.* (1998) suggest that zeaxanthin bound to CP29 is directly involved in NPQ. In contrast, Horton and co-workers (1999) have suggested that violaxanthin loosely associated with the periphery of LHCII is important in NPQ. However, significant NPQ can still be detected in the absence of LHCII as observed in the *chlorine f2* mutant of barley (Falk *et al.*, 1994) as well as in pea plants grown under an intermittent light regime (Jahns and Krause, 1994).

In addition to their roles in light harvesting and photoprotection, carotenoids are also important as structural components of LHCII (Paulsen, 1999). The superhelix formed by the thylakoid transmembrane α-helices A and B of the major LHCII polypeptides are stabilized by two carotenoid molecules (Kulhbrandt *et al.*, 1994). Since the major Chl *a/b*

LHCII as well as the minor Chl *a/b* complexes, CP24, CP26 and CP29 bind two luteins per apoprotein (Bassi *et al.*, 1993), it is presumed that the two carotenoids detected in trimeric crystals of LHCII are lutein molecules (Kuhlbrandt *et al.*, 1994). Furthermore, carotenoids are essential also for the proper folding and assembly of LHCII monomers. When reconstitution of LHCII monomers was performed *in vitro* in the presence of Chl *a* and Chl *b* but in the absence of carotenoids, the LHCII apoprotein did not fold properly and no pigment-protein was formed. Stable pigment-protein complexes did form *in vitro* only in the presence of Chl *a*, Chl *b* plus carotenoids (Plumley and Schmidt 1987; Booth and Paulsen 1996). For LHCII, pigment binding affinity appears to decrease in the following order: Chl *b* > neoxanthin > Chl *a* > lutein > zeaxanthin > violaxanthin (Horton *et al.*, 1999). *In vitro* reconstitution results are consistent with the observations that a mutant of *Scenedesmus obliquus* which lacks lutein also fails to assemble LHCII complexes (Bishop, 1996). Generally, all carotenoids including the xanthophylls appear to promote LHCII monomer assembly but the stability of the LHCII monomer varies depending upon the combination of carotenoids employed in reconstitution experiments. By expressing and mutagenizing *Lhcb* in *E. coli*, Paulsen and Hobe (1992) and Cammarata and Schmidt (1992) were able to show that the first 61 N-terminal amino acids and the last 10 C-terminal amino acids of LHCb are not required for pigment binding. However, all 3 membrane spanning regions of LHCb are required for pigment binding and formation of LHCII monomers. It was assumed that thylakoid lipids were not critical for LHCII monomer formation.

The 3.4 Å resolution of LHCII structure of crystals from pea indicated that native LHCII exists in the trimeric form (Kuhlbrandt *et al.*, 1994; Jansson, 1994; Paulsen, 1994). In antisense *Arabidopsis* plants which lack trimeric LHCII, the minor pigment-protein complex, CP26, accumulates in large amounts and also becomes organized into trimers to replace the absent trimeric LHCII (Ruban *et al.*, 2003). Horton *et al.* (1991, 1996, 1999) suggest that the xanthophylls cycle pigments, violaxanthin and zeaxanthin, loosely bound to the periphery of LHCII, may be important in stabilizing LHCII supramolecular aggregates. The seminal work of Trémolières and co-workers showed that phosphatidylglycerol (PG), the major phospholipid present in chloroplast thylakoid membranes, and its fatty acid composition play a crucial role in stabilizing the oligomeric or trimeric state of LHCII (Trémolières *et al.*, 1981; Dubacq and Trémolières, 1983). *Trans*-Δ3-hexadecenoic acid (16:1(3*t*)) of PG is a unique fatty acid for several reasons: first, this fatty acid exhibits its double bond in the *trans* configuration whereas all other fatty acids of the highly unsaturated chloroplast thylakoid membrane are in the *cis* configuration (Selstam, 1998); second, 16:1(3*t*) is always esterified specifically to the *sn*-2 position of the glycerol backbone of PG; third, although a fatty acid with a *trans* double bond is assumed to have the same physical properties as a saturated fatty acid, the phase transition temperature of PG-16:0/16:1(3*t*) is 10°C lower than that of 16:0/16:0-PG (Bishop and Kenrick, 1987); fourth, 16:1(3*t*) is the only chloroplast fatty acid in angiosperms whose biosynthesis is strictly light-dependent (Gray *et al.*, 1997; Trémolières and Siegenthaler, 1998; Siegenthaler and Trémolières, 1998) and last, 16:1(3*t*) is found exclusively in eukaryotic, Chl*a/b*-containing photoautotrophs (Selstam, 1998). It has not been detected in any other organisms.

In vitro studies by Trémolières and co-workers (Trémolières and Siegenthaler, 1998) showed that the apparent stability of oligomeric LHCII is dependent not on PG content, but rather, dependent on the molecular species composition of PG such that oligomeric LHCII was stabilized when thylakoids of exhibit high levels of PG-16:0/16:1(3*t*) relative to PG-16:0/16:0 (Bishop and Kendrick 1987; Selstam, 1998). The results of the *in vitro* experiments were confirmed by examining oligomeric LHCII stability in the *mf2* mutant of *Chlamydomonas reinhardtii* which lacks PG-16:1(3*t*) Dubertret *et al.*, 1994). Furthermore, purification of rye light harvesting complex showed that PG is specifically bound to LHCII and *in vitro* reconstitution experiments of delipidated rye LHCII showed that the conversion of monomeric LHCII to oligomeric LHCII was strictly dependent upon the presence of PG-16:0/16:1(3*t*) (Krupa *et al.*, 1987; 1992). Chloroplast biogenesis showed that LHCII is inserted into the thylakoid membrane in its monomeric form which subsequently is stabilized in its oligomeric form (Krol *et al.*, 1988; 1989; Dreyfuss and Thornber, 1994). Monitoring the kinetics of the light-dependent accumulation of PG-16:0/16:1(3*t*) indicated the maximum 16:1(3*t*) accumulation preceded the conversion of monomeric LHCII to oligomeric LHCII (Krol *et al.*, 1988;1989).

Through the elegant use of purified LHCb apoprotein over-expressed in *E. coli* combined with in vitro lipid and pigment constitution studies, Paulsen's group identified a trimerization motif (WYXXXR) found between residues 16 and 21 from N-terminus of all known sequences of LHCII apoproteins (Hobe *et al.*, 1994; 1995). This positively charged motif was identified as the site on LHCb which interacts with the negatively charged head group of PG (Trémolières and Siegenthaler, 1998). Removal of this motif prevented the in vitro trimerization of Lhcb. This indicates that the molar ratio of PG to trimer should be 3, which is consistent with PG analyses of purified rye LHCII (Krupa *et al.*, 1987). Furthermore, Nußberger *et al.* (1993) showed that crystallization of LHCII trimers was dependent upon the presence of both PG and DGDG.

In summary, it appears that carotenoids are important for the proper folding of monomeric LHCII whereas PG plays a role in stabilizing the higher order trimeric organization of LHCII. Presently, it is not known how trimeric CP26 is stabilized (Ruban *et al.*, 2003). However, it is clear that the PG-16:0/16:1(3*t*) is not required to stabilize trimeric LHCII in all plant species. Although orchids do not synthesize 16:1(3*t*) (Huner *et al.*, 1989; Selstam, 1998), LHCII is stabilized in its oligomeric form. Similarly, a fatty acid mutant of *Arabidopsis thaliana* which specifically lacks 16:1(3*t*) still exhibits oligomeric LHCII (McCourt *et al.*, 1985). Clearly, factors other than molecular species of PG are also involved in the stabilization of trimeric LHCII. Horton *et al.* (1999) suggest that violaxanthin may play such a structural role.

2.2 Cyanobacteria

PBSs are functionally but not structurally homologous to LHCII (Mimuro and Kikuchi, 2003). In contrast to the non-covalent association of pigments to LHCII, the linear, tetrapyrrole, bilin chromophores called phycobilins are covalently attached to their respective PBS apoproteins through thioether linkages (Table 2). The blue green colour typical of most cyanobacteria is due to the presence of the blue coloured phyco-

Table 2. Phycobiliproteins and their associated phycobilins and linker polypeptides

Phycobiliprotein	Phycobilin	Linker Polypeptide
APC	phycocyanobilin	$L_c^{7.8}$, $L_c^{8.9}$, L_{cm}
PC	phycocyanobilin	$L_R^{8.9}$, L_R^{34}, $L_{RC}^{29.5}$
PE	phycoerythrobilin	L_R^{35}

cyanobilin (PCB) and the red coloured phycoerythrobilin (PEB). PBSs are composed of two phycobiliprotein structural domains:

1. an allophycocyanin (APC) core that is in direct contact with the PSII core antenna.
2. 6 rods of stacked phycocyanin (PC) and phycoerythrin (PE).

Phycobiliproteins associate as heterodimers of α and β monomeric subunits that, in turn aggregate into trimeric $(\alpha\beta)_3$ and hexameric discs $(\alpha\beta)_6$. The rod structure characteristic of PBSs is stabilized by non-pigmented linker polypeptides (L) specifically associated with each type of phycobiliprotein (Table 2) to optimize their absorption and energy transfer properties.

Generally there are four types of colourless, linker polypeptide: (1) linkers within the rod structure (L_R); (2) linkers that connect the rod to the core (L_{RC}); (3) linkers within the core (L_C) and (4) linkers that connect the core to the thylakoid membrane (L_{CM}). These linkers are usually designated by a superscript indicating their molecular mass (Table 2). The small molecular mass L_R polypeptide linkers (8 – 9 kDa) are located at the distal ends of the rods whereas the larger molecular mass L_R and L_{RC} polypeptide linkers (30 – 35 kDa) are involved in rod elongation and the orientation of the hexamers within the rod. The L_{CM} is a unique polypeptide with a molecular mass of 60 – 120 kDa that anchors the rod to the membrane (Mimuro and Kikuchi, 2003). The crystal structures of the trimeric APC from *Spirulina platensis* have been resolved to 2.3 Å (Brejc *et al.*, 1995), of trimeric PC from *Fremyella diplosiphon* to 1.66 Å (Dürring *et al.*, 1991) and trimeric PE from *Polysiphonia urceolata* to 2.8 Å (Chang *et al.*, 1996). All three phycopbiliproteins exhibit similar 3-D structure (Mimuro and Kikuchi,

2003). Furthermore, the genes encoding the core and rod polypeptides as well as their associated linker polypeptides have been cloned, sequenced and their expression characterized (Grossman *et al.*, 2003).

The phycobiliprotein discs present in the rods are arranged from the distal end to the PSII core in the following manner: PE, PC, APC, PSII core. PE absorbs at the shortest wavelengths (500 – 565 nm), PC at intermediate wavelengths (595 – 640 nm) whereas APC absorbs between (650 – 655 nm). This is consistent with the energy transfer pathway of PE $\rightarrow$ PC $\rightarrow$ APC $\rightarrow$ PSII core $\rightarrow$ P_{680}. The energy gradient within a PBS is a consequence of the pigment composition as well as the influence of the linker polypeptides on the absorption properties of these chromophores. The establishment of this energy gradient ensures efficient energy transfer through the PBS rod to the reaction centre by the resonance energy transfer mechanism (Förster, 1965).

Phycobilisome composition and structure are very sensitive to light quality and light intensity as well as nutrient levels. These features of PBSs have been summarized in a series of excellent reviews (Tandeau de Marsac and Houmard, 1993; Grossman *et al.*, 1994; Grossman *et al.*, 2003). Complementary chromatic adaptation reflects the capacity of cyanobacteria to adjust phycobilin composition in response to changes in light quality (Grossman, 1990). This adjustment in pigment composition is regulated by a two-component regulatory system (Hoch and Silhavy, 1995). The protein RcaE is the sensor His kinase that perceives changes in light quality and is related to the plant photoreceptor, phytochrome (Kehoe and Grossman, 1996). Red light induces RcaE to undergo autophosphorylation after which the phosphoryl group is transferred to the response regulator, RcaF. Subsequently, RcaF transfers its phosphoryl group to RcaC which is critical in regulating phycobilin composition. In contrast, exposure to green light inhibits this phosphorelay system and consequently inhibits complementary chromatic adaptation in cyanobacteria (Grossman *et al.*, 2003).

In addition to the presence of PBSs, cyanobacteria are also distinct from eukaryotic chloroplasts with respect to respiratory and photosynthetic electron transport. Both bioenergetic processes function within the thylakoid membrane of cyanobacteria, and thus share a common PQ pool and a common Cyt b_6/f complex (Scherer, 1990;

Cooley and Vermass, 2001). Thus, electron fluxes in the intersystem electron transport chain may be affected by the supply of electrons from PSII, cytosolic NAD(P)H dehydrogenase, the succinate dehydrogenase-mediated electron transport pathway from respiratory donors, as well as PSI cyclic electron transport. Furthermore, the possibility of electron consumption by either the terminal respiratory Cyt oxidase or by PSI will also affect intersystem electron flux in cyanobacterial thylakoid membranes.

Although these data on the structure and composition of light harvesting complexes are essential to understand the function of this important supramolecular, thylakoid pigment-lipid-protein complex, they provide a static view of the light harvesting process. However, landmark research conducted by several laboratories on terrestrial plants (Berry and Björkman; 1980; Anderson, 1986; Anderson *et al.* 1995;Melis, 1991; 1998), green algae (Falkowski, 1983; Sukenik *et al.*, 1987; LaRoche *et al.*, 1991; Melis, 1998; Falkowski and Chen, 2003) and cyanobacteria (Fujita *et al.*, 1994; Grossman *et al.*, 1994; Fujita, 1997; Grossman *et al.*, 2003) on photoacclimation have illustrated the dynamic nature of the oxygenic photosynthetic apparatus with respect to its response to environmental cues such as light intensity, light quality and nutrient availability. We suggest that the dynamic modulation in the structure, composition and function of the photosynthetic apparatus reflects the imperative of photoautotrophs to maintain photostasis (Melis, 1998; Huner *et al.*, 1998; Huner *et al.*, 2003), that is, a balance between energy supply through light harvesting and photochemistry and energy consumption through the metabolic reduction of C, N, S and O. Modulation of light harvesting not only appears to reflect the broad influence of photostasis on diverse molecular, physiological and developmental processes (Anderson *et al.*, 1995; Huner *et al.*, 1998; 2003; Falkowski and Chen, 2003) but also it appears to be linked to ecological fitness as estimated by plant seed production (Kulheim *et al.*, 2002; Andersson *et al.*, 2003; Ganeteg *et al.*, 2004).

3. EXCITATION PRESSURE AND PHOTOSTASIS

The process of photosynthesis represents an integration of extremely rapid, temperature-insensitive photochemical reactions with relatively slow, temperature-dependent biochemical reactions. In the conversion of light energy into redox potential energy, the rate-limiting step is considered to be the conversion of "closed" PSII reaction centers to "open" PSII reaction centers through the oxidation of Q_A^- by PQ and the Cyt b_6/f complex of the intersystem electron transport chain (Krause and Weis, 1991).

Since the oxidation of $(PQ)_{red}$ by $(Cyt\ b_6/f)_{ox}$ is diffusion limited (Haehnel, 1984), the "closure" of PSII reaction centers through photochemistry will always be faster than the subsequent "opening" of PSII reaction centers which is dependent upon the oxidation of the PQ pool through intersystem electron transport. Thus, exposure to excess light causes the over-reduction of the PQ pool which, in turn, increases the probability that PSII reaction centers remain closed. As a consequence, any environmental condition that exacerbates the difference between the rates at which PSII reaction centers are closed through photochemistry versus the rates at which they are opened through redox biochemistry will be reflected in an increase in the relative reduction state of the PQ pool.

Chl a fluorescence is a property exhibited by all photosynthetic organisms due to the essential role of chlorophyll in the structure and function of the photosynthetic apparatus. Typically less than 3% of the absorbed light is ever re-emitted as chlorophyll fluorescence and, at room temperature, the latter primarily emanates from PSII (Krause and Jahns, 2003). Quantification of chlorophyll a fluorescence induction has proven to be an extremely useful tool to assess the structure and function of PSII and the overall process of photosynthesis (Krause and Weis, 1991; Krause and Jahns, 2003). Since reduced Q_B is in equilibrium with the reduced PQ pool, the relative reduction state of the PQ pool, also called excitation pressure, can be estimated in vivo as the relative reduction state of Q_A, that is $[Q_A^-] / ([Q_A] + [Q_A^-])$, which can be conveniently measured *in vivo* as 1-qP using pulse amplitude modulated chlorophyll fluorescence where qP is the photochemical quenching parameter (Maxwell *et al.*, 1995; Huner *et al.*, 1998).

The rate of energy absorption by PSII is proportional to $I\sigma_{PSII}$ where σ_{PSII} is the functional absorption cross-sectional area of PSII and I is the absorbed photon flux. This product represents light harvesting and is, by and large, insensitive to temperature in the biologically significant range. Under light saturating conditions, the rate of utilization of the absorbed light through temperature-sensitive photosynthetic electron transport and the ultimate use of these photosynthetic electrons to reduce C, O, N and S may be expressed as $n\tau^{-1}$ where n is the number of photosynthetic units and τ^{-1} represents their turnover rate (Durnford and Falkowski, 1997). This product represents the electron sink capacity. Accordingly, photoautrophs are exposed to excitation pressure whenever $I\sigma_{PSII} > n\tau^{-1}$ (Huner *et al.*, 1998; Huner *et al.*, 2003).

According to this inequality, excitation pressure may be induced by changes in several different environmental parameters. For example, increasing growth irradiance at a constant temperature would cause an increase in the relative reduction state of the PQ pool due to an increase in I, and thus an increase in $I\sigma_{PSII}$, assuming no changes in the capacity to utilize the absorbed energy, that is no change in $n\tau^{-1}$. Theoretically, a similar increase in the relative reduction state of the PQ pool could be created by maintaining the same photon flux but concomitantly decreasing the temperature. The lower temperature would decrease the rate of the biochemical redox reactions (τ^{-1}) that utilize the absorbed energy, that is, would decrease $n\tau^{-1}$, with no change in $I\sigma_{PSII}$. Similarly, exposure to drought or the lack of specific essential nutrients would also cause an increase in the reduction state of the PQ pool due to decrease in $n\tau^{-1}$ as a consequence of the limitations in the availability of electron acceptors such as CO_2, O_2, NO_3^- or SO_4^{2-}.

Any change in environmental conditions such as light, temperature, water and nutrient availability may modulate the photochemical reactions of photosynthesis to a different extent than the biochemical reactions involved in carbon reduction cycle, photorespiration, and nitrogen and sulphur assimilation. Consequently, these environmental changes will modulate excitation pressure. Excitation pressure reflects an imbalance between energy absorbed through photochemistry and energy utilized

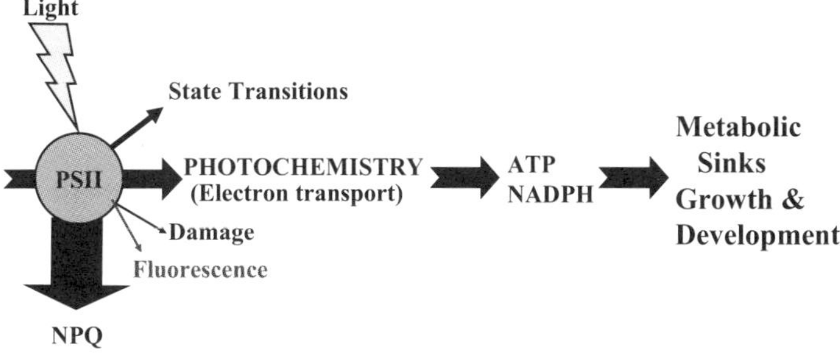

Fig. 1. Possible fates of absorbed energy in photosynthesis (Modified from Krause, 1988).

through the consumption of photosynthetically generated reducing power ($I\sigma_{PSII} > n\tau^{-1}$; Huner *et al.*, 1998). The predisposition of photosynthetic organisms to attain a balance in energy budget is defined as photostasis (Melis, 1998; Huner *et al.*, 2003). This would be attained whenever energy absorption equals energy utilization through photochemistry plus nonphotochemical dissipation of excess energy (Fig. 1). Due to the photophysical and photochemical nature of light absorption, energy transfer and charge separation, photostasis is rarely, if ever, attained under natural environmental conditions. In fact, the light absorbed usually exceeds the energy that can be consumed through metabolism.

The inequality illustrated above also provides insights into the possible mechanisms by which photosynthetic organisms may respond to the imbalance in energy budget to attain photostasis. Fig. 1 illustrates the possible fates of absorbed light energy through the photosynthetic apparatus. Energy balance could be attained by either reducing σ_{PSII}, by reducing light harvesting antenna size and/or reducing the effective absorption cross-sectional area of PSII by dissipating energy non-photochemically as heat (Krause and Weis, 1991; Horton *et al.*, 1996; Falkowski and Chen, 2003). Alternatively, photostasis also could be attained by increasing sink capacity ($n\tau^{-1}$) (Huner *et al.*, 2003). This may be accomplished by elevating the levels of Calvin cycle enzymes and enzymes involved in cytosolic sucrose biosynthesis, which would increase the capacity for CO_2 assimilation relative to the capacity for photosynthetic electron transport. Clearly, photoautotrophs may exploit any one or a combination of these mechanisms to attain photostasis in an environment which exhibits hourly, daily and seasonal changes in irradiance, temperature, water availability and nutrient status. The present discussion will focus on the capacity of photosynthetic organisms to modulate the structure and function of the major light harvesting complex of PSII in response to environmental changes in temperature, irradiance and nutrient availability.

4. ROLE OF LIGHT HARVESTING IN PHOTOPROTECTION AND PHOTOSTASIS

4.1. Short-term Response Mechanisms

4.1.1. State Transitions

The spectral distribution of the solar radiation reaching the earth is attenuated due to filtering through either aquatic environments (Falkowski, 1983), crop and forest canopies (Bjorkman and Ludlow, 1972), or through a single leaf (Vogelmann *et al.*, 1996). Such attenuation inevitably results in an imbalance in the absorption of light between PSII and PSI, which causes a decreased efficiency of linear electron transport. In terrestrial plants and green algae, light absorbed preferentially by PSII relative to PSI (state 2) leads to an over-reduction of the PQ pool whereas preferential excitation of PSI relative to PSII (state 1) results in oxidation of the PQ pool. The redox state of the PQ pool regulates a thylakoid protein kinase which controls the phosphorylation state of the peripheral Lhcb antenna polypeptides and energy transfer between PSII and PSI (Allen and Pfannschmidt, 2000). Thus, the regulation of energy transfer by the redox state of the thylakoid PQ pool reflects a photoprotective mechanism to counteract the potential for uneven absorption of light by PSI and PSII by adjustment of σ_{PSII} to maintain photostasis and to ensure maximum photosynthetic efficiency on a short-term basis. Thus, the state transition is a dynamic mechanism that enables photoautotrophs to respond rapidly to changes in illumination. Lunde *et al.* (2000) showed that the presence of the PSI subunit, PSI-H, is an absolute requirement for energy transfer from the mobile, phosphorylated LHCII to PSI in *Arabidopsis thaliana*.

It was first proposed by Fujita *et al* (1994) that modulation of photosystem stoichiometry is a response to changes in the redox state of the intersystem electron transport chain to ensure equal rates of electron flow through both PSI and PSII (Allen and Pfannschmidt, 2000). Pfannschmidt *et al.* (1999) have shown that the transcription of the chloroplast encoded *psbA* which codes for D1, and *psaAB* genes which code for the PSI reaction center polypeptides, are controlled by the redox state of the PQ pool. Over-reduction of the PQ pool by the preferential excitation of PSII not only favours energy transfer from PSII to PSI through phosphorylation of LHCII but also favours the activation of *psaAB* transcription and the concomitant repression of *psbA*. Conversely, oxidation of the PQ pool by preferential excitation of PSI not only favours de-phosphorylation of LHCII but also the activation of transcription of *psbA* and the repression of *psaAB* (Pfannschmidt *et al.*, 1999; Allen and Pfannschmidt, 2000). Thus, PQ, the redox sensor that controls state transitions, also appears to be the sensor that regulates chloroplast photosystem stoichiometry (Pfannschmidt, 2003).

4.1.2 Antenna Quenching

Photoinhibition is defined as the light-dependent decrease in photosynthetic rate which may occur whenever the photon flux is in excess of that required for photosynthesis (Long *et al.*, 1994), that is, whenever $I\sigma_{PSII} > n\tau^{-1}$. Sudden, short-term exposure to high light and/or low temperature may induce photoinhibition (Krause, 1988; Huner *et al.*, 1993). Rapidly reversible photoinhibition is a consequence of an increase in thermal energy dissipation (NPQ, non-photochemical quenching) which leads to a down-regulation of PSII activity (Öquist *et al.*, 1992). Non-photochemical quenching processes are, by and large, associated with LHCII and PSII core antennae (Horton *et al.*, 1999). This process or any other process which protects PSII from over-excitation is referred to as 'photoprotection' (Demmig-Adams and Adams, 1996; Demmig-Adams *et al.*, 1999; Demmig-Adams and Adams, 2002).

Although not always the case (Hurry *et al.*, 1997), NPQ is thought to occur *via* the xanthophylls cycle which interconverts the light harvesting xanthophyll, violaxanthin (V), to the energy quenching xanthophylls, antheraxanthin (A) and zeaxanthin (Z) (Demmig-Adams and Adams 1996; Horton *et al.* 1999; Demmig-Adams *et al.*, 1999; Gilmore 1999; Gilmore and Ball, 2000; Ort, 2001; Demmig-Adams and Adams, 2002). Two molecular mechanisms have been proposed to account for NPQ. The direct mechanism for NPQ proposes that the S_1 state of A and Z within LHCII is lower than that of Chl *a* within the antenna pigment bed. Thus, A and Z are not able to transfer energy to the S_1 state of antenna chlorophyll whereas V is able to transfer energy to antenna chlorophyll. Consequently, excited states A and Z decay to the ground state with the release of heat (Frank *et al.*, 1994). This light-dependent, reversible interconversion of V to A and Z has been called a 'molecular gear shift' regulating energy transfer within LHCII (Frank *et al.*, 1994). However, Polivka *et al.* (1999; 2002) conclude that the 'molecular gear shift' hypothesis is not likely to be valid since the S_1 states of both zeaxanthin and violaxanthin are lower than the Q_y transition of Chl *a*. In contrast, the indirect mechanism proposes that the ΔpH-dependent transthylakoid and xanthophyll cycle pigments regulate the oligomerization state of LHCII which affects the rapidly relaxing energy-dependent component (qE) of NPQ (Horton *et al.*, 1999; Ruban *et al.*, 2002; Aspinall-O'Dea *et al.*, 2002; Wentworth *et al.*, 2003). In support of the indirect mechanism, Elrad *et al.* (2002) reported that LHCII trimerization is required for antenna quenching in the *npq5* mutant of *Chlamydomonas reinhardtii*. Polivka *et al.* (1999; 2002) also suggest that the indirect mechanism through regulation of the structure of LHCII by zeaxanthin and violaxanthin is the more likely mechanism underlying antenna quenching. Although the precise molecular mechanism underlying NPQ is still equivocal, antenna quenching through the xanthophyll cycle leads to photostasis *via* a decrease in σ_{PSII} even though the physical size of LHCII remains constant.

Recently, major insights into our understanding of the molecular mechanism(s) of NPQ have occurred as a consequence the isolation of NPQ mutants of *Arabidopsis thaliana* and *Chlamydomonas reinhardtii*. The *PsbS* deletion mutant, *npq4-1* (Li *et al.*, 2000) and various *PsbS-*

defective mutants of *Arabidopsis* (Niyogi, 1999; Havaux and Kloppstech, 2001; Peterson and Havir, 2001; 2003; Grasses *et al.*, 2002) are impaired in the development of NPQ. However, despite the development of mutants specifically deficient in PsbS and NPQ, the precise function of the PsbS protein, and its specific role in NPQ, remains equivocal. Based on the original observation that PsbS binds chlorophylls and xanthophylls (Funk, 2001), and its role in the development of qE, Li *et al.* (2000) suggested that this protein is the site of ΔpH and xanthophyll-dependent NPQ. However, more detailed biochemical analyses suggest that the PsbS protein does not bind pigments (Dominici *et al.* 2002). This is in agreement with the fact that most of the highly conserved amino acids that form the ligands for chlorophyll in most of the LHC proteins (Kühlbrandt *et al.* 1994; Bassi and Caffari, 2000) are not found in PsbS. In addition, the availability of the 3-D map of the PSII supercomplex (Nield *et al.* 2000a) and the structure of the LHCII trimer (Kühlbrandt *et al.* 1994) indicate that there is not sufficient space to accommodate the PsbS protein within the LHCII-PSII supercomplex (Nield *et al.* 2000b). Through fluorescence analysis of the *npq4-1* mutant of *Arabidopsis thaliana* lacking PsbS, Peterson and Havir (2003) have suggested that the PsbS polypeptide may regulate exciton distribution within PSII. Wentworth *et al.* (2003) suggest that the role of PsbS is to regulate the oligomerization of antenna complexes involved in antenna quenching. A detailed summary of the structure and function of this intriguing protein is provided by Funk (2001).

Swiatek *et al.* (2001) have provided convincing evidence that the *ycf9* gene which encodes PsbZ, a core PSII subunit, plays a critical role in NPQ in tobacco and *Chlamydomonas reinhardtii*. PsbZ appears to stabilize the supramolecular organization of PSII core complexes with the peripheral antennae (Swiatek *et al.*, 2001). Although NPQ was significantly inhibited in Δ*ycf9* tobacco plants, which was associated with a decrease in the level of PsbZ, PsbS accumulation was unaffected in this mutant. Furthermore, PsbZ is present in phycobilisome-containing eukaryotic and prokaryotic organisms which exhibit NPQ but no xanthophylls cycle. Swiatek *et al.* (2001) suggest that PsbZ is a critical component in the regulation of NPQ in these organisms.

Early light-inducible proteins (ELIPs) are a family of proteins related to the LHC gene family (Montane and Kloppstech, 2000). ELIPs also have been shown to accumulate under conditions of high light or low temperature stress in mature leaves under controlled as well as natural field conditions (Lindahl *et al.*, 1997; Montane *et al.*, 1997; Norén *et al.*, 2003) and during chloroplast development (Meyer and Kloppstech, 1984; Krol *et al.*, 1999). Although it has been presumed that ELIPs are involved in the non-photochemical photoprotection of PSII, their precise role still remains to be elucidated. Recently, Hutin *et al.* (2003) reported that ELIPs play a significant role in the protection of *Arabidopsis thaliana* against photooxidative stress. Their results indicate that this family of LHC-related polypeptides fulfil a photoprotective function by binding free chlorophylls that may be released during turnover of pigment-binding proteins or, alternatively, ELIPs may the stabilize of the proper assembly of pigment-proteins during high-light stress. This is consistent with an earlier report which showed that both maximum xanthophyll cycle activity as well as maximum accumulation of ELIPs was transient and observed during the early stages of the light-dependent assembly of the photosynthetic apparatus (Krol *et al.*, 1999).

4.2 Long-Term Response Mechanisms

4.2.1 Photoacclimation

Adjustments to the structure and function of the photosynthetic apparatus in response to changes in growth irradiance are called photoacclimation. One mechanism of long-term photoacclimation involves the modulation of the physical size and composition of the light harvesting antennae of PSII (Sukenik *et al.*, 1988; LaRoche *et al.*, 1991; Melis, 1998; Falkowski and Chen, 2003). Generally, there is an inverse relationship between growth irradiance and light harvesting antenna size. Thus, low growth light promotes large PSI and PSII light harvesting antenna size whereas growth at high light generates a small photosynthetic unit size. Recently, it was shown that the modulation of the size of LHCII is the consequence of regulation of the nuclear *Lhcb* gene family by the redox state of the PQ pool in the chloroplast thylakoid membrane in *Dunaliella tertiolecta*

(Escoubas *et al.*, 1995) and *Chlorella vulgaris* (Wilson and Huner, 2000; Wilson *et al.*, 2003). In the presence of the PSII-specific inhibitor, DCMU, the PQ pool remains oxidized and *Lhcb* expression and Lhcb accumulation is maximum and cells exhibit a low light phenotype. In contrast, when these algal cells are exposed to sub-lethal doses of DBMIB, the PQ pool remains reduced and *Lhcb* expression and Lhcb accumulation are significantly depressed and the cells exhibit a high light phenotype.(Escoubas *et al.*,1995; Wilson and Huner, 2000; Wilson *et al.*, 2003). Thus, when the PQ pool is reduced by exposure to high light, the transcription of the *Lhcb* genes is down-regulated, which results in a decrease in the size of LHCII. This photoprotective mechanism is consistent with the notion that photostasis in response to high light may be attained through modulation of σ_{PSII}.

The persistent retention of Z and A in overwintering plants led to the development of the concept of sustained xanthophyll-dependent energy dissipation, which involved sustained thylakoid lumen acidification, even in the dark (Demmig-Adams *et al.* 1999). This form of persistent energy dissipation through the antenna has been suggested as an important protective mechanism enabling evergreen plants to maintain their leaves during the winter through reorganization of the LHCII into xanthophyll-containing aggregates, when a combination of low temperature and high light usually occurs (Gilmore and Ball, 2000; Öquist and Huner, 2003; Gilmore *et al.*, 2003). However, in addition to its role in non-photochemical quenching, Z also appears to act as an anti-oxidant to protect against photooxidative stress (Havaux and Niyogi, 1999; Baroli *et al.*, 2003).

4.2.2 Cold Acclimation

Photosynthetic adjustment during cold acclimation of the unicellular green algae *Chlorella vulgaris* and *Dunaliella salina* by growth at low temperature and moderate irradiance $5°C/150 \ \mu mol \ m^{-2} \ s^{-1}$ (5/150) mimics photoacclimation of these algal species grown at high light and moderate temperatures (27/2200) (Huner *et al.*, 1998). Cells grown at 5/150 are indistinguishable from those grown at 27/2200 with respect to photosynthetic efficiency, photosynthetic capacity, pigmentation, Lhcb content and sensitivity to photoinhibition. These results are explained on the basis that cultures grown at either 5/150 or 27/2200 indeed exhibit comparable excitation pressure measured as 1-qP (Huner *et al.*, 1998). Similar conclusions regarding the role of excitation pressure have been reported for thermal and photoacclimation of *Laminaria saccharina* (Machalek *et al.*, 1996) and cold acclimation of the filamentous cyanobacterium, *Plectonema boryanum* (Miskiewicz *et al.*, 2000; 2002). *Plectonema boryanum* exposed to either high light or low temperature not only exhibited a decrease in PBS rod length but also exhibited a concomitant uncoupling between the PBS and the PSII reaction centres. These alterations in PBS organization drastically decreased the efficiency of energy transfer from the PBS to PSII (Miskiewicz *et al.*, 2002). These results are consistent with the thesis that exposure to low temperature creates a similar imbalance in energy budget as exposure to high light.

These green algal and cyanobacterial species are unable to up-regulate carbon metabolism and thus are unable to adjust electron-consuming sink capacity during growth and development at low temperature (Savitch *et al.*, 1996; Miskiewicz *et al.*, 2000). As a consequence, these organisms are unable to adjust $n\tau^{-1}$ significantly with respect to changes in growth temperature. Thus, to attain photostasis, these organisms adjust $I\sigma_{PSII}$ through a reduction in the size of PSII light-harvesting complex coupled with an increased capacity for NPQ which result in a decrease in σ_{PSII}.

Cold temperate conifers such as Lodgepole pine (*Pinus contorta* L.) and herbaceous cereals such winter wheat (*Triticum aestivum* L.) and winter rye (*Secale cereale* L.) are representative of some of the most cold-tolerant plants that retain their foliage during the autumn and winter (Öquist *et al.*, 2001). This capacity to cold acclimate is an essential requirement for the development of maximum freezing tolerance, which allows them to survive the freezing temperatures during the winter. Although neither pine nor wheat and rye exhibit significant decreases in Chl contents, these two groups of plants exhibit quite different strategies for the utilization of light energy during growth and cold acclimation (Öquist *et al.*, 2001; Savitch *et al.*, 2002; Öquist and Huner, 2003). Cold acclimation of conifers induces the cessation of primary growth in contrast to winter cereals, which require continued growth and development during the cold acclimation period to attain maximum freezing tolerance (Fowler and Carles, 1979; Gray *et al.*, 1997). In the context of these different growth and cold acclimation strategies, the requirement for photosynthetic assimilates also differs considerably. Conifers exhibit a decreased requirement for photosynthetic assimilates upon the induction of dormancy and cold acclimation. In contrast, overwintering cereals maintain a high demand for photoassimilates during cold acclimation.

As a consequence of the decreased sink demand for photoassimilates, that is, a decrease in $n\tau^{-1}$, conifers exhibit feedback inhibition of CO_2 assimilation (Savitch *et al.*, 2002). To attain photostasis under these conditions, conifers adjust their capacity to transform light energy photochemically by decreasing the content of PSII reaction centers and by decreasing their efficiency for light harvesting by increasing the capacity for NPQ through the up-regulation of *PsbS* and the xanthophyll cycle. Concomitantly, the major and minor LHCII polypeptides associate into large supramolecular aggregates which results in a highly quenched energetic state (Ottander *et al.*, 1995; Savitch *et al.*, 2002). This phenomenon is consistent with the recent reports that the overwintering evergreens, snow gum and mistletoe, exhibit a distinctive 'cold-hard-band' (CHB) in their 77K fluorescence emission spectrum, which is associated Chl aggregation, and dissipate excess energy as heat separate from PSII while simultaneously decreasing the quantum yield of PSII (Gilmore and Ball, 2000; Gilmore *et al.*, 2003). Since conifers exhibit the capacity to recover fully from this quenched state with the onset of spring (Ottander *et al.*, 1995), this capacity to down-regulate photosynthesis during cold acclimation is considered an important mechanism for the successful establishment of evergreen conifers in cold temperate and sub-arctic climates (Öquist and Huner, 2003).

5. NUTRIENT LIMITATIONS

Since photosynthesis is integrated into virtually all metabolic pathways in photoautrophs, through either the requirement of carbon skeletons, reducing power or ATP, nutrient limitations may induce a feedback inhibition of photosynthetic electron transport due to decreases in sink capacity ($n\tau^{-1}$). Growth of higher plants such as sugar beet under Fe deficiency induces a co-ordinated decrease in the content of LHCII, electron transport components and Rubisco (Terry, 1983; Winder and Nishio, 1995). In addition to the decrease in the apparent size of LHCII, growth under Fe deficient conditions increases the capacity for NPQ (Abadía *et al.*, 2000). Thus, these observations are consistent with the thesis that plants grown under Fe deficient conditions attempt to attain photostasis by decreasing σ_{PSII} by lowering the LHCII content and increasing the xanthophyll cycle activity.

Cyanobacteria exhibit an impressive capacity to alter the composition, structure and function of the photosynthetic apparatus in response to iron limitation (Straus, 1994). A novel chlorophyll-protein complex, denoted CP43', is encoded by the *isiA* gene in *Synechococcus* and *Synechocystis* which has close sequence similarity to the *psbC* gene that encodes the core antenna polypeptide, CP43. Iron stress induces the expression of isiA and the accumulation of CP43' with the concomitant disappearance of CP43 (Straus, 1994). Initially it was proposed that CP43' may simply replace CP43 during iron stress in these cyanobacteria (Burnap *et al.*, 1993). However, picosecond Chl *a* fluorescence lifetime measurements indicate inefficient energy transfer from CP43' to PSII reaction centres (Falk *et al.*, 1995). Park *et al.* (1999) concluded that CP43' acts as a quencher of excess energy to photoprotect PSII during conditions of iron limitation. However, recent structural evidence indicates that CP43' is primarily associated with PSI rather than with PSII (Bibby *et al.*, 2001; Boekema *et al.*, 2001; Huner *et al.*, 2001). It has been suggested that the ring structure formed around PSI by CP43' may function as an antenna for PSI (Bibby *et al.*, 2001; Boekema *et al.*, 2001). Thus, any role of CP43' in photoprotection during iron stress requires further investigation.

In the green alga, *Chlamydomonas reinhardtii*, both phosphate and sulfate limitations independently result in a decrease in photosynthetic efficiency (Wykoff *et al.*, 1998; Davies and Grossman, 1998). This is due to the combined effects of increased capacity for xanthophyll cycle-dependent NPQ plus an enhanced energy transfer from PSII to PSI during exposure to nutrient-limited conditions. Thus, the photoacclimation response induced by nutrient limitations in this green alga, at least in part, appears to involve an adjustment of σ_{PSII}.

The effects of altered electron sink capacity on excitation pressure was illustrated elegantly by Holtgrefe *et al.* (2003). Transgenic potato plants were generated that exhibited levels of leaf Fd that varied between 40% and 80% of that observed in wild type (WT) potato plants. Fd is the terminal photosynthetic electron acceptor for linear photosynthetic electron transport. The reducing power of Fd is consumed in various metabolic processes including CO_2 assimilation, NO_3^- assimilation in plants. Thus, a decrease in the level of Fd should increase excitation pressure because of a lower capacity for energy consumption ($n\tau^{-1}$). Holtgrefe *et al.* (2003) reported that the transgenic potato plants with low levels of Fd mimicked plants grown at high light since they exhibited lower total Chl contents, increased Chl *a/b* ratios, as well as an increase in the reduction state of Q_A compared to WT plants. These results are consistent with the thesis that transgenic potato plants with low Fd contents maintain photostasis in response to high excitation pressure by modulating σ_{PSII} through a decrease in the physical size of LHCII.

Cyanobacteria have proven to be an excellent model system to elucidate the molecular mechanism regulating light harvesting in response to nutrient limitations. Seminal research by Grossman and co-workers (1994; 2003) has provided important insights into the global regulation of PBS turnover in response to N- and S-limitations in *Synechococcus*. The most apparent affect of growth under nutrient deprivation in *Synechococcus* is a decrease in pigment content per cell. This change in pigmentation was associated with an ordered degradation of PBSs during N- and S-limited growth (Collier and Grossman, 1992; 1994). It has been suggested that the degradation of PBSs during S-starvation could provide the cell with S in the form of the amino acids, cysteine and methionine (Grossman *et al.*, 2003). However, limitation of macronutrients such as phosphorus does not result in the degradation of PBSs (Collier and Grossman, 1992). The ordered reduction in the size of PBSs would help to maintain photostasis by a reduction in the absorption of excess energy by decreasing light harvesting efficiency ($I\sigma_{PSII}$).

To elucidate the molecular mechanism regulating PBS degradation during nutrient limitation, Grossman and co-workers isolated non-bleaching mutants (*nbl*) of *Synechococcus* that do not degrade PBSs during either N- or S-deprivation (Grossman *et al.*, 1994; 2003). The expression of the *nblS* gene is sensitive to the absence of a variety of nutrients including N, S and inorganic C. NblS is thought to interact with another protein, NblR, to regulate PBS degradation in response to nutrient stress as well as high light stress. Grossman et al (2003) suggest that NblS and NblR represent a two-component, histidine sensor kinase-response regulator couple. NblS expression appears to be sensitive to the redox state of the photosynthetic intersystem electron transport chain, which is consistent with the possibility that *nblS* expression is in fact regulated by excitation pressure. The redox sensitivity of *nblS* expression may explain how a single redox sensing protein such as NblS may act as a global regulator to balance energy absorption through light harvesting and subsequent photochemical transformation with energy consumption through cellular metabolism (van Waasbergen *et al.*, 2002; Grossman *et al.*, 2003).

6. SENSING CELLULAR ENERGY IMBALANCE AND REGULATION OF LIGHT HARVESTING

Sensing of light quality and its associated signal transduction pathways in relation to photomorphogenesis and plant development are well established and involve photoreceptors such as phytochrome and crytochrome (Chory 1997; Ballaré, 1999). However, we are unaware of any unequivocal evidence to show that these photoreceptors regulate gene expression during steady-state vegetative growth. Walters *et al.* (1993) reported that *Arabidopsis* mutants lacking phytochrome were still capable of photoacclimation to changes in light intensity even though these mutants did not exhibit the expected changes in photo-morphogenesis in response to light quality.

As suggested by Falkowski and co-workers (Durnford and Falkowski, 1997; Falkowski and Chen, 2003), the transduction of the excitation pressure signal is mediated by a nested set of responses as a result of cellular integration over various timescales. On the biologically short timescale of seconds to minutes, the reduction state of the PQ pool or some other component of the electron transport chain may increase due to a sudden increase in irradiance or a sudden decrease in temperature or nutrient availability. The nested signal hypothesis (Falkowski and Chen, 2003) predicts that such short-term, abiotic stresses will result in a rapid increase in NPQ through the induction of the xanthophyll cycle and/or state transitions which will reduce the effective absorptive cross-section of PSII and maintain photostasis. However, if excitation pressure continues to increase over longer time periods of hours to days, the these short-term photoprotective processes will not be able to keep the PQ pool sufficiently oxidized, and as a consequence, will be superseded by longer term adjustments involving the down regulation of *Lhcb* expression in eukaryotes and the induction of the *nblS-nblR* two-component regulatory system in cyanobacteria. Furthermore, exposure to chronically high excitation pressure over extended periods of time in certain species will induce the accumulation of non-photosynthetically active pigments that may act as natural sunscreens to protect the photosynthetic apparatus from excess light. Specific examples include the accumulation of the carotenoid, myxoxanthophyll, in the outer membranes of *Plectonema boryanum* upon exposure to either high light or low temperature (Miskiewicz *et al.*, 2000), the accumulation of anthocyanin localized to the epidermal cells of sun exposed needles during cold acclimation of *Pinus banksiana* (Krol *et al.*, 1995; Huner *et al.*, 1998) and the accumulation of the carotenoid, rhodoxanthin, during cold acclimation in western red cedar (Weger *et al.*, 1993). The responses at all three timescales reflect mechanisms to reduce light harvesting ($I\sigma_{PSII}$) under conditions whereby the energy absorbed exceeds the cellular capacity to utilize this energy through metabolism ($n\tau^{-1}$).

Over the past decade, research has provided significant breakthroughs in our understanding of the mechanisms by which photoautrophic organisms sense cellular energy imbalances and respond by adjusting photosynthetic light harvesting. The consensus is that the redox state of the PQ pool acts as an important energy sensor regulating the expression of genes coding for the polypeptide components of eukaryotic and prokaryotic light harvesting systems (Falkowski and Chen, 2003; Huner *et al.*, 2002; 2003). When light harvesting ($I\sigma_{PSII}$) exceeds the capacity to utilize the absorbed energy through metabolism

($n\tau^{-1}$) in *Dunaliella* and *Chorella*, the PQ pool becomes reduced, *Lhcb* expression is repressed and Lhcb accumulation is inhibited. This is associated with an increase in the capacity for NPQ due to zeaxanthin accumulation and the induction of the nuclear encoded carotenoid binding protein, Cbr (Krol *et al.*, 1997). These changes lower the efficiency for light harvesting, reduce excitation pressure and protect the photosynthetic apparatus from photodamage (Escoubas *et al.* 1995; Maxwell *et al.*, 1995; Wilson and Huner, 2000). It has been suggested that changes in excitation pressure are sensed by modulation of redox state of the PQ pool which, in turn, regulates the expression of nuclear encoded *Lhcb* and *Cbr* genes. In addition, changes in excitation pressure are also sensed through changes in transthylakoid ΔpH which regulate xanthophyll cycle activity and hence NPQ (Wilson and Huner, 2000). In addition to regulating nuclear photosynthetic genes, the redox state of the PQ pool also regulates chloroplast translation (Bruick and Mayfield, 1999), the expression of the chloroplast encoded reaction centre polypeptides, psbA and psaA/B as well as the accumulation of Cyt b_6f (Fujita 1997; Pfannschmidt, 2003; Wilson *et al.*, 2003). The retrograde signal transduction pathway between the chloroplast and the nucleus presently is unknown. However, inhibitors of protein phosphatases prevent the repression of *Lhcb* by high light, indicating that a protein phosphorylation cascade may be involved in linking the redox state of the PQ pool to nuclear gene expression (Escoubas *et al.*, 1995). Recently, Mg-protoporphyrin, an intermediate of the chlorophyll biosynthetic pathway, has been implicated in the signalling pathway between the chloroplast and the nuclear (Kropat *et al.*, 1997; 2000; Strand *et al.*, 2003).

In cyanobacteria, NblS and NblR are considered to act as a two-component sensing/signalling system regulating the degradation of PBSs in response to nutrient limitations (Grossman *et al.*, 2003). The regulation of *nblS* and *nblR* by DCMU and DBMIB is consistent with the model whereby high light as well as nutrient deprivation alter the redox state of PQ pool and thereby regulate this two-component sensing/signalling pathway. However, the redox sensor controlling changes in pigmentation in *Plectonema boryanum* in response to either light or low temperature resides somewhere downstream of the DBMIB binding site in the Cyt b_6f complex state of the PQ pool (Miskiewicz *et al.*, 2000). Clearly, further experimentation is required to elucidate the precise nature of the redox sensor(s) within the photosynthetic electron transport chain of plants, green algae and cyanobacteria (Durnford and Falkowski, 1997; Pfannschmidt *et al.*, 2001; Grossman *et al.*, 2003). Nevertheless, since photosynthetic redox sensing/signalling appears to have global impacts as indicated by its effect on photosynthetic gene expression, cold acclimation, freezing tolerance and plant morphology (Gray *et al.*, 1997), systemic acclimation to excessive light (Karpinski *et al.*, 1999) and cyanobacterial differentiation (Campbell *et al.* 1993), we suggest that the photosynthetic apparatus has a dual role. Not only does it act to harvest and transform light energy, it also acts a primary sensor of environmental change in photoautrophs by sensing imbalances in energy budget (Huner *et al.*, 1998; Huner *et al.*, 2003). This is consistent with the notion of a 'grand design' for photosynthesis proposed initially by Arnon (1982).

ACKNOWLEDGEMENTS

NPAH, DPM and GRG are grateful for financial support through their individual NSERCC Discovery Grants.

References

1. Abadía J, Morales F, and Abadía A. Photosystem II efficiency in low chlorophyll, iron deficient leaves. *Plant and Soil* 2000; **215**: 183-192
2. Allen JF and Pfannschmidt T. Balancing the two photosystems: photosynthetic electron transfer governs transcription of reaction center genes in chloroplasts. *Phil Trans Roy Soc Lond B* 2000; **355**: 1351-1359.
3. Anderson JM. Photoregulation of the composition, function and structure of thylakoid membranes. *Ann Rev Plant Physiol* 1986; **37**: 93-136.
4. Anderson JM, Chow WS, and Park Y-I. The grand design of photosynthesis: acclimation of the photosynthetic apparatus to environmental cues. *Photosynth. Res* 1995; **46**: 129-139.
5. Andersson U, Heddad M, and Adamska I. Light Stress-Induced One-Helix Protein of the Chlorophyll a/b-Binding Family Associated with Photosystem I. *Plant Physiol* 2003; **132**: 811-820.

6. Arnon DI. Sunlight, earth and life: the grand design of photosynthesis. *The Sciences* 1982; **22**: 22-27.

7. Aspinall-O'Dea, Wentworth M, Pascal A, Robert B, Ruban AV, and Horton P. *In vitro* reconstitution of the activated zeaxanthin state associated with energy dissipation in plants. *Proc Natl Acad Sci USA.* 2002; **99**: 16331-16335.

8. Ballaré CL. Keeping up with the neighbours: phytochrome sensing and other signalling mechanisms. *Trends Plant Sci* 1999; **4**: 97-102.

9. Barbato R, Race HL, Friso G, and Barber J. Chlorophyll levels in the pigment-binding proteins of photosystem II. *FEBS Lett* 1991; **86**: 86-90.

10. Baroli I, Do AD, Yamane T, and Niyogi KK. Zeaxanthin Accumulation in the Absence of a Functional Xanthophyll Cycle Protects *Chlamydomonas reinhardtii* from Photooxidative Stress. *Plant Cell* 2003; **4**: 992-1008.

11. Bassi R, and Caffari S. Lhc proteins and the regulation of photosynthetic light harvesting function by xanthophylls. *Photosynth Res* 2000; **64**: 243-256.

12. Bassi R, Pineau B, Dainese P, and Marquardt J. Carotenoid-binding proteins of photosystem II. *Eur J Biochem* 1993; **121**: 297-303.

13. Berry JA and Björkman O. Photosynthetic response and adaptation to temperature in higher plants. *Ann Rev Plant Physiol* 1980; **31**: 491-543.

14. Bibby TS, Nield J, and Barber J. A photosystem II-like protein, induced under iron stress, forms and antenna ring around the photosystem I trimer in cyanobacteria. *Nature* 2001; **412**: 743-745.

15. Bishop DG, and Kendrick JR. Thermal properties of 1-hexadecanoyl-2-trans-3-hexadecenoyl phosphatidylglycerol. *Phytochem* 1987; **26**: 3065-3067.

16. Bishop NI. The carotenoid, lutein, is specifically required for the formation of the oligomeric forms of the light harvesting complex in green the green alga, *Scenedesmus obliquus*. *J Photochem Photobiol* 1996; **36**: 279-283.

17. Björkman O, and Ludlow MM. (1972) Characterization of the light climate on the floor of a Queensland rainforest. *Carnegie Inst Washington Year Book* 1972; **71**, 85-94.

18. Blankenship RE. *Molecular Mechanisms of Photosynthesis.* 2002. Blackwell Science, Oxford, UK, p321.

19. Boekema EJ, Hifney A, Yakushevska AE, Poitrowski M, Keegstra W, Berry S, Michel, K P, Pistorius, E K, and Kruip J. A giant chlorophyll-protein complex indiced by iron deficiency in cyanobacteria. *Nature* **412**; 745-748.

20. Booth PJ, and Paulsen H. Assembly of light harvesting chlorophyll a/b complex in vitro. Time-resolved fluorescence measurements. *Biochem* 1996; **35**: 5103-5108.

21. Brejc K, Ficner R, Huber R, and Steinbacher S. Isolation, crystallization, crystal structure analysis and refinement of allophycocyanin from the cyanobacterium Spirulina platenis at 2.3 Å resolution. *J Mol Biol.* 1995; **249**: 424-440.

22. Bruick RK, and Mayfield SP. Light-activated translation of chloroplast mRNAs. *Trends Plant Sci* 1999; **4**: 190-195.

23. Burnap RL, Troyan T and Sherman LA. The highly abundant chlorophyll-protein complex of iron-deficient Synechococcus sp PCC7942 (CP43') is encoded by the isiA gene. *Plant Physiol.* 1993; **103**: 893-902.

24. Butler WL. Energy distribution in the photochemical apparatus of photosynthesis. *Ann Rev Plant Physiol* 1978; **29**: 345-378.

25. Cammarata KV, and Schmidt GW. In-vitro reconstitution of a light harvesting gene product-deletion mutagenesis and analyses of pigment binding. *Biochem* 1992; **31**: 2779-2789.

26. Campbell D, Houmard J, and Tandeau de Marsac N. Electron transport regulates cellular differentiation in the filamentous cyanobacterium Calothrix. *Plant Cell* 1993; **5**: 451-463.

27. Chang WR, Jiang T, Wan ZL, Zhang JP, Yang ZX, and Liang DC. Crystal structure of R-phycoerythrin from Polysiphonia urceolata at 2.8Å resolution. *J Mol Biol.* 1996; **262**: 721-731.

28. Chory J. Light modulation of vegetative development. *Plant Cell* 1997; **9**: 1225-1234.

29. Collier JL, and Grossman AR. Chlorosis induced by nutrient deprivation in Synechococcus sp strain PCC 7942: not all bleaching is the same. *J Bacteriol.* 1992; **174**: 4718-4726.

30. Collier JL, and Grossman AR. A small polypeptide triggers complete degradation of light harvesting phycobilisomes in nutrient deprived cyanobacteria. *EMBO J* 1994; **13**: 1039-1047.

31. Cooley JW, Hewitt CA, and Vermaas WJF. Succinate:quinol oxidoreductases in the cyanobacterium *Synchocystis* sp. strain PCC6803: presence and function in metabolism and electron transport. *J Bacteriol.* 2000; **182**: 714-722.

32. Cooley JW, and Vermaas WJF. Succinate dehydrogenase and other respiratory pathways in thylakoid membranes of *Synechocystis* sp. strain

PCC6803: capacity comparisons and physiological function. *J Bacteriol.* 2001; **183**: 4251-4248.

33. Davies J, and Grossman AR. Responses to deficiencies in macronutrients. *Advances in Photosynthesis and Respiration. The Molecular Biology of Chloroplasts and Mitochondria in Chlamydomonas* (Rochaix J-D, Goldschmidt-Clermont M, Merchant S, eds). 1998; **7**: 613-635. Kluwer Academic Publishers, Dordrecht.

34. Demmig-Adams B, and Adams WW. The role of xanthophyll cycle carotenoids in the protection of photosynthesis. *Trends Plant Sci.* 1996; **1**: 21-26.

35. Demmig-Adams B, and Adams WW. Antioxidants in Photosynthesis and Human Nutrition. *Science.* 2002; **298**: 2149-2153.

36. Demmig-Adams B, Adams WW, Ebbert V, and Logan BA. Ecophysiology of the xanthophyll cycle. *Advances in Photosynthesis . The Photochemistry of Carotenoids* (Frank HA, Young AJ, Britton G, Cogdell RJ, eds). 1999; **8**: 245-269. Kluwer Academic Publishers, Dordrecht.

37. Dominici P, Caffari S, Armenante F, Ceoldo S, Crimi M, and Bassi R. Biochemical properties of the PsbS subunit of photosystem II either purified from chloroplasts or recombinant. *J Biol Chem.* 2002; **277**: 22750-22758.

38. Dreyfuss BW, and Thornber JP. Assembly of the light-harvesting complexes (LHCs) of photosystem II. Monomeric LHCIIb complexes are intermediates in the formation of oligomeric LHCIIb complexes. *Plant Physiol.* 1994; **106**: 829-839.

39. Dubacq JP, and Trémolières A. Occurence and function of phosphatidylglycerol containing delta3-trans-hexadecenoic acid in photosynthetic lamellae. *Physiol Vég.* 1983; **21**: 293-312.

40. Dubertret G, Mirshahi A, Mirshahi M, Gerard-Hirne C, and Trémolières A. Evidence from *in vivo* manipulations of lipid composition in mutants that the Δ^3-*trans*- hexadecenoic acid-containing phosphatidylglycerol is involved in the biogenesis of the light harvesting chlorophyll a/b-protein complex of *Chlamydomonas reinhardtii. Eur J Biochem.* 1994; **226**: 473-482.

41. Düring M, Schmidt GB and Huber R. Isolation, crystallization, crystal structure analysis and refinement of the constitutive C-phycocyanin from the chromatically adapting cyanobacterium *Fremyella dipolysiphon* at 1.66Å resolution. *J Mol Biol.* 1991; **217**: 577-592.

42. Durnford DG, and Falkowski PG. Chloroplast redox regulation of nuclear gene transcription during photoacclimation. *Photosynth Res.* 1997; **53**, 229-241.

43. Elrad D, Niyogi, KK, and Grossman AR. A Major Light-Harvesting Polypeptide of Photosystem II Functions in Thermal Dissipation. *Plant Cell.* 2002; **14**: 1801-1816.

44. Escoubas J-M, Lomas M, LaRoche J, and Falkowski, PG. Light intensity regulates cab gene transcription via the redox state of the plastoquinone pool in the green alga, *Dunaliella tertiolecta. Proc Natl Acad Sci USA.* 1995; **92**, 10237-10241.

45. Falk S, Samson G, Bruce D, Huner NPA, and Laudenbach DE. Functional analysis of the iron-stress induced CP43' polypeptide of PSII in the cyanobacterium Synechococcus sp. PCCC 7942. *Photosynth Res.* 1995; **45**: 51-60.

46. Falk S, Krol M, Maxwell DP, Rezansoff DA, Gray GR, and Huner NPA. Changes in *in vivo* fluorescence quenching in rye and barley as a function of reduced PSII light harvesting antenna size. *Physiol Plant.* 1994; **91**: 551-558.

47. Falkowski PG. Light-shade adaptation and vertical mixing of marine phytoplankton: a comparative field study. *J Mar Res.* 1983: **41**, 215-237.

48. Falkowski PG, and Chen Y-B. Photoacclimation of light harvesting systems in eukaryotic algae. *Advances in Photosynthesis and Respration. Light Harvesting Systems in Photosynthesis* (Green BR, Parson WW, eds). 2003; **13**: 423-447. Kluwer Academic Publishers, Dordrecht.

49. Förster T. Delocalized excitation and excitation transfer. *Modern Quantum Chemistry. Istanbul Lectures* (Sinanoglu O, ed), **3**: 93-137. Academic Press, New York.

50. Fowler DB, and Carles RJ. Growth, development and cold tolerance of fall acclimated cereal grains. *Crop Sci.* 1979; **19**: 915-922.

51. Frank HA, Cua A, Chynwat V, Young A, Gosztola D, and Wasielewski MR. Photophysics of the carotenoids associated with the xanthophyll cycle in photosynthesis. *Photosynth Res.* 1994; **41**: 389-395.

52. Fujita Y. A study on the dynamic features of photosystem stoichiometry - accomplishments and problems for future studies. *Photosynth Res.* 1997; **53**: 83-93.

53. Fujita Y, Murakami A, Aizawa K, and Ohki K. Short-term and long-term adaptation of the photosynthetic apparatus: homeostatic properties of thylakoids. *Advances in Photosynthesis. The Molecular Biology of*

Cyanobacteria (Bryant, DA, ed.). 1994; **1**: 677-692. Kluwer Academic, Dordrecht.

54. Funk C. The PsbS protein: a Cab-protein with a function of its own. *Advances in Photosynthesis and Respiration. Regulation of Photosynthesis* (Aro E-M, Andersson B, eds). 2001; **11**: 453-467. Kluwer Academic Publishers, Dordrecht

55. Ganeteg U, Kulheim C, Andersson J, and Jansson S. Is each light-harvesting complex protein important for plant fitness? *Plant Physiol.* 2004; **134**: 502-509.

56. Gilmore AM. Mechanistic aspects of xanthophyll cycle-dependent photoprotection in higher plant chloroplasts and leaves. *Physiol Plant.* 1997; 99: 197-209.

57. Gilmore AM, and Ball MC. Photoprotection and storage of chlorophyll in overwintering evergreens. *Proc Natl Acad Sci USA.* 2002; **97**: 11098-11101.

58. Gilmore A M, Matsubara S, Ball MC, Barker DH, and Itoh S. Excitation energy flow at 77K in the photosynthetic apparatus of overwintering evergreens. *Plant Cell Environ.* 2003; **26**: 1021-1034.

59. Glazer AN. Adaptive variations in phycobilisome structure. *Adv Mol Cell Biol* (Barber J, ed.). 1994; **10**: 119-149. JAI Press, London.

60. Gradinaru CC, van Stokkum IHM, van Grondelle R, and van Amerongen H. Identifying the pathways of energy transfer between carotenoids and chlorophylls in LHCII and CP29. A multicolor, femtosecond pump-probe study. *J Phys Chem.* 2000; **104**: 9330-9342.

61. Grasses T, Pcsarcsi P, Schiavon F, Varotto C, Salamini F, Jahns P, and Leister D. The role of ΔpH-dependent dissipation of excitation energy in protecting photosystem II against light-induced damage in *Arabidopsis thaliana. Plant Physiol Biochem.* 2002; **40**: 41-49.

62. Gray GR, Krol M, Khan MU, Williams JP, and Huner NPA. Growth temperature and irradiance modulate trans-Δ^3-hexadecenoic acid content and photosynthetic light harvesting complex organization. *Physiology, Biochemistry and Molecular Biology of Plant Lipids* (Williams JP, Khan MU, Lem NW, eds). 1997; 206-208. Kluwer Academic, Dordrecht.

63. Gray GR, Chauvin L-P, Sarhan F, and Huner NPA. Cold acclimation and freezing tolerance. A complex interaction of light and temperature. *Plant Physiol.* 1997; **114**: 467-474.

64. Green BR, and Durnford DG. The chlorophyll-carotenoid proteins of oxygenic photosynthesis. *Ann. Rev. Plant Physiol. Plant Mol. Biol.* 1996; **47**, 685-714.

65. Green BR, and Parson WW. *Advances in Photosynthesis and Respiration. Light-Harvesting Antennas in Photosynthesis* (Green BR, Parson WW, eds). 2003; **13**; 495-513. Kluwer Academic Publishers, Dordrecht.

66. Grossman AR, van Waasbergen LG, and Kehoe D. Environmental regulation of phycobilisome biosynthesis. *Advances in Photosynthesis and Respiration. Light-Harvesting Antennas in Photosynthesis* (Green BR, Parson WW, eds). 2003; **13**: 471-493. Kluwer Academic Publishers, Dordrecht.

67. Grossman AR, Schaefer MR, Chiang GG, and Collier JL. The responses of cyanobacteria to environmental conditions: light and nutrients. *Advances in Photosynthesis. The Molecular Biology of Cyanobacteria* (Bryant DA, ed.). 1994; **1**: 641-675. Kluwer Academic, Dordrecht.

68. Grossman AR. Chromatic adaptation and the events involved in phycobilisome biosynthesis. *Plant Cell Environ.* 1990; **13**: 651-666.

69. Haehnel W. Photosynthetic electron transport in higher plants. *Ann. Rev. Plant Physiol.* 1984; **35**: 659-693.

70. Havaux M, and Kloppstech K. The protective functions of carotenoids and flavonoid pigments against excess visible radiation at chilling temperature investigated in Arabidopsis npq and tt mutants. *Planta* 2001; **213**: 953-966.

71. Havaux M, and Niyogi KK. The violaxanthin cycle protects plants from photooxidative damage by more than one mechanism. *Proc Natl Acad Sci USA.* 1999; **96**: 8762-8767.

72. Hobe S, Forster R, Klingler J, and Paulsen H. N-Proximal sequence motif in light-harvesting chlorophyll *a*/*b*-binding protein is essential for the trimerization of light-harvesting chlorophyll *a*/*b* complex. *Biochem.* 1995; **34**: 10224-10228.

73. Hobe S, Prytulla S, Kulhlbrandt W, and Paulsen H. Trimerization and crystallization of reconstituted light-harvesting chlorophyll a/b complex. *EMBO* 1994; **13**: 3423-3429.

74. Hoch, JA, and Silhavy TJ. *Two-component signal transduction.* 1995. p486. ASM Press, Washington, D.C.

75. Holtgrefe S, Bader KP, Horton P, Scheibe R, vonSchaewen A, and Backhausen JE. Decreased content of leaf ferredoxin changes electron distribution and limits photosynthesis in transgenic potato plants. *Plant Physiol.* 2003; **133**: 1768-1778.

76. Horton P, Ruban AV, Rees D, Pascal AA, Noctor G, and Young AJ. Control of light harvesting function of chloroplast membranes by aggregation of the LHCII chlorophyll-protein complex: a hypothesis. *FEBS Lett.* 1991; **292**: 1-4.

77. Horton P, Ruban AV, and Walters RG. Regulation of light harvesting in green plants. *Ann.Rev. Plant Physiol. Plant Mol.Biol.* 1996; **47**, 655-684.

78. Horton P, Ruban AV, and Young AJ. Regulation of the structure and function of the light harvesting complexes of photosystem II by the xanthophyll cycle. *Advances in Photosynthesis . The Photochemistry of Carotenoids* (Frank HA, Young AJ, Britton G, Cogdell RJ, eds). 1999; **8**: 271-291. Kluwer Academic Publishers, Dordrecht.

79. Huner NPA, Öquist G, Hurry VM, Krol M, Falk S, and Griffith M. Photosynthesis, photoinhibition and low temperature acclimation in cold tolerant plants. *Photosynth Res.* 1993; **37**: 19-39.

80. Huner NPA, Krol M, Ivanov AG, Sveshnikov D, and Öquist G. CP43' induced under Fe-stress in *Synechococcus* sp PCC 7942 is associated with PSI. *Proc 12th Intern Congr Photosynth.* 2001; **S3**: 061. CSIRO Publishing, Melbourne.

81. Huner N P A, Öquist G, and Melis A. Photostasis in plants, green algae and cyanobacteria: the role of light harvesting antenna complexes. *Advances in Photosynthesis and Respiration. Light Harvesting Antennas in Photosynthesis* (Green BR, Parson WW, eds). 2003; **13**: 401-421. Kluwer Academic Publishers, Dordrecht.

82. Huner NPA, Öquist G, and Sarhan F. Energy balance and acclimation to light and cold. *Trends Plant Sci.* 1998; **3**: 224-230.

83. Huner N P A, Williams JP, Maissan EE, Myscich EG, Krol M, Laroche A, and Singh J. Low temperature-induced decrease in trans-delta3-hexadecenoic acid content is correlated with freezing tolerance in cereals. *Plant Physiol.* 1989; **89**: 144-150.

84. Hurry VM, Anderson JM, Chow WS, and Osmond CB. Accumulation of zeaxanthin in abscisic acid-deficient mutants of Arabidopsis does not affect chlorophyll fluorescence quenching or sensitivity to photoinhibition in vivo. *Plant Physiol.* 1997; **113**: 639-648.

85. Hutin C, Nussaume L, Moise N, Moya I, Kloppstech K, and Havaux M. Early light-induced proteins protect Arabidopsis from photooxidative stress. *Proc Natl Acad Sci USA.* 2003; **100**: 4921-4926.

86. Jahns P, and Krause GH. Xanthophyll cycle and energy-dependent fluorescence quenching in leaves from pea plants grown under intermittent light. *Planta* 1994; **192**: 176-182.

87. Jansson S. The light-harvesting chlorophyll a/b-binding proteins. *Biochim Biophys Acta* 1994; **1184**: 1-19.

88. Jordan P, Fromme P, Klukas O, Witt HT, Saenger W, and Krauss N. Three dimensional structure of cyanobacterial PSI at 2.5A resolution. *Nature* 2001; **411**: 909-917.

89. Karpinski S, Reynolds H, Karpinska B, Wingsle G, Creissen G, and Mullineaux P. Systemic signaling and acclimation in response to excess excitation energy in Arabidopsis. *Science* 1999; **284**: 654-657.

90. Kehoe DM, Grossman AR. Similarity of a chromatic adaptation sensor to phytochrome and ethylene receptors. *Science* 1996; **273**: 1409-1412.

91. Krause GH. Photoinhibition of photosynthesis. An evaluation of damaging and protective mechanisms. *Physiol Plant.* 1988; **74**: 566-574.

92. Krause GH, and Jahns P. Pulse amplitude modulated chlorophyll fluorometry and its application in plant science. *Advances in Photosynthesis and Respiration. Light Harvesting Antennas in Photosynthesis* (Green BR, Parson WW eds). 2003; **13**: 373-399. Kluwer Academic Publishers, Dordrecht.

93. Krause GH, and Weis E. Chlorophyll fluorescence and photosynthesis: the basics. *Ann Rev Plant Physiol Plant Mol Biol.* 1991; **42**: 313-349.

94. Krol M, Gray GR, Hurry VM, Öquist G, and Huner NPA. Low temperature stress and photoperiod affect an increased tolerance to photoinhibition in *Pinus banksiana* seedlings. *Can J Bot.* 1995; **73**: 1119-1127.

95. Krol M, Huner NPA, Williams JP, and Maissan E. Chloroplast biogenesis at cold hardening temperatures. Kinetics of trans-3-hexadecenoic acid accumulation and the assembly of LHCII. *Photosynth Res.* 1988; **15**: 115-132.

96. Krol M, Huner NPA, Williams JP, and Maissan EE. Prior accumulation of phosphatidylglycerol high in trans-Δ^3-hexadecenoic acid enhances the in vitro stability of oligomeric light harvesting complex II. *J Plant Physiol.* 1989; **135**: 75-90.

97. Krol M, Ivanov AG, Jansson S, Kloppstech K, and Huner NPA. Greening under high light or cold temperature affects the level of xanthophyll-cycle pigments, early light-inducible proteins, and light-harvesting polypeptides

in wild-type barley and the chlorina f2 mutant. *Plant Physiol.* 1999; **120**: 193-203.

98. Krol M, Maxwell DP, and Huner NPA. Exposure of *Dunaliella salina* to low temperature mimics the high light-induced accumulation of carotenoids and the carotenoid binding protein (Cbr). *Plant Cell Physiol.* 1997; **38**: 213-216.

99. Krol M, Spangfort MD, Huner NPA, Öquist G, Gustafsson P, Jansson S. Chlorophyll *a/b*-binding proteins, pigment conversions and early light-induced proteins in a chlb-less barley mutant. *Plant Physiol.* 1995; **107**: 873-883.

100. Kropat J, Oster U, Rudiger W, Beck CF. Chlorophyll precursors are signals of chloroplast origin involved in light induction of nuclear heat-shock genes. *Proc Natl Acad Sci USA.* 1997; **94**: 14168-14172.

101. Kropat J, Oster U, Rudiger W, and Beck CF. Chloroplast signalling in the light induction of nuclear HSP70 genes requires the accumulation of chlorophyll precursors and their accessibility to cytoplasm/nucleus. *Plant J.* 2000; **24**: 523-531.

102. Krupa Z, Huner NPA, Williams JP, Maissan E, and James DR. Development at cold-hardening temperatures. The structure and composition of purified rye light harvesting complex II. *Plant Physiol.* 1987; **84**: 19-24.

103. Krupa Z, Williams JP, and Huner, NPA. The role of acyl lipids in reconstitution of lipid-depleted light harvesting complex II from cold hardened and nonhardened rye. *Plant Physiol.* 1992; **100**: 931-938.

104. Kuhlbrandt W, Wang DN, and Fuijyoshi Y. Atomic model of plant light-harvesting determined by electron crystallography. *Nature* 1994; **367**: 614-621.

105. Kulheim C, Agren J, and Jansson S. Rapid regulation of light harvesting and plant fitness in the field. *Science* 2002; **297**: 91-93.

106. Kurisu G, Zhang H, Smith J, and Cramer WA. Structure of the cytochrome b6f complex of oxygenic photosynthesis: tuning the cavity. *Science* 2003; **302**: 1009-1014.

107. Laroche J, Mortain-Bertrand A, Falkowski PG. Light intensity-induced changes in cab mRNA and light harvesting complex II apoprotein levels in the unicellular chlorophyte *Dunaliella tertiolecta*. *Plant Physiol.* 1991; **97**:147-153.

108.Li X P, Björkman O, Shih C, Grossman AR, Rosenquist M, Jansson S, and Niyogi KK. A pigment-binding protein essential for regulation of photosynthetic light harvesting. *Nature* 2000; **403**: 391-395.

109.Lindahl M, Funk C, Webster J, Bingsmark S, Adamska I, and Andersson B. Expression of ELIPs and PSII-S protein in spinach during acclimative reduction of the photosystem II in response to increased light intensities. *Photosynth Res.* 1997; **54**: 227-236.

110.Long SP, Humphries S, and Falkowski PG. Photoinhibtion of photosynthesis in nature. *Ann Rev Plant Physiol Plant Mol Biol.* 1994; **45**: 633-662.

111.Lunde C, Jensen PE, Haldrup A, Knoetzel J, and Scheller H V. The PSI-H subunit of photosystem I is essential for state transitions in plant photosynthesis. *Nature* 2000; **408**: 613-615.

112.Machalek KM, Davison IR, and Falkowski PG. Thermal acclimation and photoacclimation of photosynthesis in the brown alga *Laminaria saccharina. Plant CellEnviron.*1996; **19**:1005-1016.

113.Maxwell DP, Laudenbach DE, and Huner NPA. Redox regulation of light-harvesting complex II and *cab* mRNA abundance in *Dunaliella salina. Plant Physiol.* 1995; **109**: 787-795.

114.McCourt P, Browse J, Watson J, Arntzen CJ, and Somerville CR. Analysis of photosynthetic antenna function in a mutant of Arabidopsis thaliana lacking trans-hexadecenoic acid. *Plant Physiol.* 1985; **78**: 853-858.

115.Melis A. Dynamics of photosynthetic membrane composition and function. *Biochim Biophys Acta* 1991; **1058**: 87-106.

116.Melis A. Photostasis in plants. *Photostasis and Related Phenomena.* (Williams E and Thistle R, eds). 1998, 207-220. Plenum Press, New York.

117.Melis A. Photosystem-II damage and repair cycle in chloroplasts: what modulates the rate of photodamage *in vivo? Trends Plant Sci.* 1999; **4**: 130-135.

118.Meyer G, and Kloppstech K. A rapidly light-induced chloroplast protein with high turnover coded for by pea nuclear DNA. *Eur J Biochem.* 1984; **138**: 201-207.

119.Mimuro M, Kikuchi H. Antenna systems and energy transfer in cyanophyta and rhodaphyta. *Advances in Photosynthesis and Respiration. Light-Harvesting Antennas in Photosynthesis* (Green BR, Parson WW, eds). 2003; **13**: 281-306. Kluwer Academic Publishers, Dordrecht.

120.Miskiewicz E, Ivanov AG, and Huner NPA. Stoichiometry of the photosynthetic apparatus and phycobilisome structure of the cyanobacterium *Plectonema boryanum* UTEX 485 are regulated by both light and temperature. *Plant Physiol.* 2002; **130**: 1414-1425.

121.Miskiewicz E, Ivanov AG, Williams JP, Khan MU, Falk S, and Huner NPA. Photosynthetic acclimation of the filamentous cyanobacterium, *Plectonema boryanum* UTEX 485, to temperature and light. *Plant Cell Physiol.* 2000; **41**: 767-775.

122.Montane MH, Dreyer S, Triantaphylides C, and Kloppstech K. Early light-inducible proteins during long-term acclimation of barley to photooxidative stress caused by light and cold - high level of accumulation by posttranscriptional regulation. *Planta* 1997; **202**: 293-302.

123.Montane M-H, and Kloppstech K. The family of light-harvesting-related proteins (LHCs, ELIPs, HLIPs): was the harvesting of light their primary function? *Gene* 2000; **258**: 1-8.

124.Nield J, Funk C, and Barber J. Supermolecular structure of photosystem II and location of the PsbS protein. *Phil Trans Royal Soc London Series B* 2000a; **355**: 1337-1343.

125.Nield J, Orlova E, Morris E, Cowen B, van Heel M, and Barber J. 3D map of the plant photosystem two supercomplex obtained by cryoelectron microscopy and single particle analysis. *Nature Struct Biol.* 2000b; **7**:44-47.

126.Niyogi KK. Photoprotection revisited: Genetic and Molecular Approaches. *Ann Rev Plant Physiol Plant Mol Biol.* 1999; **50**: 333-359.

127.Norén H, Svensson P, Stegmark R, Funk C, Adamska I, and Andersson B. Expression of the early light-induced protein but not the PsbS protein is influenced by low temperature and depends on the developmental stage of the plant in field-grown pea cultivars. *Plant Cell Environ.* 2003; **26**: 245-253.

128.Nußberger S, Dorr K, Wang DN, and Kuhlbrandt W. Lipid-protein interactions in crystals of plant light-harvesting complex. *J Mol Biol.* 1993; **234**: 347-356.

129.Öquist G, Chow WS, and Anderson JM. Photoinhibition of photosynthesis represents a mechanism for long term regulation of photosystem II. *Planta* 1992; **186**, 450-460.

130.Öquist G, Gardeström P, Huner N P A. Metabolic changes during cold acclimation and subsequent freezing and thawing. *Conifer Cold Hardiness*

(Bigras FJ, Colombo SJ, eds). 2001; pp.137-163. Kluwer Academic, Dordrecht.

131. Öquist G, and Huner NPA. Photosynthesis of overwintering evergreens. *Ann Rev Plant Biol.* 2003; **54**: 329-355.

132. Ort DR. When there is too much light. *Plant Physiol.* 2001; **125**: 29-32.

133. Ottander C, Campbell D and Öquist G. Seasonal changes in photosystem II organization and pigment composition in *Pinus sylvestris. Planta* 1995; **197**, 176-183.

134. Park Y-I, Sandstrom S, Gustafsson P, and Oquist G. Expression of the isiA gene is essential for the survival of the cyanobacterium *Synechococcus* sp. PCC 7942 by protecting photosystem II from excess light under iron limitation. *Mol Microbiol.* 1999; **32**: 123-129.

135. Paulsen H. Carotenoids and the assembly of light harvesting complexes. *Advances in Photosynthesis and Respiration. The Photochemistry of Carotenoids* (Frank HA, Young AJ, Britton G, Cogdell RJ, eds).1999; **8**:123-136. Kluwer Academic Publishers, Dordrecht.

136. Paulsen H, Hobe S. Pigment binding properties of mutant light harvesting chlorophyll a/b-binding protein. *Eur J Biochem.* 1992; **205**: 71-76.

137. Peterson RB, and Havir EA. Photosynthetic properties of an *Arabidopsis thaliana* mutant possessing a defective PsbS gene. *Planta* 2001; **214**: 142-152.

138. Peterson RB, and Havir EA. Contrasting modes of regulation of PSII light utilization with changing irradiance in normal and PsbS mutant leaves of Arabidopsis thaliana. *Photosynth Res.* 2003; **75**: 57-70.

139. Pfannschmidt T. Chloroplast redox signals: how photosynthesis controls its own genes. *Trends Plant Sci.* 2003; 8: 33-41.

140. Pfannschmidt T, Nilsson A, and Allen JF. Photosynthetic control of chloroplast gene expression. *Nature* 1999; **397**, 625-628.

141. Plumley FG, and Schmidt GW. Reconstitution of chlorophyll a/b light harvesting complexes: xanthophyll-dependent assembly and energy transfer. *Proc Natl Acad Sci USA.* 1987; **84**: 146-150.

142. Polivka T, Herek JL, Zigmantas D, Akerlund H-E, and Sundström V. Direct observation of the (forbidden) S1 state in carotenoids. *Proc Natl Acad Sci USA.* 1999; **96**: 4912-4917.

143. Polivka T, Zigmantis D, and Sundström V. Carotenoid S1state in a recombinant light harvesting complex of photosystem II. *Biochem.* 2002, **41**: 439-450.

144.Ruban AV, Pascal AA, Robert B, and Horton P. Activation of zeaxanthin is an obligatory event in the regulation of photosynthetic light harvesting. *J Biol Chem.* 2002; **277**: 7785-7789.

145.Ruban AV, Wentworth M, Yakushevska AE, Andersson J, Lee PJ, Keegstra W, Dekker JP, Boekema EJ, Jansson S, and Horton P. Plants lacking the main light harvesting complex retain photosystem II macro-organization. *Nature* 2003; **421**: 648-652.

146.Sandona D, Croce R, Pagano A, Crimi M, and Bassi R. Higher plant light harvesting proteins. Structure and function as revealed by mutation analysis of either protein or chromophore moieties. *Biochim Biophys Acta* 1998; **1365**: 201-214.

147.Savitch LV, Leonardos ED, Krol M, Jansson S, Grodzinski B, Huner NPA and Öquist G. Two different strategies for light utilization in photosynthesis in relation to growth and cold acclimation. *Plant Cell Environ.* 2002; **25**: 761-771.

148.Savitch LV, Maxwell DP and Huner NPA. Photosystem II excitation pressure and photosynthetic carbon metabolism in *Chlorella vulgaris.* *Plant Physiol.* 1996; **111**: 127-136.

149.Schatz GH, Brock H, and Holzwarth AR. Kinetic and energetic model for the primary processes in photosystem II. *Biophys J.* 1988; **54**: 397-405.

150.Scherer S. Do photosynthetic and respiratory electron transport chains share redox proteins? *Trends Biochem Sci.* 1990; **15**: 458-462.

151.Selstam E. Development of thylakoid membranes with respect to lipids. *Advances in Photosynthesis and Respiration. Lipids in Photosynthesis: Structure, Function and Genetics* (Siegenthaler P-A, ed). 1998; **6**: 209-224. Kluwer Academic Publishers, Dordrecht.

152.Sidler WA. Phycobilisome and phycobiliprotein structures. *Advances in Photosynthesis. Molecular Biology of Cyanobacteria* (Bryant DA, ed.). 1994; **1**: 139-216. Kluwer Academic, Dordrecht.

153.Siegenthaler P-A, and Trémolières A. Role of acyl lipids in the function of photosynthetic membranes in higher plants. *Advances in Photosynthesis and Respiration. Lipids in Photosynthesis: Structure, Function and Genetics* (Siegenthaler P-A, ed). 1998; **6**: 145-173. Kluwer Academic Publishers, Dordrecht.

154.Strand A, Asami T, Alonso J, Ecker JR, and Chory J. Chloroplast to nucleus communication trigered by accumulation of Mg-protoporphyrin IX. *Nature* 2003; **421**: 79-83.

155. Straus NA. Iron deprivation: physiology and gene regulation. *Advances in Photosynthesis. The Molecular Biology of Cyanobacteria* (Bryant DA, ed.) 1994; **1**: 731-750. Kluwer Academic, Dordrecht.

156. Stroebel D, Choquet Y, Popot J-L, and Picot D. An atypical haem in the cytochrome b_6f complex. *Nature* 2003; **426**: 413-418.

157. Sukenik A, Wyman KD, Bennett J, Falkowski PG. A novel mechanism for regulating the excitation of photosystem II in a green alga. *Nature* 1987; **327**: 704-707.

158. Sukenik A, Bennett J, and Falkowski PG. Changes in the abundance of individual apoproteins of light-harvesting chlorophyll *a/b* complexes of photosystem I and II with growth irradiance in the marine chlorophyte *Dunaliella teriolecta. Biochim Biophys Acta* 1988; **932**: 206-215.

159. Swiatek M, Kuras R, Sokolenko A, Higgs D, Olive J, Cinque G, Muller B, Eichacker LA, Stern DB, Bassi R, Herrmann R G, and Wollman F-A. The chloroplast gene ycf9 encodes a photosystem II (PSII) core subunit, PsbZ, that participates in PSII supramolecular architecture. *Plant Cell* 2001; **13**: 1347-1368.

160. Tandeau de Marsac N, and Houmard J. Adaptation of cyanobacteria to environmental stimuli: new steps towards molecular mechanisms. *FEMS Microbiol Rev.* 1993; **104**: 119-190.

161. Terry N. Limiting factors in photosynthesis. IV. Iron stress mediated changes in light harvesting and electron transport capacity and its effects on photosynthesis in vivo. *Plant Physiol.* 1983; **71**, 855-860.

162. Trémolières A, Dubacq JP, Ambard-Bretteville F, and Remy R. Lipid composition of chlorophyll-protein complexes. Specific enrichment in trans-hexadecenoic acid of an oligomeric form of light harvesting chlorophyll a/b protein. *FEBS Lett.* 1981; **130**: 27-31.

163. Trémolières A, and Siegenthaler P-A. Reconstitution of photosynthetic structures and activities with lipids. *Advances in Photosynthesis and Respiration. Lipids in Photosynthesis: Structure, Function and Genetics* (Siegenthaler P-A, ed). 1998; **6**: 175-189. Kluwer Academic Publishers, Dordrecht.

164. van Amerongen H and Dekker JP. Light harvesting in photosystem II. *Advances in Photosynthesis and Respiration. Light Harvesting Antennas in Photosystem II* (Green BR, Parson WW, eds) 2003; **13**: 219-251, Kluwer Academic Publishers, Dordrecht

165. van Waasbergen LG, Dolganov N, and Grossman AR. nblS, a gene involved in controlling photosynthesis-related gene expression during high

light and nutrient stress in *Synechococcus elongatus* PCC 7942. *J Bacteriol.* 2002; **184**: 2481-2490.

166. Vogelmann TC, Nishio JN, and Smith WK. Leaves and light capture: light propagation and gradients of carbon fixation within leaves. *Trends Plant Sci.* 1996; **1**: 65-70.

167. Walters RG, and Horton, P. Theoretical assessment of alternative mechanisms for non-photochemical quenching of PSII fluorescence in barley leaves. *Photosynth Res.* 1993; **36**: 119-139.

168. Weger HG, Silim SN, and Guy RD. Photosynthetic acclimation to low temperature by Western red cedar seedlings. *Plant Cell Environ.* 1993; **16**: 711-717.

169. Wentworth M, Ruban AV, and Horton P. Thermodynamic Investigation into the Mechanism of the Chlorophyll Fluorescence Quenching in Isolated Photosystem II Light-harvesting Complexes. *J Biol Chem.* 2003; **278**: 21845-21850.

170. Wilson KE and Huner NPA. The role of growth rate, redox-state of the plastoquinone pool and the trans-thylakoid ΔpH in photoacclimation of *Chlorella vulgaris* to growth irradiance and temperature. *Planta* 2000; **212**, 93-102.

171. Wilson KE, Krol M, and Huner NPA. Temperature-induced greening of *Chlorella vulgaris*. The role of the cellular energy balance and zeaxanthin-dependent nonphotochemical quenching. *Planta* 2003; **217**: 616-627.

172. Winder, T.L. & Nishio, J. Early iron deficiency stress response in leaves of sugar beet. *Plant Physiol.* 1995; **108**, 1487-1494.

173. Wykoff, D.D., Davies, J.P., Melis, A. & Grossman, A.R. The regulation of photosynthetic electron transport during deprivation in *Chlamydomonas reinhardtii*. *Plant Physiol.* 1998; **117**, 129-139.

174. Zouni A, Witt HT, Kern J, Fromme P, Krauss N, Saenger W, Orth P. Crystal structure of photosystem II from *Synechococcus elongatus* at 3.8 angstrom resolution. *Nature* 2001; **409**: 739-743.

Chapter 5

FROM BIOLOGICAL TO SYNTHETIC LIGHT-HARVESTING MATERIALS – THE ELEMENTARY STEPS

Tõnu Pullerits and Villy Sundström

Light harvesting in photosynthetic purple bacteria is reviewed in detail based on recent advances in structure determination and ultrafast laser spectroscopy. Knowledge obtained from photosynthesis research forms a solid ground for studies of various artificial light harvesting systems. We present our recent research of excitation transfer and charge separation in conjugated polymers. We continue by summarizing our studies of ultrafast photoinduced processes in dye-sensitized nanocrystalline large band-gap semiconductor films – a key part of the Grätzel solar cell. Finally, our recent studies of energy transfer in transition metal supramolecular complexes, a kind of artificial antenna, are presented.

Keywords: Photosynthesis, purple bacteria, excitons, polarons, ultrafast dynamics, conducting polymers, semiconductor films.

1. INTRODUCTION

The energy sources we are using today are associated with unacceptable impact on our environment and are in addition rapidly becoming depleted. There is therefore a strong need for renewable and clean energy sources. Our sun is delivering much more energy to the planet than we presently consume – we 'just' have to find efficient ways to convert it to forms of energy that we can conveniently use. Presently, two options appear feasible; conversion to electricity with the help of solar cells based on some material for light to charge conversion, or conversion to some kind of fuel in a process mimicking photosynthesis. In both processes, light energy is harvested by some molecular entity and then converted to charge – which is used directly as electrical current, or used to drive chemical reactions producing energy-rich molecules. Nature has in photosynthesis, over billions of years, developed sophisticated molecular structures to achieve this. In designing artificial systems for light energy conversion, Nature's concepts are probably a good starting point. In this chapter we are discussing the nature of the very first light-driven processes in photosynthesis, and also comparing light to charge reactions in some synthetic supramolecular and polymeric systems that could form the active materials in future solar energy converting systems.

2. PHOTOSYNTHETIC LIGHT-HARVESTING – ENERGY TRANSFER AND TRAPPING

Primary energy transfer and trapping in photosynthesis occurs in sophisticated machinery consisting of a so-called antenna and a reaction centre (RC) together forming a photosynthetic unit (PSU). The antenna is an array of pigment molecules which absorbs light energy and transfers

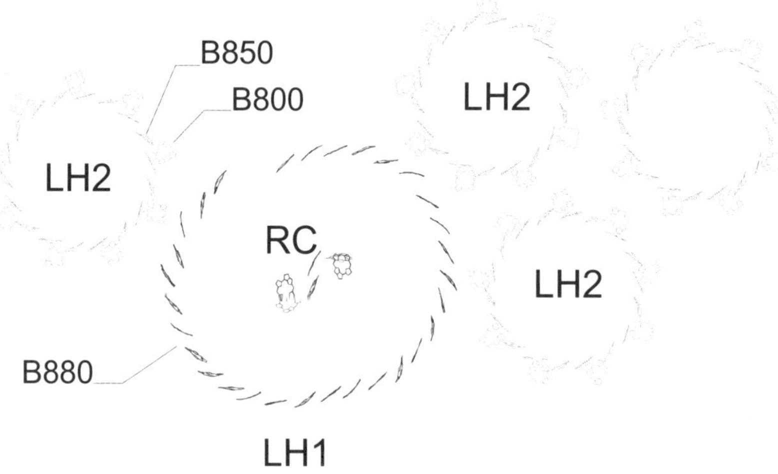

Fig. 1. Model of the photosynthetic unit of purple bacteria based on the known structural data of LH2[1] and LH1+RC.[2] B800, B850 and B880 correspond to bacteriochlorophylls absorbing at 800, 850 and 880 nm, respectively. The diameter of the PSU is about 250 Å.

the energy in a form of molecular electronic excitation to a trap, the RC, where primary charge separation takes place. In Nature many different types of RCs and even more antennas exist. In the current section we will review some fundamental details of photosynthetic light harvesting, using the PSU of photosynthetic purple bacteria (Figure 1) as a working model. Compared to the complexity of green plants,[3,4,5] the bacterial light harvesting may seem far too primitive. However, the basic principles of light harvesting, relevant from the point of view of applications, are also likely to be very similar in other more complex photosynthetic systems. Furthermore, light harvesting in purple bacteria is among the most studied and best understood processes of primary photosynthesis.[6]

The PSU of purple bacteria generally consists of more than one antenna complex. The peripheral antenna (LH2) is in touch with the core antenna (LH1), which surrounds the RC. In constructing Figure 1 we have used the crystal structures of LH2 from *Rhodopseudomonas* (*Rps.*) *acidophila*[1] and RC-LH1 from *Rps. palustris*.[2] Even if there are slight differences in structural details of various species, the PSUs from different purple bacteria are expected to be almost the same. The peripheral antenna LH2 consists of two concentric rings of BChl molecules, named B800 and B850 according to their characteristic Q_y absorption maxima at 800 nm and 850 nm. In LH1 the B880 BChls are similar to the B850s. One remarkable difference is that the B880s do not form a full ring – one pair of BChls is missing, making it possible for the reduced ubiquinone to transport electrons out from the RC. In most photosynthetic systems, besides the chlorophyll-type pigments there are many different carotenoid molecules (not shown in Figure 1) serving as light harvesters and protectors against photo-degradation. The photophysics of carotenoids is currently an active research field of its own and goes beyond the scope of the current article. Instead we refer the interested readers to a recent review by Polivka and Sundström.[7]

2.1. B800

The BChls of B800 are well separated from each other and from the B850s and thereby have mainly monomeric spectroscopic properties. Excitation absorbed in the B800 ring is first transferred among B800

BChls.[8] Calculations based on the Förster theory agree remarkably well with the measured pairwise transfer time of 300 fs suggesting that for the B800 ring the point-dipole approximation is applicable.[9] Also low temperature transient absorption anisotropy kinetics measured at different wavelengths inside the B800 band were successfully simulated by a model of Förster hopping in a spectrally inhomogeneous ring of BChl molecules.[9]

B800 to B850 transfer occurs with a time-constant of 0.7 ps at room temperature[10,11,12] and it slows down upon lowering the temperature to 1.2 ps at 77 K and to 1.5 ps at 4 K.[10] B800 BChls can be exchanged to other similar pigments.[13] A series of systems where the B800 band was blue-shifted was studied and a significant increase of transfer time was observed.[14] The most blue-shifted pigment, Chl at 670 nm, gave a transfer time of 8.3 ps. All these trends of the transfer time are in qualitative agreement with what one would expect based on Förster spectral overlap. However, quantitative Förster theory calculations of the transfer time based on the dipole-dipole interaction between B800 and nearby B850s failed to reproduce the observed lifetimes by almost an order of magnitude. Since the spectral overlap did describe the qualitative trend it was suggested that the source of the discrepancy is the electronic coupling term.[14] It has been pointed out that the carotenoid molecule may contribute to the electronic coupling between B800 and B850.[15] Alternatively the non-diagonal[16] and/or diagonal[17] electron phonon coupling may facilitate efficient excitation transfer to optically forbidden exciton levels. In the spirit of the same ideas a modified Förster theory adapted for transfer to the collective exciton states with spectral inhomogeneity has been used for describing B800 to B850 transfer, apparently leading to quantitative agreement between theory and experiment.[18]

2.2. Excitons and Polarons in B850

The B850 ring forms a densely packed excitonically coupled aggregate. Shortly after the structure of the LH2 became available, the question of the extent of exciton delocalisation in B850 became a hot research subject. Different experimental techniques and theoretical methods lead to diverging, sometimes even contradicting conclusions.[6] Based on nonlinear absorption experiments[19] it was concluded that the excitation is delocalised over almost the whole B850 ring, whereas analyses of the transient absorption spectra suggested quite limited exciton delocalisation on the picosecond timescale – about 4 BChl molecules.[20] Many following studies supported one or the other point of view.[21-23] The apparent controversy was resolved by considering that the exciton, and thereby also exciton delocalisation, is time-dependent.[24] Furthermore, different definitions of exciton delocalisation length led to rather different numerical values for the quantity. We showed that at the moment of excitation, certain measures of exciton delocalisation indeed gave numerical values suggesting full-ring excitons (see Figure 2.). After a few hundreds of femtoseconds most of the measures gave the value ~4 BChls.[24]

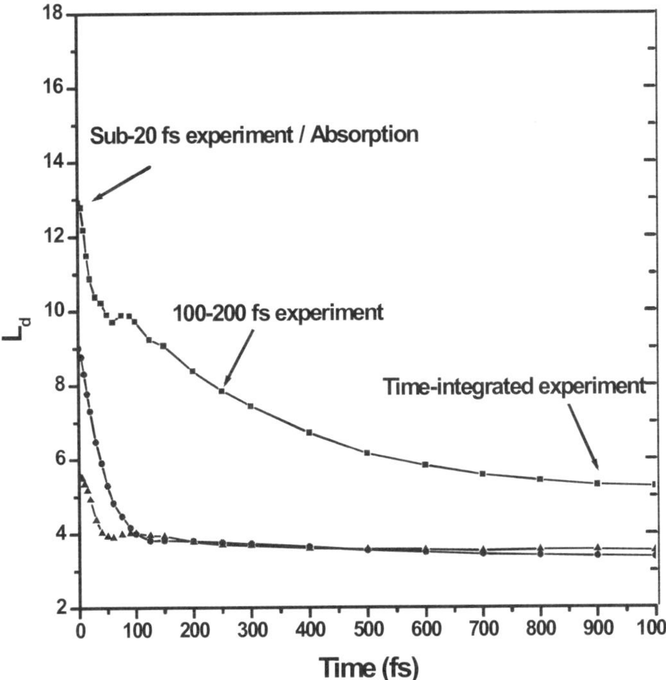

Fig. 2. Time evolution of three different numerical characteristics of exciton delocalisation length in B850. For more detailed description of the characteristics see ref. 24.

At high light intensities it may happen that two or more excitations are simultaneously present in a set of molecules exchanging excitation, like B850. In this case excitation annihilation can take place.[25,26] The annihilation process is usually described as excitation transfer to an already excited molecule, where it produces a higher (doubly) excited electronic state. By very fast internal conversion the molecule relaxes to the lowest excited molecular state and one excitation is lost. The same process has been also described in terms of collective excitations *i.e.*, excitons, by coupling the one- and two-exciton manifolds.[27] Simultaneous analysis of excitation annihilation dynamics and transient absorption anisotropy decay in well-separated rings of LH2 was recently used for obtaining detailed information about excited state properties and dynamics in a B850 ring.[28] It was confirmed that excitons in B850 are delocalised over 3-4 BChls.

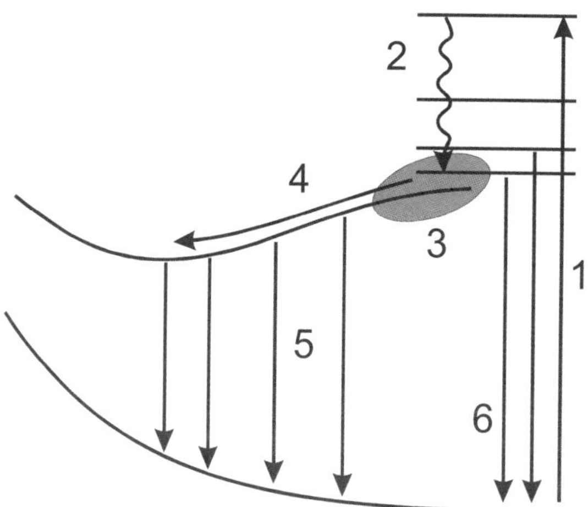

Fig. 3. Schematic picture of the different processes occurring in B850 at low temperature: excitation into the B850 band creates an initial, nonselective population on exciton levels (1); this population relaxes through the exciton band in about 100 fs (2); the lowest exciton states are mixed with charge-transfer states and these states are populated by means of a slower process occurring within 0.6 ps (3); due to polaron formation, slow motion along the relaxation coordinate takes place on a time scale of about 10 ps (4); stimulated emission from the polaron states is seen as a new band in the red part of TAS (5). At room temperature, thermal excitations do not allow population of charge-transfer states for a sufficient time to relax the population along the relaxation coordinate; hence, only stimulated emission from the lowest exciton states is observed (6).

At low temperatures transient absorption dynamics reveal a new, significantly red-shifted stimulated emission band in B850 on a picosecond timescale.[29] It was suggested that the band appears due to polaron formation in B850 (see Figure 3). Later, a similar interpretation was given to strongly Stokes-shifted emission from B850.[30] In that work the process was called exciton self-trapping. Exciton self-trapping and polaron formation are two names for the same physical process where electron-phonon coupling leads to a significant change in the excited state nuclear configuration accompanied by lowering of the energy on the excited site. The latter causes exciton localization.[31,32] We point out that it has been argued that the above red stimulated emission band may reflect interring excitation transfer among inhomogeneously distributed B850 rings.[33] In the following section we show that the interring transfer takes place on a considerably slower timescale.

2.3. Inter-Complex Excitation Transfer

Recently we studied the transfer between aggregated B850 rings using time-resolved excitation annihilation together with transmission electron microscopy.[34] Figure 4 shows transient absorption kinetics after direct excitation of the B850 band, as obtained for various surfactant (N,N-dimethyl dodecylamine oxide, LDAO, was used in this study) concentrations in the case of an LH2 solution at 850 nm, corresponding to 0.06 μM LH2 complexes. In all curves a subpicosecond component is present, reflecting the initial *intra*-ring annihilation for the B850 rings that are excited by more than one excitation by the laser pulse. The slower dynamics contain two different components: the single-excitation decay and the *inter*-ring annihilation. These dynamics depend strongly on the aggregation state of LH2. At higher LDAO concentrations ($c_{LDAO} = 1.5$-15 mM), after the initial *intra*-ring annihilation only the exponential single-excitation decay occurs, with a time constant $\tau \approx 750$ ps corresponding well to the previously reported excitation lifetime ~1 ns.[21] This means that conditions are achieved where the LH2 complexes are well separated.[28] For lower LDAO concentrations ($c_{LDAO} = 0.15$ and 0.5 mM) the decay of the signal is faster and non-exponential due to *inter*-ring annihilation, indicating aggregation of the LH2 protein complexes.

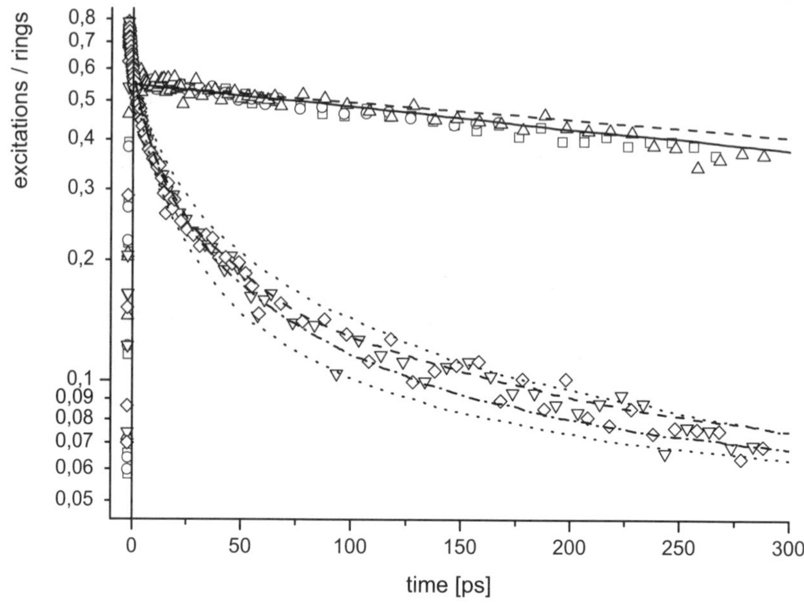

Fig. 4. Kinetics for samples with optical density (OD) of 0.02 (c_{LH2}= 0.06 µM) at different surfactant concentrations: c_{LDAO}= 15 mM (□), 5 mM (○), 1.5 mM (Δ), 0.5 mM (∇), and 0.15 mM (◊). Furthermore, simulations are shown for aggregate sizes N = 1 (*solid line*), 10 (*dashed line*), and 12 (*dash-dotted line*) both with a ring-to-ring hopping rate k = 30 ns^{-1}. For N = 12 simulations with k = 20 ns^{-1} (*upper dotted line*) and 40 ns^{-1} (*lower dotted line*) mark the confidence band.

To establish the initial conditions of excitation, we first analyzed the fast *intra*-ring annihilation at the beginning of each kinetic trace. This enabled us to determine the fraction of initially excited B850-rings.[28] The initial signal intensity before *intra*-ring annihilation is proportional to the fraction f of excited BChl molecules. According to the binominal distribution, for 18 BChls per B850-ring the fraction of rings which contains *no* excitation is $p_0 = (1-f)^{18}$. Consequently, the fraction of rings which carry initially *one or more* excitations is given by $p_{\geq 1} = 1-p_0$. After completion of the initial *intra*-ring annihilation, all these rings carry only *one* excitation. Correspondingly the fraction of excited BChls is now $p_{\geq 1}/18$. The signal intensity after *intra*-ring annihilation is proportional to that number. Comparing the amplitude of the transient absorption signal after completion of the *intra*-ring annihilation with the initial amplitude we can obtain $p_{\geq 1}/(18f)$ directly from the experimental curves. For example from the uppermost curve in Figure 4 ('□') we estimate that the remaining signal after the *intra*-ring annihilation is $\sim 80\ \%$, giving

$$\frac{p_{\geq 1}}{18f} = \frac{1-\left(1-f\right)^{18}}{18f} = \left(80 \pm 5\right)\% \ ,\tag{1}$$

which results in $f = (2.7\pm0.8)\%$ and subsequently in $p_{\geq 1} = (40\pm10)\%$. The latter is the initial occupation for the simulations of *inter*-ring annihilation after completion of the initial *intra*-ring annihilation. We have also estimated f using the absorption cross-section of the Bchl, the energy of the laser pulse and the diameter of the beam. The results are similar but the error of such an estimation is significantly larger. Hence, in what follows we will only use values of the initial population as obtained from the *intra*-ring annihilation analysis for each set of curves separately.

The population kinetics of the *inter*-ring annihilation process are modeled by random walk of the hopping excitations on a two-dimensional hexagonal lattice of N nodes, each representing one B850-ring. The hexagonal coordination of the lattice is strongly suggested by the TEM images.[34] It is also the most favorable structure with respect to the surface polarity distribution. For the same reason the N nodes are arranged for a minimum perimeter of the aggregate. At the beginning of each simulation track, some of the N lattice-nodes are randomly occupied by single excitations with probability $p_{\geq 1}$ as obtained from the previous analysis of the *intra*-ring annihilation. Thus the initial state of the simulation describes the situation immediately after the fast *intra*-ring annihilation has been completed. The subsequent random-walk simulation by means of the Monte Carlo method has been performed in short time-steps Δt (here 0.1 ps) as follows:

1. During a time-step each excitation has a probability $\Delta t/\tau$ to decay due to the single-excitation decay, $\tau = 750\ \text{ps}$. The decision to remove the excitation is taken from a uniform random distribution via the Monte Carlo method.

2. For the excitations that remain after step (1) we need to decide whether they make a jump or stay where they are during the time step. The probability to jump is given as $P = nk\Delta t$, where n is the number of nearest neighbors ($n = 6$ for the rings inside the aggregate but $n < 6$ for the rings on the edges) and k is the ring-to-ring hopping rate for the *inter*-ring excitation transfer. The decision to jump is taken using the same Monte Carlo method as in step (1).

3. If the excitation is to jump, the acceptor node is chosen from a uniform distribution of the n nearest neighbors. This condition means that k does not depend on the acceptor state but is the same for the acceptors with and without excitation. This assumption is justified by the fact that the B850 ring contains 18 BChl molecules and even if the excitation is delocalized over 2-4 BChl molecules[20,28] the accepting BChls remain most of the time unaffected.

4. If the jump is made, the source node occupation is set to zero and that of the target node is set to one, independent of whether the latter is already occupied or not. This means that if the acceptor already had an excitation, one of the two initial excitations is annihilated.

5. The procedure is carried out for all remaining excitations.

6. The time t is increased by one time step Δt and the algorithm is repeated until $t = 300$ ps.

7. If $t = 300$ ps a new track is started for a new random initial occupation of the same lattice.

For good statistics, 1000 simulations are accumulated for each combination of fitting parameters k and N. In the case of no aggregation i.e., $N = 1$, the only kinetic component after *intra*-ring annihilation has occurred (see Figure 4 solid line) is the single-exponential quenching with $\tau = 750$ ps. In LH2 aggregates where the ring-to-ring hopping rate k is much higher than the single-excitation decay rate τ^{-1}, the calculated kinetics depend on the two fitting parameters in qualitatively different ways. The parameter k mainly determines the initial time-profile of the annihilation part of the decay e.g., for $k = 20$ and 40 ns^{-1} in Figure 4 (dotted line), whereas the relative amplitude of the asymptotic exponential decay at long times determines the aggregate size N e.g., for $N = 10$ and 12 in Figure 4 (dashed line). The latter can be understood as follows: from the $p_{\geq 1} \times N$ nodes, which are excited immediately after *intra*-ring annihilation, only one carries an excitation after *inter*-ring annihilation. Hence, the signal amplitude after completing the *intra*-ring annihilation is $p_{\geq 1} \times N$ times larger than the signal corresponding to the last excitation when interpolated to $t = 0$ for the single excitation decay. Thus, the parameters k and N are independently determinable.

For LDAO concentrations below the critical micellar concentration (CMC) ($c_{LDAO} < 1.2$ mM) the aggregate size is $N = 14 \pm 4$ and ring-to-ring hopping rate is $k = 30 \pm 15$ ns^{-1} (Figure 4 dashed and dash-dotted lines). These aggregate sizes fit remarkably well to the sizes of the spots observed in the TEM images.[34]

The ring-to-ring hopping rate $k = 30 \pm 15$ ns^{-1} (corresponding time is ~30 ps) found for the cluster-like aggregates at LDAO concentrations below the CMC means that for a B850-ring surrounded by six neighboring rings the excitation residence time is ~5 ps. This number is remarkably close to the low-temperature dephasing time of 6.6 ps measured by spectral hole-burning at the red edge of the B850 band in native LH2 membranes.[35] Direct time-resolved measurement of the interring excitation transfer in B850 is not possible because of the overlap of signals. Even for transfer from the peripheral antenna LH2 to the core antenna LH1 it is not a trivial issue. In native membranes a range of times for LH2-LH1 transfer has been obtained: 3.3 ps,[36] 4.5 ps,[37] and 8 ps.[38]

Comparing these results with those in the previous section we can conclude that interring energy transfer can hardly explain the red stimulated emission band seen in B850[29] – one would need at least 1-2 steps of random walk for an appreciable downward shift of the stimulated emission. These hops would take about 10 ps whereas the band appears within 1 ps – a discrepancy of an order of magnitude.

The last step in excitation transfer is trapping by the RC. Already early studies indicated that the trapping by RC is slow.[39] The rate of the antenna to RC transfer was first determined by Timpmann *et al.*[40] to be 30 - 40 ps. The trapping of excitation energy by photosynthetic RC has often been discussed in terms of a trap-limited or diffusion-limited process. The new situation after revealing the elementary rate was termed transfer-to-trap-limited.[6] The structure of LH1-RC (Figure 1) explains the result – the distance between RC and LH1 BChls is the decisive factor.

An issue sometimes raised in literature is the role of the collective excited states, excitons, in photosynthetic light harvesting. Since the detailed structural information has not been available for a sufficiently long time, no rigorous work concerning the whole PSU has appeared so far. The qualitative discussion given by us quite some time ago[41] still holds and accordingly it is the last energy transfer step – transfer to the RC – which is the most influenced by collective excitations. The overall trapping and thereby efficiency of photosynthesis is significantly improved via speeding up this transfer step.

3. ORGANIC CONDUCTING POLYMERS: LIGHT HARVESTING AND GENERATION

Organic conjugated polymers are remarkable in the sense that they may be designed for the dual functions of light-generation and light-energy conversion. The first function relies on the fact that these materials are electroluminescent, *i.e.* upon leading electric current through the material, electron and holes recombine to form light-emitting excited states. The second function rests on the property of polymer excited states formed by light absorption to dissociate into electrons and positive holes. Thus, we see that both functions involve very much the same species and processes, excited states (excitons) and charges that are interconverted and transported through the material in the course of action. As an example, in a conjugated polymer used as a solar cell material, light absorption creates an excited state, which migrates as a neutral exciton along and between polymer chains until it dissociates into an electron-hole pair In an efficient solar cell material the exciton has a low binding energy and the charges easily escape the mutual Coulomb attraction to form free mobile charges that contribute to photocurrent. In a light-emitting material (light-emitting diode or display) on the other hand, charges are injected into the polymer and migrate until they recombine to form emitting excited states. From these considerations we can see that it is necessary to understand the processes of excited state and charge generation and transport, as well as their participation in various quenching processes reducing the efficiency of the material, for the design and production of optimized materials. Here we will discuss processes involving both neutral excited states and charges. We start by briefly considering properties and processes involving neutral excited states and continue with a more detailed discussion of photoinduced charge generation.

3.1. Excited State Dynamics

The nature of the excited state, relaxation processes and excited state energy migration and quenching are related processes connected to the temporal evolution of the excited state, and determining *e.g.* if it has strong emission or is more prone to charge separation. Energy migration is a basic property of a material and is strongly related to inhomogeneity and the presence of traps. Energy transfer has been studied by many groups for many different polymers, both in thin solid films and for polymers in solution.[42] This work shows that energy transfer occurs along polymer chains, as well as between polymer chains when they are closely packed in a solid film. Depending on the nature of the polymer spectroscopic unit and distances between polymer chains, the site-to-site energy transfer is on the sub-picosecond to few picosecond time-scale.[43,44] Also faster processes ($<$ 100 fs), assigned to exciton relaxation or localization within a single spectroscopic unit, have been observed.[45] In our work we have used energy migration, probed *e.g.* by fluorescence depolarization, as a tool to study polymer conformations. Our recent work on this topic resulted in detailed insights into the mechanism of photoluminescence quenching – it was concluded that in polymer thin films with densely packed chains having strong intermolecular interactions, aggregates formed by chains in close approach act as non-luminescent traps.[46,47] As excitation energy migrates along and between polymer chains, the measured fluorescence or transient absorption signal depolarizes because different segments have different orientations. Thus, time-resolved polarized measurements can be used to probe the degree of conformational disorder of a polymer. By comparing the results of ultrafast measurements and Monte Carlo simulations on computer-generated polymer chains, we could obtain detailed information regarding the conformational properties of several polymers.[44,45,48]

3.2. Light-Induced Charge Generation

As mentioned above, the conversion of light to charge and *vice versa* are the two basic processes determining the performance of conjugated polymers as electrooptic materials. Two very different concepts have been advanced to describe charge photogeneration. The semiconductor approach describes the polymer as a one-dimensional semiconductor, assuming direct photogeneration of free charge carriers via optical interband transitons.[49,50] The molecular approach, on the other hand, considers charge generation as a two-step process, in which neutral excitons are created by photon absorption followed by exciton splitting into charge carriers. Support for the latter notion comes from a number of publications that have shown that in derivatives of polyphenylene vinylenes[51,52,53,54,55,56] and ladder type polyphenylene (MeLPPP)[56,57,58,59,60] field or temperature assisted break-up of singlet excitons creates charge carriers.

A disputed question within the molecular picture is whether the dissociation proceeds from a Franck Condon state prior to vibrational cooling, or from the vibrationally cold singlet exciton. Based upon energy-selective photoluminescence studies of the PPV family of polymers, one idea is that there is branching into charge carriers and relaxed, i.e. fluorescent, singlet excitons before the primary excitation looses its excess energy.[61] On the other hand Scherf *et al.*[62] presented evidence that the relaxed exciton can also dissociate. Kersting *et al.* investigated the dynamic field-induced fluorescence quenching and observed that the main quenching occurs within several ps after excitation, followed by a minor process that continues on a timescale of several hundred ps.[56] They proposed that quenching of the fluorescence dynamics is related to the interplay of exciton breaking and spectral relaxation of excitons within an inhomogeneously broadened density of states. Observation of rapid onset of both fluorescence quenching and formation of transient absorption due to charged species (polarons) in an electric field[57] supported the notion that those processes do occur on a timescale of a few ps, suggesting that it is not the relaxed singlet exciton, which dissociates. It was suggested that the exciton breaking occurs on 'dissociation sites' reached during the exciton migration towards lower energy sites.[57] Recently, we showed that field-assisted charge pair generation in a ladder type-PPP polymer occurs during the entire exciton lifetime, and the exciton breaking dynamics is typical for random systems[60] and may be well approximated by a Poole-Frenkel type of field dependence, implying field-assisted energy barrier crossing. By measuring photocurrent in ladder type-PPP using two femtosecond laser pulses, Muller *et al.*[59] also concluded that exciton dissociation from relaxed exciton states is the only relevant channel for photocurrent generation and that excitons are formed during their entire lifetime. These authors also identified a minor channel (10 %) of intra-chain charge pair formation, not contributing to the photocurrent. Ultrafast spectroscopy results on a polythiophene polymer suggested that charge generation may also occur in the absence of an external electric field.[63]

Many different theoretical models have been proposed over the years to describe charge photo-generation in conjugated polymers. According to the semiconductor model, carriers are directly generated by the optical excitation and thermalize to form charged polarons, which later bind into polaron excitons due to the Coulomb interaction. The initial generation step in this model is field-independent, while the field dependence of the generation efficiency comes from the competition between the Coulomb interaction and the external field preventing exciton formation, and/or from breaking of already formed excitons. Onsager-type models are similar, in the sense that they describe exciton dissociation in terms of charge pair dissociation and consider primary exciton dissociation as being field-independent. Arkhipov *et al.*[64] took into consideration the excess vibrational energy stored in a polymer segment after excitation with a high-energy photon and proposed a model where photo-generation of free mobile charges occurs in two steps involving three species, a neutral exciton, a geminately bound *e-h* pair and separated charges. These three species are separated by electric field-dependent potential barriers, which may be overcome by excess thermal energy and/or by applying an external electric field that lowers the barriers. Thus, if a neutral exciton is created close to the bottom of its potential well a substantial electric field is required for its dissociation. If, on the other hand, sufficient excess energy is delivered to the polymer segment via the optical excitation, the exciton dissociation is facilitated by the excess energy delivered to the phonon bath and is predicted to have a significantly lower electric field-dependence than the dissociation originating from the thermalized exciton.[64] According to this model, hot exciton dissociation is expected to occur within a time window set by vibrational cooling of a polymer segment, *i.e.* the few picosecond time-scale. Fully dissociated mobile charges contributing to photocurrent are formed by breaking the geminate *e-h* pairs held together by the mutual Coulombic attraction energy. Charge photo-generation could also be thought of as autoionization of a hot Franck-Condon state, which after thermalization forms a charge pair. This process was first discovered by Geacintov and Pope[65] and theoretically explained by Jortner.[66] The time-scale of exciton dissociation in this case is set by redistribution of excess vibrational energy into the heath bath, *i.e.* ~100 fs.

The experimental results summarized above, supporting one or other model, were obtained for different polymers, and different experimental techniques and conditions were used. It is therefore difficult to obtain a general picture of the charge generation process. In order to contribute towards a general picture of charge photo-generation in conjugated poly-mers we have performed systematic studies of two different polymer types that, on the basis of previous studies, appear to represent two quite different limits of charge generation behavior, a ladder type poly- (para-phenylene) (MeLPPP) and poly(3-(2'-methoxy-5'octylphenyl)thiophene (POMeOPT). Exciton dissociation was investigated as a function of excitation photon energy, electric field strength and excitation power density, with the aim to obtain exciton binding energies, dependence of exciton breaking on excess vibrational energy and excitation density.

Using electric field modulated femtosecond transient absorption spectroscopy[60] we have studied charge pair generation in thin films of MeLPPP and POMeOPT. We will briefly summarize the results and then present a picture of photo-induced charge generation that emerges. Upon exciting MeLPPP with a femtosecond pulse into the $S_0 \rightarrow S_1$ transition (3.1 eV) with low excess vibrational energy, an external electric field is required for the formation of a discernable number of charge pairs. The temporal evolution of the electron-hole concentration is characterized by a rise time of approx 30 ps, implying that charge pairs are formed throughout the exciton lifetime from thermally relaxed excitons (see

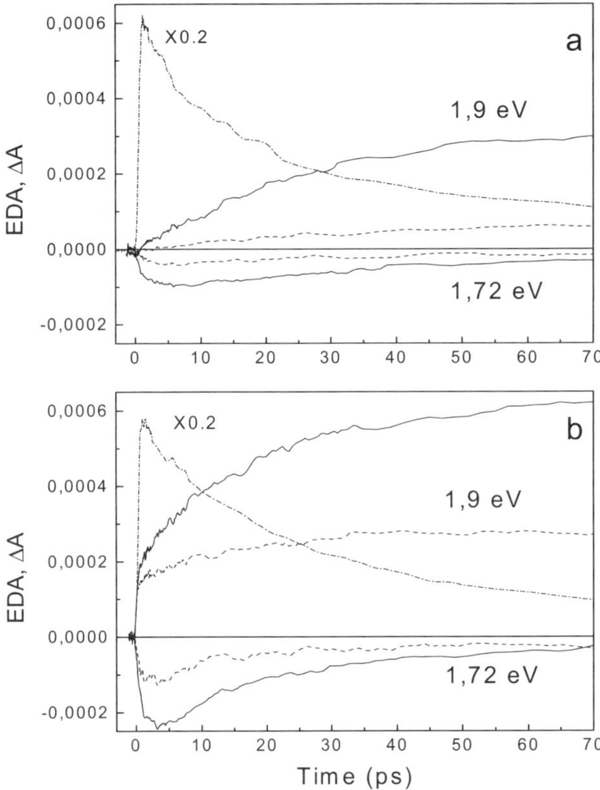

Fig. 5. ΔA and EDA kinetics measured at two different probe energies with (a) 3.1 eV and (b) 4.66 eV excitation. Solid lines show EDA kinetics at $2.2*10^6$ V/cm applied field (22 V) and dashed lines those at $1.5*10^6$ V/cm (15 V). Dash-dot lines show ΔA kinetics measured at 1,72 eV without applied voltage. The ΔA kinetics are divided by a factor of 5.

Figure 5a. The charge generation exhibits a threshold-like electric field dependence with an onset at $\sim 0.7 \times 10^6$ V/cm, showing that there is a substantial potential barrier (several tenths of an eV) between the exciton and charge pair states, which is reduced by the electric field.[60] The charge pair generation rate is not constant in time, but decreases with time as $\gamma(t) \propto (t_0/t)^{0.4}$, typical for a random system, as a result of a distribution of inter-chain distances and segment orientations leading to a distribution of charge transfer rates (dotted line in Figure 6).

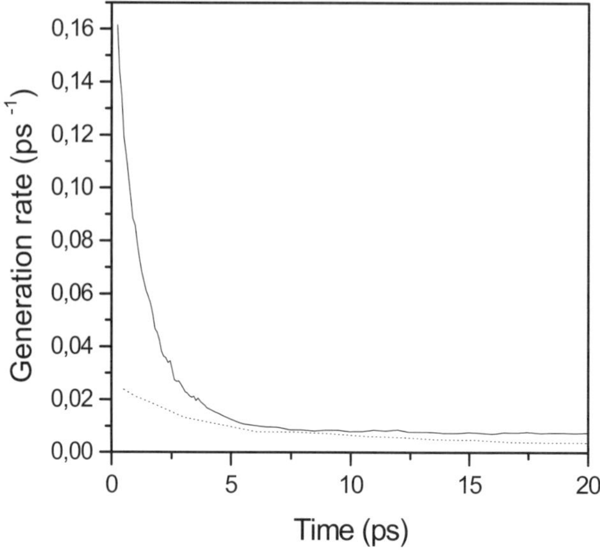

Fig. 6. Time dependence of the charge pair generation rate with 4.66 eV (solid line) and 3.1 eV (dotted line) excitation.

Upon exciting the polymer at 4.66 eV, *i.e.* with almost 2 eV of excess energy, an instantaneous generation of charge pairs superimposed on a much slower more gradual response, reminiscent of that with 3.1-eV excitation, is observed (Figure 5b). The instantaneous charge pair generation has a weaker electric field dependence, without the threshold behavior observed for lower (3.1 eV) photon energy excitation, and the time dependence of the slow (> 500 fs) charge generation rate is much more pronounced (solid line in Figure 6) – the rate decreases by a factor of ~8, as compared to a factor of 2 with low excess energy excitation during the first 5 ps after excitation.[67]

These results suggest that, with ~2 eV excess energy for excitation of MeLPPP, photogeneration occurs from a hot Franck-Condon exciton state, prior to complete cooling. Using the potential energy surface diagram of Fig 7 the following picture of charge generation in MeLPPP can be given. The threshold-type voltage dependence of the charge pair generation efficiency with 3.1-eV excitation shows that a barrier exists between the exciton and the charge pair states, and that this barrier is reduced by the electric field. With 4.66-eV excitation, the charge pair

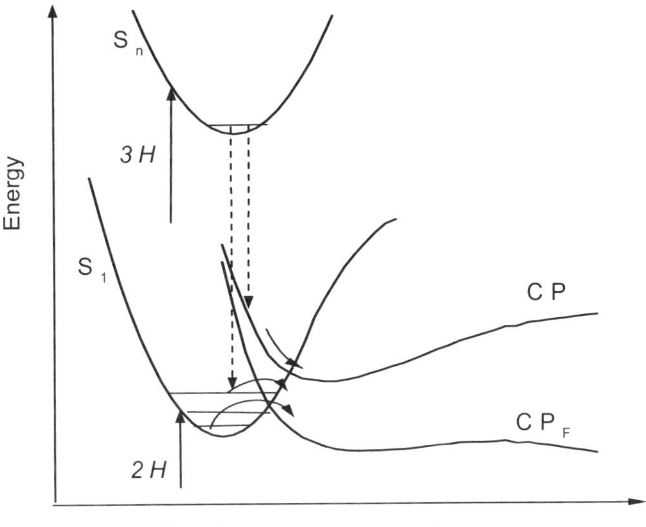

Fig. 7. Charge pair generation model. S_1 and S_n are electronic states localized within a single conjugated polymer segment. CP is the charge pair state and CP_F is that modified by the applied electric field.

states may be populated by three different processes: a) branching during the relaxation of the higher energy excited state to the charge pair and the S_1 exciton state; b) thermally activated barrier crossing from the hot S_1 state, and c) barrier crossing from the thermalized S_1 state. The process a) accounts for the very fast <100 fs charge pair formation observed with 4.66 eV excitation and would correspond to the autoionization proposed by Geacintov and Pope.[65] Process b) corresponds to the strongly time-dependent charge pair generation and would correspond to the vibrational energy assisted charge generation discussed by Arkhipov *et al.*[64] Finally, process c) is the slow, almost time-independent process observed with low excess energy (3.1 eV) excitation. The model is significantly simplified, since it neither accounts for different carrier separation directions relative to the electric field, nor effects related to the structural disorder and polymer morphology. Nevertheless, we believe that it explains the basic features of photodissociation in MeLPPP and conjugated polymers with similar properties.

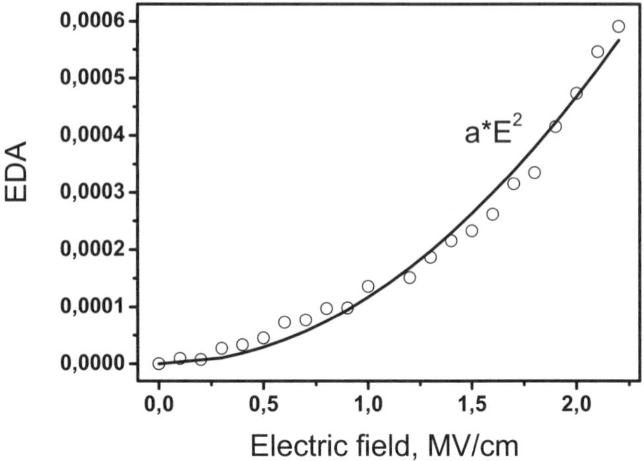

Fig. 8. Voltage dependence of charge pair formation (EDA = Electric field modulated Differential Absorption signal) in POMeOPT (circles). Solid line is a parabolic fit: EDA=aE^2, where $a = 1.17 \times 10^{-4}$ (MV/cm)$^{-2}$.

For the polythiophene POMeOPT, the charge photogeneration behavior is quite different from that of MeLPPP. Although the timescale for charge generation from singlet excitons is similar for the two polymers, there are several differences in the charge formation dynamics. As revealed by the electric field dependence of the charge generation yield, the exciton breaking in POMeOPT is barrierless – already at zero applied external field there is appreciable charge pair formation at low excitation photon energy, and the electric field dependence of the charge formation yield is much more gradual than for MeLPPP (Figure 8). Despite the barrierless nature of exciton breaking in POMeOPT, excess vibrational energy deposited in the polymer through direct optical excitation or excitation annihilation does not change the charge generation yield or dynamics. Comparison of charge generation and energy transfer times (the latter obtained from time-resolved absorption and fluorescence anisotropy) shows that the two processes occur on a similar timescale, suggesting that charge generation is controlled by energy migration to sites with particularly low exciton binding energy.

Combining these results with those already available in the literature (see summary above), a general picture of charge generation is beginning to emerge. The resulting scenario includes differences in polymer properties as well as different experimental conditions. The different charge generation behavior of MeLPPP and POMeOPT can be understood as being a result of the presence of sites with low exciton binding energy in POMeOPT.[68] In POMeOPT, being a flexible and disordered polymer, strong interaction between chains may give rise to aggregates or donor-acceptor structures, while this tendency is lower in the stiff ladder structure of MeLPPP. Thus we suggest that, in POMeOPT, sites with particularly low energy barriers towards charge separation are formed where there is strong interaction between polymer chains, while in MeLPPP the exciton binding energy has a finite value (several tenths of an eV) with a distribution characteristic of the random nature of the polymer. This will give rise to the following dynamical characteristics of charge photogeneration.

1. In a polymer with high exciton binding energy, the energy barrier may be overcome and charge pairs formed by applying an external electric field or by depositing sufficient (> 1 eV) excess vibrational energy in the polymer, by direct optical excitation or by exciton annihilation. Charge pair generation has a threshold type of electric field dependence, signaling that a substantial energy barrier has to be overcome for breaking the exciton.
2. For a polymer with high exciton binding energy, the field-assisted charge pair generation is slow and occurs throughout the exciton lifetime with a time-dependent rate characteristic of a disordered material, with a distribution of exciton binding energies and electron transfer times. With substantial excess vibrational energy a component of ultrafast (<100 fs) charge generation appears in competition with vibrational relaxation and cooling. Charge generation from hot S_1 states occurs with a strongly time-dependent rate.
3. In a polymer with sites having low (~0 eV) exciton binding energy, charge pair generation occurs in the absence of an external electric field and applying a field increases the charge pair yield.
4. Dynamics of charge pair formation in a polymer with low barrier sites is controlled by energy migration, since the excitons have to move to a dissociation site for charge pair generation. A similar picture has previously been proposed,[57] but with the difference that

the dominating part of the charge generation was thought to occur during downhill energy migration. This would place charge generation on the ~1 ps timescale, clearly different from the slow charge generation from thermalized exciton states observed here.

5.   In a polymer with low barrier dissociation sites, excess vibrational energy may not increase the charge pair yield. This puzzling behavior may be understood as a result of weak excess energy dependence of the barrierless exciton dissociation and dissipation of excess energy during energy migration to dissociation sites.

It should once more be emphasized that the above picture is in an early stage of development and more results including several more polymers will certainly add other features and more details to this picture.

4. DYE-SENSITIZED NANOSTRUCTURED SEMI-CONDUCTORS – ENERGY CONVERSION BY ULTRAFAST ELECTRON TRANSFER

Sensitized wide band-gap semiconductors have long been considered as materials for photovoltaics. However, until the discovery by Grätzel and coworkers,[69] that a material with high conversion efficiency could be produced by sensitising a nanostructured thin film of a wide band-gap semiconductor, no real progress towards a competitive solar cell device was made. A so-called dye-sensitized solar cell (DSSC) typically consists of a thin (~1 μm) TiO_2 film of nanometer-sized (~10 nm) particles sintered together for electrical contact. This thin film is deposited on a conducting glass (ITO) electrode and brought into electrical contact wit a counter electrode via an electrolyte. Light energy conversion in such a solar cell starts when light is absorbed by the sensitising dye and electrons are injected from the excited state of the dye into the conduction band of the semiconductor. Electrons then migrate between semiconductor particles until they reach the back contact and the external circuit, where they can perform work. Finally, electrons are lead back to a redox couple (often I^-/I_3^-), which regenerates the oxidized dye back to the neutral ground state and the dye can absorb another photon and start a new conversion cycle.

High efficiency of the electron injection step and a low yield of electron recombination between electrons in the semiconductor conduction band and oxidized dye are essential for an efficient material. The electron injection efficiency can be optimised by ensuring very fast injection, much faster than all competing processes deactivating the excited state of the sensitizer. The fraction of injected electrons that recombine with the oxidized may be influenced in several different ways. One possibility is to choose sensitizer and redox couple in such a way that their electrochemical properties maximize the ratio of rates for oxidized dye reduction by the redox couple and electron recombination from the semiconductor (k_{redox}/k_{recomb}). Another possibility is to decrease the rate of back-electron transfer from the semiconductor to the oxidized dye, by increasing the distance between the semiconductor and sensitizer.[70] Still another approach that has been explored is to use a secondary electron donor attached to the sensitizer and hence move the positive hole on the sensitizer further away from the semiconductor surface.[71] The most efficient DSSC so far constructed has an overall light-to-electricity conversion efficiency of 10.4 %[72] and relies on TiO_2 anatase for the semiconductor and a transition metal complex $(Ru(dcbpy)_2(NCS)_2$ (dcbpy = 4,4'-dicarboxylate-2,2'-bipyridine, RuN3 for short) as sensitizer.

It was early realized that electron injection in many dye-semiconductor systems is very fast and highly efficient. For the TiO_2-RuN3 couple used in the most efficient solar cell, a great deal of spectroscopy and time-resolved work have been devoted to elucidating the mechanisms and pathways of photoinduced electron injection. In what follows we will give an account of our contribution towards this goal. Until approximately two years ago, the generally accepted picture of the electron injection in this system was that a dominating part of it was very fast (< 100 fs) and efficient (> 90 %), with a minor part occurring on a much slower timescale (1-100 ps). The fact that electron injection proceeds over three orders of magnitude in time (< 0.1 – ~100 ps) was explained by highly varying electron transfer rates, due to a wide distribution of sensitizer-semiconductor couplings caused by varying binding distances and geometries.[73] This picture of electron injection did not include any details of the excited electronic states of the sensitizer, or specific pathways of the injection. Neither was it known how processes like vibrational relaxation, intersystem crossing and intramolecular electron transfer within the sensitizer are coupled to and influence the electron injection.

Before we start discussing these matters, we will give a brief description of the photophysics of transition metal complexes necessary to understand the light-induced processes in a sensitizer-semiconductor system. The strong visible absorption of those molecules is due to singlet and triplet excited states. For the Ru-based molecules to be discussed here, the lowest excited state is a triplet metal to ligand charge transfer state (^{3}MLCT) and there is a ^{1}MLCT state at higher energy responsible for the main absorption band. Light absorption into this band therefore generates the excited ^{1}MLCT state, but within very short time (< 100 fs) the molecule has relaxed into the lowest ^{3}MLCT state, because of efficient intersystem crossing caused by the heavy metal atom. For efficient electron injection and energy conversion in the sensitised semi-

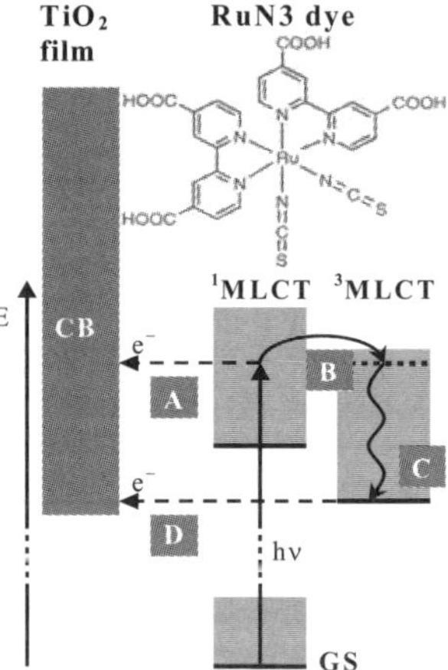

Fig 9. Schematic model for two-state electron injection and structure of RuN3. Following MLCT excitation (at 530 nm) of the RuN3-sensitized TiO$_2$ film an electron is promoted from a mixed ruthenium NCS state to an excited π^* state of the dcbpy-ligand and injected into the conduction band (CB) of the semiconductor. GS: ground state of RuN3. Channel A: electron injection from the non-thermalized, singlet ^{1}MLCT excited state. Channel B and C: ISC followed by internal vibrational relaxation in the triplet ^{3}MLCT excited state. Channel D: electron injection from the thermalized, triplet ^{3}MLCT excited state.

semiconductor system, the energy of the lowest ^{3}MLCT state has to be above the conduction band edge of the semiconductor. The resulting scenario is illustrated by Figure 9, showing the valence and conduction bands of the semiconductor, as well as the ground and excited states of the RuN3 sensitizer. Our work shows that, following light absorption to the ^{1}MLCT state, ~60 % of the molecules inject electrons directly from this state into the semiconductor conduction band with a characteristic time constant of 50 fs. Upon excitation to higher-lying vibrational levels of the ^{1}MLCT state, even faster injection occurs (< 20 fs), in competition with vibrational energy relaxation and redistribution. The residual ~40 % of the excited sensitizers relax to the triplet state, from which they inject electrons much more slowly on the 1-100 ps timescale. Below, we show how this picture was obtained by using ultrafast spectroscopy to probe the various species involved in the light energy conversion process.

To distinguish the early-time involvement of the singlet channel in the ET, we used ~30 fs laser pulses to monitor the optical population of

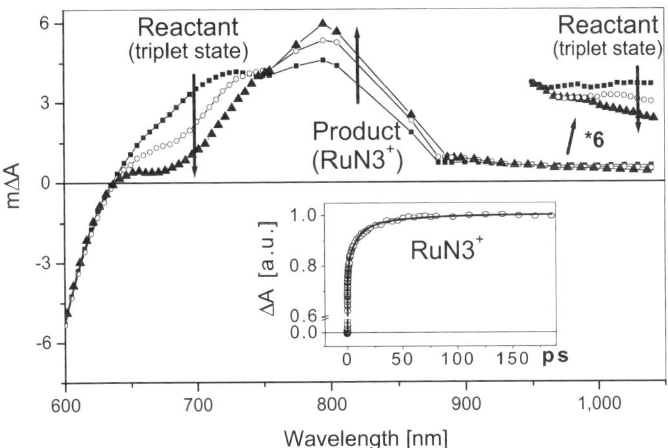

Fig. 10. Visible and near-IR transient absorption spectra of RuN3-sensitized TiO$_2$ film. At time delays of 0.5 ps (——■——), 10 ps (——○——), and 150 ps (——▲——) between the pump and probe pulses we observe characteristic dynamics of the differential spectra with two isobestic points (the wavelength where $\Delta A_{reactant} = \Delta A_{product}$) at 760 and 940 nm, and one nearly time-independent $\Delta A=0$ point at 630 nm. Inset, transient absorption kinetics of oxidized RuN3 cation-TiO$_2$ measured at 860 nm. Symbols are measured data, while the curve is a fit with the following time constants and amplitudes: rise within the laser pulse (20±5%), 28±3 fs (50%), 1±0.1 ps (11%), 9.5±1 ps (12%), and 50±5 ps (7%).

the singlet state, and its decay into the triplet state intermediate and oxidized dye product (channels A, B and C of Figure 9). The transient spectra of Figure 10 indicate the key spectral regions for probing these species – the appearance and decay of the triplet state is monitored around 700 nm and 1000 nm, while the formation of the oxidized dye is probed at ~850 nm. The decay of the initially excited singlet state is monitored by stimulated emission, and the kinetics measured at 600 nm is shown in Figure 11a. The initial fast decay of the stimulated emission is characterized by a 30 fs time constant to a time-independent (on this timescale) signal level, due to ground state bleaching. The stimulated emission decay of RuN3 in ethanol solution due to intersystem crossing is 70 fs,[74] showing that electron injection effectively competes with the very fast intersystem crossing. Ultrafast dynamics matching the decay of the singlet state can also be observed in other spectral regions. At 690 nm (Figure 11b), the instantaneous rise followed by the 30±3 fs decay again reflects the evolution of the singlet state, while the 80±5 fs rise corres-

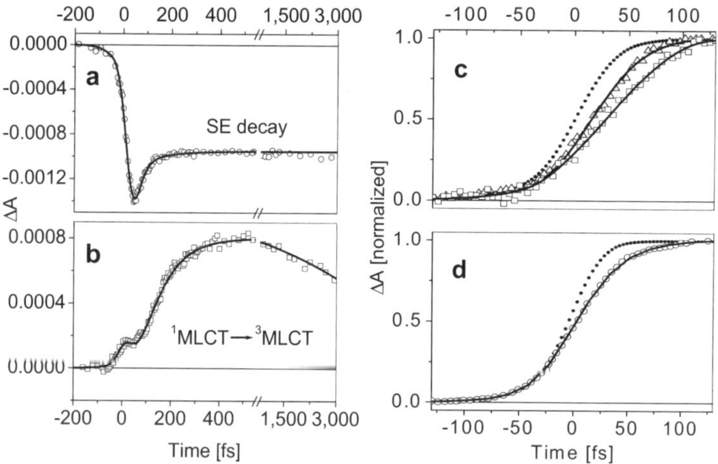

Fig. 11. Transient absorption kinetics. Open symbols are measured data, curves are fits, and instrument responses are represented by dotted curves: a) SE decay probed at 600 nm; b) excited state evolution (^{1}MLCT→^{3}MLCT) probed at 690 nm; c) formation of the triplet state at 1050 nm; the data are well fitted with rise times of 30±5 fs (for RuN3-TiO$_2$; triangles) and 70±15 fs (for RuN3-EtOH; squares); d) early-time transient absorption kinetics of oxidized RuN3 cation-TiO$_2$ measured at 860 nm; parameters of the fit are given in the caption of Figure 10.

ponds to the formation and thermalization of the triplet state (channels B and C of Figure 9). It is intriguing that the decay of the singlet proceeds in ~30 fs, but the rise of the triplet state occurs in ~100 fs. However, the probe signal centered at 690 nm in the blue wing of the triplet absorption band (Figure 10) is sensitive to the spectral blue-shift caused by the triplet equilibration process. Thus, the ~100 fs time constant encompasses both the ISC process and ensuing thermalization of the triplet state. The kinetics at 1050 nm (Figure 11c) provide one more possibility to monitor the formation of the triplet state. In this spectral region the absorption of the triplet is very broad and featureless and therefore not sensitive to spectral shifts that may be caused by the thermalization processes. Hence, a ~30fs rise time is observed, corresponding directly to the decay of the singlet state monitored by SE. In Figure 11 c, the formation of the triplet state for RuN3 in solution is compared to that of RuN3-sensitized TiO_2 film and found to be considerably slower in solution (~70 fs). The faster formation of the triplet in RuN3-TiO_2 reflects the very fast and efficient electron injection from the singlet state prior to IVR, IC and ISC. Finally, if the ~30 fs decay corresponds to electron injection from the RuN3 singlet state, then we expect to observe this time constant also for the formation of the oxidized RuN3. Indeed, the fit to the data at 860 nm (Figure 11d) confirms the presence of the ~30 fs time constant with high amplitude (>50%). Hence, we see that the optically excited singlet state decays with the same time constant of ~30 fs as the triplet state and oxidized dye products appear, thus coupling these species in a reactant - product relationship.

In a simplified reaction model, the rate constant related to the observed ~30-fs decay of the singlet state reflects the sum of the rates of electron injection from the ^{1}MLCT state (k_A) and ISC (k_B): $1/30$ fs^{-1} = $k_A + k_B$. The ratio between k_A and k_B of 1.5 is calculated from the amplitude ratio of femtosecond (~60%) and picosecond (~40%) electron injection contributions to the signal at 860 nm, yielding $k_A = 1/50$ fs^{-1} and $k_B = 1/75$ fs^{-1}. The value determined for $1/k_B$ is in good agreement with previous measurements of ISC for this class of molecule[75-77] and is in excellent agreement with the formation of the triplet state measured here for RuN3 in ethanol solution (Figure 11c). We conclude that after excitation of the sample at 530 nm, ~60% of the RuN3 molecules inject electrons from the singlet state into TiO_2 (channel A of Figure 9) and the rest undergoes ISC (ch. B). After relaxation to the bottom of the triplet state (ch. C), electrons are again injected into the semiconductor, but now the reaction occurs on the picosecond time scale (ch. D).

In order to examine the role of vibrational relaxation in the singlet injection, electron transfer rates were measured as a function of the excitation wavelength. When exciting the RuN3-sensitized TiO_2 film with pulses of central wavelength at 455 nm, and probing the formation of the oxidized dye at 860 nm, the singlet injection is faster than 20 fs and the triplet injection very similar to that observed with 530 nm excitation. This finding implies that the femtosecond ET either precedes or occurs in concert with IVR in this system. We conclude that the kinetics of electron injection from RuN3 to TiO_2 is dependent on the initially populated vibronic state: the higher the state, the faster the electron injection. This result implies that dyes with an excited-state redox potential below the conduction band edge of the semiconductor are also capable of electron injection, when they are excited by sufficiently energetic photons.

The highly non-exponential slow electron injection from the triplet state was in the past suggested to be a result of weak and highly inhomogeneous coupling between the sensitizer and semiconductor. However, with this explanation it is difficult to understand how triplet injection can be independent of semiconductor (the same ps injection rates were observed for RuN3 attached to TiO_2 and SnO_2[78]), but dependent on the solvent in contact with the dye-semiconductor film (a slower injection was observed in ethanol than in acetonitrile[79]). Also with the injection scheme of Figure 9, one would expect to observe excitation wavelength-dependent (hot) injection from the triplet state, similarly to what was observed for the singlet state. No evidence for hot triplet injection was however found.[74,79] These observations suggest that perhaps some intramolecular process within the sensitizer is the rate-limiting step of the triplet injection.

In what follows we will summarize the results that lead to the conclusion that electron transfer between the pyridyl ligands of the RuN3 sensitizer (interligand electron transfer, ILET) is the rate-limiting process controlling electron injection from the triplet MLCT state to the semiconductor. With this explanation of the triplet electron injection dynamics, the mode of binding of the RuN3 sensitizer to the semiconductor surface is a key feature. Various results in the literature have been interpreted to suggest that RuN3 is either attached with one bipyridyl ring,[80-82] or both bipyridyl rings[83,84] to the semiconductor surface (the one-ring attachment is illustrated in Figure 12). With the one-ring attachment the two ligands become non-equivalent and it can be expected that injection from the attached and non-attached ligand is different. By comparing a number of spectroscopic properties, solar cell performance, and electron injection characteristics for RuN3 and an analog (Ru520DN) only capable of attachment to the semiconductor with one bipyridyl ring, we concluded that RuN3 most likely is attached to the semiconductor with only one bipyridyl ring. This results in a dramatically different picture of the mechanism of electron injection from the triplet state, illustrated by the schematic diagram of Figure 13. The resulting scenario is that the majority of the excited molecules inject electrons from the initially excited delocalized ^{1}MLCT state into the conduction band of the TiO_2 (Figure 13, pathway A_1) on the <100 fs time-

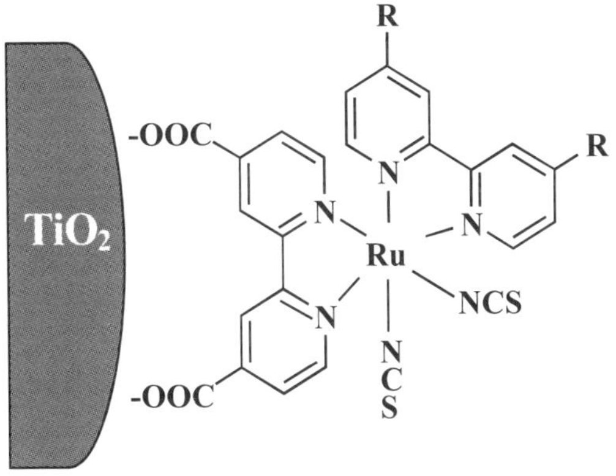

Fig. 12. Molecular structures of the Ru(II)-polypyridyl complexes. The carboxylate groups ensure efficient adsorption of the dye on the TiO_2 nanoparticle surface.

scale. In concert with this, the remaining population undergoes intersystem crossing (ISC), and localizes in the ^{3}MLCT state (pathway A$_2$). After relaxation to the bottom of the triplet state (pathway C), and interligand electron transfer (ILET, pathway D), the electrons are injected from the ^{3}MLCT state of the attached ligand (pathway E). This pathway of electron injection is controlled by ILET, which proceeds with different picosecond time constants depending on the dye molecule-TiO$_2$ particle interaction and solvent environment. In total, up to ~70% of the ET takes place on the femtosecond timescale.

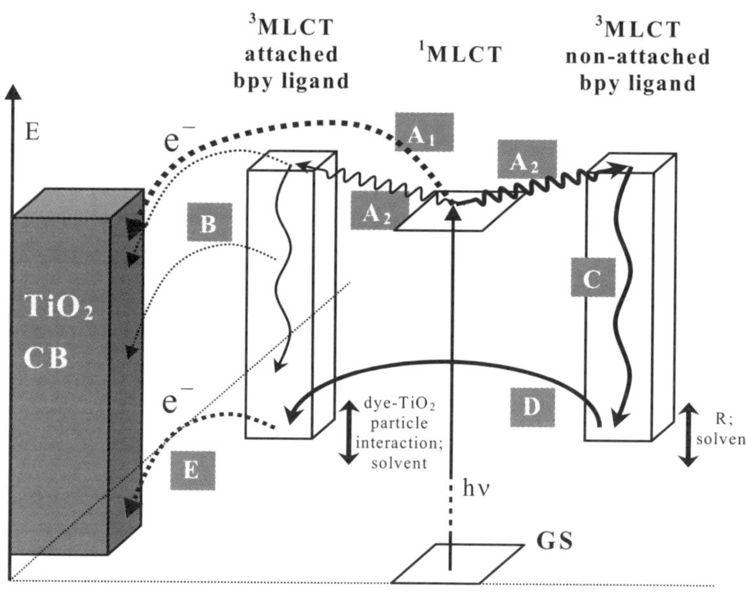

Fig. 13. Schematic model for electron injection of RuN3. Following MLCT excitation of the RuN3-sensitized TiO$_2$ nanocrystalline film, electron injection occurs from both excited states of the dye, ^{1}MLCT and ^{3}MLCT, into the conduction band (CB) of the semiconductor. GS: ground state of RuN3. Pathway A$_1$: electron injection from the initially excited delocalized ^{1}MLCT excited state. Pathway A$_2$: ISC and localization in the ^{3}MLCT excited state. Pathway B: electron injection from the hot ^{3}MLCT excited state of the attached bipyridine ligand (not observed in the present study). Pathway C: internal vibrational relaxation in the ^{3}MLCT excited state of the non-attached bipyridine ligand. Pathways D and E: ILET between the bipyridine ligands and ensuing electron injection.

This picture of the triplet injection was obtained from a comparison of ILET dynamics of RuN3 in solution and attached to the nanostructured TiO_2 film. When the RuN3 molecule is in solution, both of its bipyridine ligands are in contact with solvent and have similar energies. Following MLCT excitation and intramolecular energy relaxation processes in the molecule, the charge in the thermalized ^{3}MLCT excited state is localized on one of the ligands on the sub-picosecond time scale, similarly to what was observed for a similar Ru-complex.[77] After localization takes place, equilibration of population between the two bipyridine ligands of the molecule, *i.e.* ILET, may occur.[85,86] Since ILET only induces solvent reorganization of the structurally equivalent ligands, it results in no spectral changes, but its dynamics can be revealed using time-resolved absorption polarization spectroscopy. ILET dynamics of Ru- and Os-complexes in solution was shown to occur on the ~5-20 ps timescale.[86]

The time dependence of the transient absorption anisotropy of RuN3 in ethanol solution (RuN3-EtOH) is shown in Figure 14. The initial fast decay from a negative value (~ -0.15) reflects ^{1}MLCT $\rightarrow$ ^{3}MLCT intersystem crossing.[74,87] This is evidenced by the fact that the decay agrees very well with the intersystem crossing time measured from stimulated emission decay or triplet state formation. Another piece of evidence is that the anisotropy decay becomes faster when RuN3 is attached to TiO_2 and again agrees very well with the electron injection time from the ^{1}MLCT state in RuN3-TiO_2. At longer delay times (>1 ps), when the molecule has already relaxed to its triplet state, the amplitude of the anisotropy signal first increases with a time constant of ~20 ps to more negative values and later starts decaying (Figure 14). This last process is very slow compared to the previous ones, and reflects in a relaxation to lower lying excited states, similar to what has been observed for other transition metal complexes.[87,88]

The process occurring with the time constant of ~20 ps we attribute to ILET, because it is:

1. An exponential process occurring on the <100 picosecond timescale, as recorded previously for RuN3 and for similar complexes.[86]

2. Observable in time-resolved absorption anisotropy measurements, and not as spectral shift[89] or population decay.
3. Sensitive to surface attachment, because the ligand symmetry of the molecule is broken – precisely what we expect from previous studies.[90]
4. Not caused by rotational reorientation of the molecule because it is not decaying to lower anisotropy values.[85,90]

Now, let us consider how ILET influences electron injection dynamis of RuN3 attached to nanostructured TiO_2. As a result of MLCT excitation of RuN3-TiO_2 the triplet excited state will be either localized on the surface-attached ligand or on the non-surface-attached one. When the first case is realized, ILET does not influence the ET and the electrons can be injected at any moment from the excited state into the TiO_2 (pathway B in Figure 13). Because of the rapid decrease in density of conduction band states towards the bottom of the band a faster electron injection rate is expected for the ET from the surface-attached

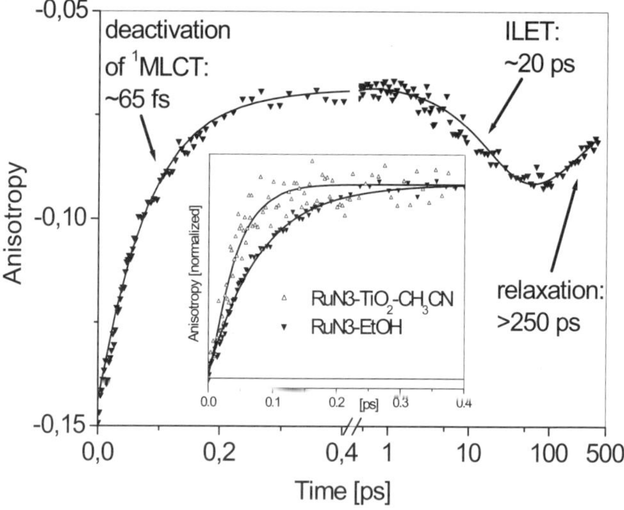

Fig. 14. Time dependence of the absorption anisotropy of RuN3 in EtOH solution at 850 nm. Inset: the same data at early times together with the scaled absorption anisotropy of the RuN3-sensitized TiO_2 film. Symbols are measured data, curves are fits to the signals with the following time constants: RuN3-EtOH: 65±10 fs, 19±2 ps, and >250 ps; RuN3-TiO_2-CH_3CN: 30±10 fs.

non-thermalized triplet state than for the thermalized triplet state. Therefore, this pathway of injection will have a decreasing contribution with time to the overall amplitude of ET. However, in our experiments we do not observe any electron injection from the hot triplet excited state localized on the attached ligand (pathway B), distinct from singlet injection. Therefore, we conclude that if this pathway is at all active, it occurs on a timescale similar to singlet ET. In the case where the ^{3}MLCT excited state is localized on the non-surface-attached ligand after photoexcitation, the electron may undergo ILET and be injected into the semiconductor if the relative energies of the two bipyridine ligands

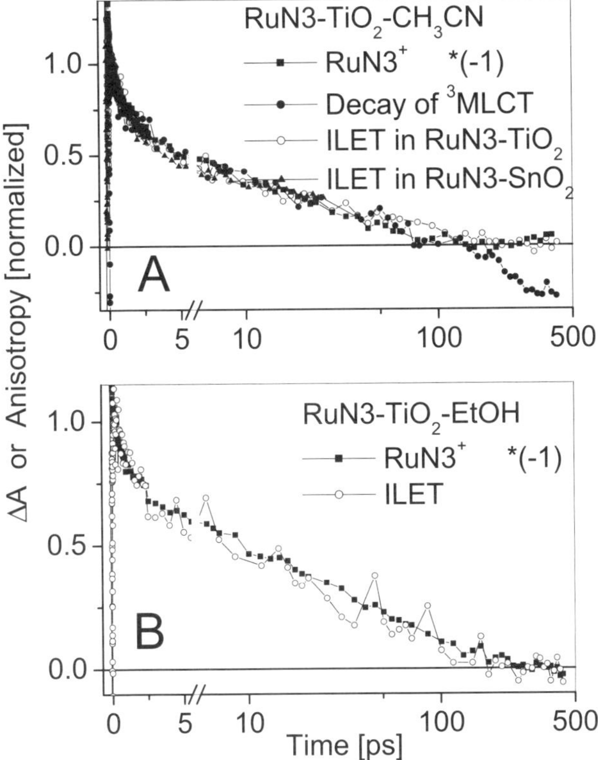

Fig. 15. Comparison between the transient absorption kinetics of picosecond electron injection at 850 nm (RuN3$^+$, data from Figure 2), decay of the ^{3}MLCT state at 1050 nm, and ILET in RuN3-sensitized TiO$_2$ and SnO$_2$ films (anisotropy at 850 nm), in the presence of CH$_3$CN (panel A) and EtOH (panel B).

permit (pathways D and E). Compared to ILET in RuN3-EtOH the rate of ILET in RuN3-TiO$_2$, regardless of solvent, is different because the ligand symmetry of the molecule is broken by the attachment to the semiconductor. ILET is not only faster in RuN3-TiO$_2$ than in RuN3-EtOH, but the time dependence is also non-exponential. While ILET in solution is well fitted with a single exponential with a time constant of ~20 ps, for RuN3-TiO$_2$ the fit requires at least three time constants ranging from ~1 to >50 ps. As ILET depends on the relative energies of the two bipyridine ligands, and since one of the ligands is in interaction with the semiconductor, the non-exponential behavior of ILET is a result of the heterogeneous interaction between RuN3 molecules and TiO$_2$ nanoparticles.[73,91,92]

The kinetics of RuN3$^+$ formation and ILET in acetonitrile and ethanol are compared in Figure 15 A and B, respectively. Also shown in Figure 15 A are the ILET kinetics of RuN3-SnO$_2$-CH$_3$CN (absorption anisotropy at 850 nm) and the kinetics of RuN3-TiO$_2$-CH$_3$CN recorded at 1050 nm, the latter directly monitoring the population of the triplet state on the non-attached ligand.[74,79] The electron injection and ILET kinetics are superimposable for both solvents and both semiconductors, showing that the population leaving the dye molecule for the semiconductor is the same as that transferred from the non-attached ligand to the attached one. The excellent agreement between the temporal evolution of the decay of the non-attached triplet state, ILET, and picosecond electron injection from the dye molecule implies that ILET efficiently controls the triplet channel of electron injection from RuN3.

In conclusion, photoinduced electron transfer from the sensitizer RuN3 to nanostructured TiO$_2$ occurs along two pathways. The major part of the molecules (~60 %) inject directly from the non-thermalized ^{1}MLCT state prior to IVR and intersystem crossing, while the remaining part inject from the thermalized triplet state of the non-attached ligand, in a process controlled by interligand electron transfer.

5. TRANSITION METAL SUPRAMOLECULAR COMPLEXES – ENERGY TRANSFER IN ARTIFICIAL ANTENNAS

In the previous section we discussed the application of transition metal complexes to a new type of photovoltaics. This kind of molecule is attracting a great deal of attention also for other applications. Transition metals complexed with polypyridine organic ligands are interesting for their multitude of properties (e.g. optical and redox), which can be combined to yield interesting functions, such as: photoinduced energy- and electron transfer; chemi- and electro-luminescence; light-activated molecular motion, and conformational changes.[93,94] Consequently, these molecules are considered for applications as light-harvesting antennas, for solar energy conversion, optical data storage and molecular machines and devices.[72,95,96] The spectral and electrochemical characteristics can be widely tuned by using a combination of different metals and ligands. By coupling more than one metal center together in a dendritic structure, supramolecules can be constructed with light-harvesting and energy funneling functions, or photoinduced electron transfer functions. A pigment system for solar energy conversion, e.g. artificial photosynthesis, could for instance consist of a multi-chromophore transition metal complex serving the function of a light-harvesting antenna, coupled to another transition metal complex carrying the reaction center function. The light-harvesting part would absorb the light energy and funnel it by directional energy transfer to the photochemical reaction center, where the energy is converted and stabilized by electron transfer. For a long-lived stable charge separation, multiple charge transfer steps are required, preventing the electron and positive hole from recombining. In particular, many Ru and Os and mixed Ru/Os complexes with varying number of metal centers have recently been designed and synthesized, and remarkable photochemical properties have been demonstrated.[97,98]

In designing and predicting the photochemical properties of a multi-centered system, consideration of the excited state energy level structure of the used building blocks is the starting point. For Ru and Os polypyridine complexes the strong visible absorption is due to ^{1}MLCT

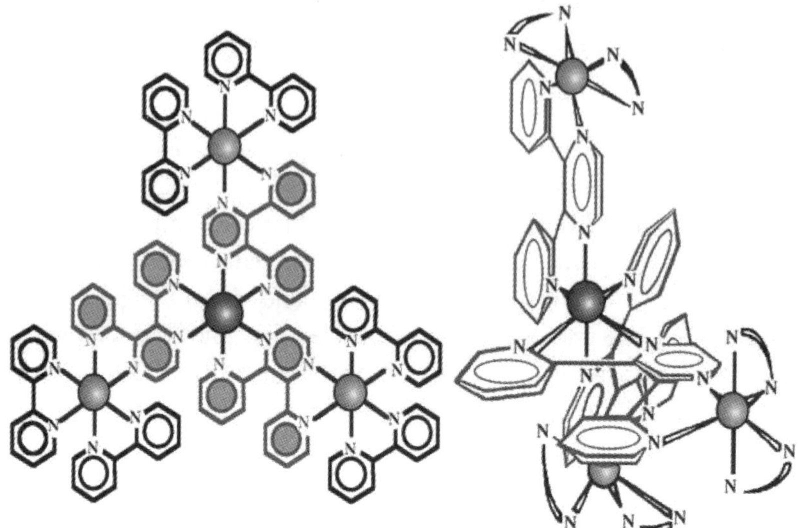

Fig 16. Representations of the OsRu3 complex, charges omitted. Left: dpp ligands shaded to distinguish from bpy ligands. Right: Exhibiting dpp ligand orientation.

states and red-shifted absorption tails or shoulders observed for many molecules are often assigned to lower-lying ^{3}MLCT states. Luminescence originates from the ^{3}MLCT states since, following optical excitation to a higher-lying singlet state, ultrafast (~ 100 fs) intersystem crossing to the lower ^{3}MLCT states occurs (section 4). This feature is the origin to the generally accepted picture of transition metal complex photochemistry – all function emanates from triplet states because the singlet states initially prepared by light absorption are depopulated via intersystem crossing on the 100 fs timescale, before they are significantly involved in chemical reactions. Hence, in designing complexes with energy funneling or electron transfer functions, the energies of the triplet states are considered for estimating directionality and driving forces of the processes. That this picture is not always true was demonstrated in section 4, for photoinduced electron transfer from a Ru-complex to a nanostructured semiconductor. In this section we are analyzing the corresponding situation for energy transfer in a tetranuclear mixed-metal complex $\{Os((\mu\text{-}2,3\text{-dpp})Ru(bpy)_2)_3\}^{8+}$ (2,3-dpp = 2,3-bis(2'-pyridyl)-pyrazine; bpy = 2,2'-bipyridine), depicted in Figure 16, and hereafter denoted as OsRu3. Steady-state and time-resolved (nano- and microsecond time scales) measurements have shown that this complex behaves as an artificial antenna system – all the light energy absorbed by the peripheral Ru-based chromophores is quantitatively transferred to the central Os-based core, which acts as a terminal emitter.[98] Using ultrafast spectroscopy we have shown that both singlet and triplet states are actively involved in this process, and that a situation analogous to that for electron transfer to a nanostructured semiconductor holds here for energy transfer. Thus, upon exciting the peripheral Ru-chromophores to their triplet state, efficient energy transfer to the Os-triplet state occurs with a 600 fs time constant. When the wavelength of the excitation is tuned to the singlet states of the Ru-moieties another much faster ($\leq$ 60 fs) channel of energy transfer between the singlet states of the Ru- and

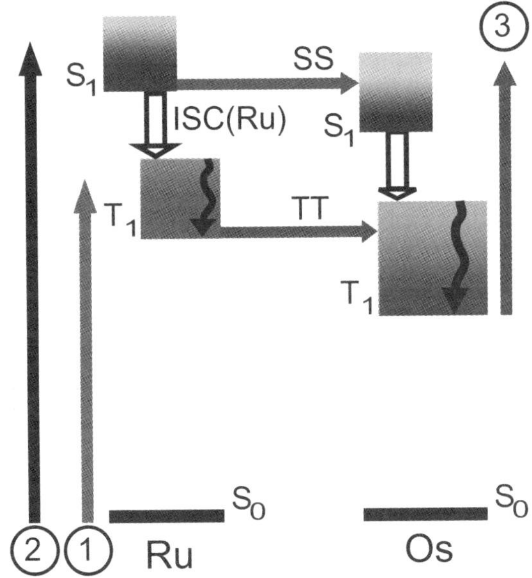

Fig. 17. Scheme of energy transfer and relaxation processes within the OsRu3 complex. Excitation of the singlet state (process 2) leads to ultrafast singlet-singlet energy transfer (SS) towards the Os core that competes with ISC crossing denoted by double arrows. Excitation of the triplet state (process 1) initiates slower triplet-triplet (TT) energy transfer. All processes are probed by means of excited state absorption from the Os triplet (process 3). Wavy arrows denote vibrational relaxation and cooling in triplet states of both metal centers.

Os-chromophores opens up in competition with singlet-triplet intersystem crossing in the Ru-unit. This scenario is illustrated by the energy level diagram in Figure 17. Details of this work can be found in two recent papers.[99,100]

Electronic energy transfer is generally discussed in terms of two different mechanisms, Förster energy transfer[101] mediated by dipole-dipole interactions (or in the general case Coulombic interactions also including higher multipoles) between donor and acceptor chromophores, and Dexter energy transfer,[102] promoted by electron exchange. The Förster mechanism is long-range and requires the involvement of strong transitions to be efficient. The Dexter mechanism on the other hand is short-range, since it effectively requires orbital overlap for efficient electron exchange. Based on the results discussed above, it is too early to draw definite conclusions regarding the mechanisms of the observed energy transfer pathways. Nevertheless, the Förster mechanism appears less likely for the T-T pathway involving the low lying MLCT states of mainly triplet character. Note that in the discussion above we have termed these states as ^{3}MLCT states, but due to the spin-orbit coupling induced by the heavy metal elements, the excited states are not of pure multiplicity but contain some singlet character which accounts for the weak red-tail absorbance. For the S-S transfer the situation is different; both the Ru- and Os-^{1}MLCT transitions carry substantial oscillator strength. Therefore, both the Förster and Dexter mechanisms could be active here. For more elaborate mechanistic considerations, more detailed spectral characterization of in particular the singlet states is required.

Together with the observations of ultrafast electron transfer processes from transition metal complexes to a semiconductor (see section 4), the present results show that such molecules can be designed to perform extremely fast (among the fastest chemical processes observed) and efficient energy-harvesting and converting functions. This opens the possibility to construct very large systems with maintained efficiency and maximum utilization of the photon energy. The very short residence time of excitation energy on a metal center will in addition minimize destructive photochemical processes in an extended energy-converting system.

ACKNOWLEDGEMENTS

We thank all our coworkers at the Department of Chemical Physics, Lund University for many fruitful collaborations and discussions.

References

1. McDermott G, Prince SM, Freer AA, Hawthornthwaite-Lawless AM, Papiz MZ, Cogdell RJ, and Isaacs NW. *Nature* 1995; **374**: 517.
2. Roszak AW, Howard TD, Southall J, Gardiner AT, Law CJ, Isaacs NW, and Cogdell RJ. *Science* 2003; **302**: 1969.
3. Ferreira KN, Iverson TM, Maghlaoui K, Barber J, and Iwata S. *Science* 2004; **303**: 1831.
4. Jordan P, Fromme P, Witt HT, Klukas O, Saenger W, and Krauß N. *Nature* 2001; **411**: 909.
5. Liu Z, Yan H, Wang K, Kuang T, Zhang J, Gui L, An X, and Chang W. *Nature* 2004; **428**: 287.
6. Sundström V, Pullerits T, and van Grondelle R. *J. Phys. Chem B.* 1999; **103**: 2327.
7. Polívka T, and Sundström V. *Chem. Rev.* 2004; **104**: *in press*.
8. Hess S, Feldchtein F, Babin A, Nurgaleev I, Pullerits T, Sergeev A, and Sundström V. *Chem. Phys. Lett.* 1993; **216**: 247.
9. Hess S, Åkesson E, Cogdell RJ, Pullerits T, and Sundström V. *Biophys. J.* 1995; **69**: 2211.
10. Shreve AP, Trautman JK, Frank HA, Owens TG, and Albrecht, AC. *Biochim. Biophys. Acta* 1991; **1058**: 280.
11. Pullerits T, Hess S, Herek JL, and Sundström V. *J. Phys. Chem.* 1997; **101**: 10560.
12. Monshouwer R, de Zarate IO, van Mourik F, and van Grondelle R. *Chem. Phys. Lett.* 1995; **246**: 341.
13. Fraser NJ, Dominy PJ, Ücker B, Simonin I, Scheer H, and Cogdell RJ. *Biochemistry* 1999; **38**: 9684.
14. Herek JL, Fraser N, Pullerits T, Martinsson P, Polivka T, Scheer H, Cogdell RJ, and Sundström V. *Biophys. J.* 2000; **78**: 2590.
15. Krueger BP, Scholes GD, Gould IR, and Fleming GR. *PhysChemComm* 1999; **8** (http://www.rsc.org/ej/qu/1999/c9903172).
16. Kühn O, and Sundström V. *J. Phys. Chem. B.* 1997; **101**: 3432.

17. Sumi H. *J. Phys. Chem. B.* 1999; **103**: 252.

18. Scholes GD, and Fleming GR. *J. Phys. Chem. B.* 2000; **104**: 1854.

19. Leupold D, Stiel H, Teuchner K, Nowak F, Sandner W, Ücker B, and Scheer H. *Phys. Rev. Lett.* 1996; **77**: 4675.

20. Pullerits T, Chachisvilis M, and Sundström V. *J. Phys. Chem.* 1996; **100**: 10787.

21. Monshouwer R, Abrahamsson M, van Mourik F, and van Grondelle R. *J. Phys. Chem. B.* 1997; **101**: 7241.

22. Chachisvilis M, Kühn O, Pullerits T, and Sundström V. *J. Phys. Chem. B.* 1997; **101**: 7275.

23. Meier T, Chernyak V, and Mukamel S. *J. Phys. Chem.* 1997; **101**: 7332.

24. Dahlbom M, Pullerits T, Mukamel S, and Sundström V. *J. Phys. Chem. B.* 2001; **105**: 5515.

25. Mauzerall D. *Biophys. J.* 1976; **16**: 87.

26. Den Hollander WTF, Bakker JGC, and van Grondelle R. *Biochim. Biophys. Acta* 1983; **725**: 492.

27. Brüggemann B, Herek JL, Sundström V, Pullerits T, and May V. *J. Phys. Chem. B.* 2001; **105**: 11391.

28. Trinkunas G, Herek JL, Polivka T, Sundström V, and Pullerits T. *Phys. Rev. Lett.* 2001; **86**: 4167.

29. Polivka T, Pullerits T, Herek JL, and Sundström V. *J. Phys. Chem. B.* 2000; **104**: 1088.

30. Timpmann K, Katiliene Z, Woodbury NW, and Freiberg A. *J. Phys. Chem. B.* 2001; **105**: 12223.

31. Dahlbom M, Beenken W, Sundström V, and Pullerits T. *Chem. Phys. Lett.* 2002; **364**: 556.

32. Beenken WJD, Dahlbom M, Kjellberg P, and Pullerits T. *J. Chem. Phys.* 2002; **117**: 5810.

33. Van Mourik F, Frese RN, van der Zwan G, Cogdell RJ, and van Grondelle R. *J. Phys. Chem. B.* 2003; **107**: 2156.

34. Schubert A, Stenstam A, Beenken WJ, Herek JL, Cogdell R, Pullerits T, and Sundström V. *Biophys. J.* 2004; **86**: 2363.

35. Reddy NRS, Picorel R, and Small GJ. *J. Phys. Chem.* 1992; **96**: 6458.

36. Hess S, Chachisvilis M, Timpmann K, Jones MR, Fowler GJS, Hunter CN, and Sundström V. *Proc. Natl. Acad. Sci. USA* 1995; **92**: 12333.

37. Nagarajan V, and Parson WW. *Biochemistry* 1997; **36**: 2300.

38. Freiberg A, Godik VI, Pullerits T, and Timpmann K. *Biochim. Biophys. Acta* 1989; **973**: 93.

39. Visscher KJ, Bergström H, Sundström V, Hunter CN, and van Grondelle R. *Photosynth. Res.* 1989; **22**: 211.

40. Timpmann K, Zhang FG, Freiberg A, and Sundström V. *Biochim. Biophys. Acta* 1993; **1183**: 185.

41. Pullerits T, and Sundström V. *Accounts Chem. Res.* 1996; **29**: 381.

42. See, *Photochem. Photobiol.* 2001; **144**:, and references therein.

43. Ruseckas A, Theander M, Valkunas L, Andersson MR, Inganäs O, and Sundström V. *J. Luminescence* 1998; **76&77**: 474.

44. Grage MML, Wood P, Ruseckas A, Pullerits T, Mitchell W, Burn PL, Samuel IDW, and Sundström V. *J. Chem. Phys.* 2003; **118**: 7644.

45. Grage MML, Zaushitsyn Y, Yartsev AP, Chachisvilis M, Sundström V, and Pullerits T. *Phys. Rev. B.* 2003; **67**: 5207.

46. Ruseckas A, Namdas E, Ganguly T, Theander M, Svensson M, Andersson MR, Inganäs O, and Sundström V. *J. Phys. Chem.* 2001; **105**: 7624.

47. Ruseckas A, Namdas E, Theander M, Svensson M, Zigmantas D, Yartsev AP, Andersson MR, Inganäs O, and Sundström V. *Photochem. Photobiol.* 2001; **144**: 3.

48. Grage MML, Pullerits T, Ruseckas A, Theander M, Inganäs O, and Sundström V. *Chem. Phys. Lett.* 2001; **339**: 96.

49. Yu G, Philips SD, Tomozawa H, and Heeger AJ. *Phys. Rev. B.* 1990; **42**: 3004.

50. Moses D, Dogariu A, and Heeger AJ. *Phys. Rev. B.* 2000; **61**: 9373.

51. Barth S, Bässler H, Wehrmeister T, and Müllen K. *J. Chem. Phys.* 1997; **106**: 321.

52. Frankevich EL, Lymarev AA, Sokolik I, Karasz FE, Blumstengel S, Baughman RH, and Hörhold HH. *Phys. Rev. B.* 1992; **46**: 9320.

53. Esteghamatian M, Popovic ZD, and Xu G. *J. Phys. Chem.* 1996; **100**: 13716.

54. Khan MI, Bazan GC, and Popovic ZD. *Chem. Phys. Lett.* 1998; **298**: 309.

55. Pfeffer N, Neher D, Remmers M, Poga C, Hopmeier M, and Mahrt R. *Chem. Phys.* 1998; **227**: 167.

56. Kersting R, Lemmer U, Deussen M, Bakker HJ, Marth RF, Kurz H, Arkhipov VI, Bässler H, and Göbel EO. *Phys. Rev. Lett.* 1994; **73**: 1440.

57. Graupner W, Cerullo C, Lanzani G, Nisoli M, List EJW, Leising G, and De Silvestri S. *Phys Rev. Lett.* 1998; **81**: 3259.

58. Arkhipov VI, Wolf U, and Bässler H. *Phys. Rev. B.* 1999; **59**: 7514.

59. Muller JG, Lemmer U, Feldmann J, and Scherf U. *Phys. Rev. Lett.* 2002; **88**: 147401.

60. Gulbinas V, Zaushitsyn Y, Sundström V, Hertel D, Bässler H, and Yartsev AP. *Phys.Rev.Lett.* 2002; **89**: 107401.

61. Yan M, Rothberg LJ, Papadimitrakopoulos F, Galvin ME, and Miller TM. *Phys. Rev. Lett.* 1994; **72**: 1104.

62. Sherf U, Bohnen A, and Müllen K. *Macromol. Chem. Phys.* 1992; **193**: 1127.

63. Ruseckas A, Theander M, Andersson MR, Svensson M, Prato M, Inganäs O, and Sundström V. *Chem. Phys. Lett.* 2000; **322**: 136.

64. Arkhipov VI, Emelianova EV, and Bässler H. *Phys. Rev. Lett.* 1999: **82**: 1321.

65. Geacintov NE, and Pope M. *J. Chem. Phys.* 1965; **47**: 1194.

66. Jortner J. *Phys. Rev. Lett.* 1968; **20**: 244.

67. Gulbinas V, Zaushitsyn Y, Bässler H, Yartsev AP, and Sundström V. *Phys. Rev. B.* 2004; **70**: 5215.

68. Zaushitsyn Y, Gulbinas V, Zigmantas D, Zhang F, Inganäs O, Sundström V, and Yartsev AP. *Phys Rev. B. In press.*

69. O'Regan B, and Grätzel M. *Nature* 1991; **353**: 737.

70. Burfeindt B, Zimmermann C, Ramakrishna S, Hannappel T, Meissner B, Storck W, and Willig F. *Z. Physikalische Chemie.* Part 1 1999; **212**: 67.

71. Ghanem R, Xu Y, Pan J, Hoffmann T, Andersson J, Polivka T, Pascher T, Styring S, Sun L, and Sundström V. *J. Inorg. Chem.* 2002; **41**: 6258.

72. Hagfeldt A, and Grätzel M. *Acc. Chem. Res.* 2000; *33*: 269.

73. Asbury JB, Hao E, Wang Y, Ghosh HH, and Lian T. *J. Phys. Chem. B.* 2001; **105**: 4545.

74. Benkö G, Kallioinen J, Korppi-Tommola JEI, Yartsev AP, and Sundström V. *J. Am. Chem. Soc.* 2002; **124**: 489.

75. Asbury JB, Ellingson RJ, Ghosh HN, Ferrere S, Nozik AJ, and Lian TQ. *J. Phys. Chem. B.* 1999; **103**: 3110.

76. Damrauer NH, Cerullo G, Yeh A, Boussie TR, Shank CV, and McCusker JK. *Science* 1997; **275**: 54.

77. Yeh AT, Shank CV, and McCusker JK. *Science.* 2000; **289**: 935.

78. Benkö G, Myllyperkiö P, Pan J, Yartsev AP, and Sundström V. *J. Am. Chem. Soc.* 2003; **125**: 1118.

79. Kallioinen J, Benkö G, Sundström V, Korppi-Tommola JEI, and Yartsev AP. *J. Phys. Chem. B.* 2002; **106**: 4396-4404.

80. Rensmo H, Westermark K, Södergren S, Kohle O, Persson P, Lunell S, and Siegbahn H. *J. Chem. Phys.* 1999; **111**: 2744-2750.

81. Persson P, and Lunell S. *Sol. Mater. & Sol. Cells.* 2000; **63**: 139-148.

82. Haukka M, and Hirva P. *Surf. Sci.* 2002; **511**: 373-378.

83. Shklover V, Ovchinnikov YE, Braginsky LS, Zakeeruddin SM, and Grätzel M. *Chem. Mater.* 1998; **10**: 2533-2541.

84. Fillinger A, Soltz D, and Parkinson BA. *J. Electrochem. Soc.* 2002; **149**: A1146-A1156.

85. Waterland MR, and Kelley DF. *J. Phys. Chem. B.* 2001; **105**: 4019-4028.

86. Olsen CM, Waterland MR, and Kelley DF. *J. Phys. Chem. B.* 2002; **106**: 6211-6219.

87. Benkö B, Kallioinen J, Myllyperkiö P, Trif F, Korppi-Tommola JEI, Yartsev AP, and Sundström V. *J. Phys. Chem. B.* In press kolla

88. Thompson DW, Wishart JF, Brunschwig BS, and Sutin NJ. *Phys. Chem. A.* 2001; **105**: 8117-8122.

89. Myllyperkiö P, Benkö G, Kallioinen J, Yartsev AP, and Sundström V. *J. Phys. Chem. B* In press

90. Shaw GB, Brown CL, and Papanikolas JM. *J. Phys. Chem. A* 2002; **106**: 1483-1495.

91. Tachibana Y, Rubtsov I V, Montanari I, Yoshihara K, Klug DR, and Durrant JR. *J. Photochem. Photobiol. A* 2001; **142**: 215-220.

92. Benkö G, Skårman B, Wallenberg R, Hagfeldt A, Sundström V, and Yartsev, A P. *J. Phys. Chem. B.* 2003; **107**: 1370-1375.

93. Crosby GA. *Acc. Chem. Res.* 1975; **8**: 231.

94. Balzani V, Juris A, Venturi M, Campagna S, and Serroni S. *Chem. Rev.* 1996; **96**: 759, and refs. therein.

95. Welter S, Brunner K, Hofstraat JW, and De Cola L. *Nature* 2003; **421**: 54, and refs. therein.

96. Balzani V, Credi A, and Venturi M. *Moleular Devices and Machines*, Wiley-VCH, Weinheim, 2003, and refs. therein.

97. Balzani V, and Scandola F. *Supramolecular Photochemistry*: Horwood, Chichester; 1991.

98. Campagna S, Serroni S, Puntoriero F, Di Pietro C, and Ricevuto V, in Balzani V (Ed.), *Electron Transfer in Chemistry*: VCH-Wiley, Weinheim; 2001, Vol. 5, p. 186, and refs. therein.

99. Andersson J, Polivka T, Puntoriero F, Campagna S, and Sundström V. *Chem. Phys. Lett.* In press

100. Andersson J, Polivka T, Puntoriero F, Campagna S, and Sundström V. *Faraday Discuss.,* 2004;127 (2004) 000

101. Förster Th. In *Modern Quantum Chemistry*, ed. Sinanoglu O: Academic Press, New York; 1965.

102. Dexter DL. *J. Chem. Phys.* 1953; **21**: 836

Chapter 6

CONTROLLING EXCITATION ENERGY AND ELECTRON TRANSFER BY TUNING THE ELECTRONIC COUPLING

Bo Albinsson and Jerker Mårtensson

Factors that control electronic coupling, as manifested through electron transfer and singlet and triplet excitation energy transfer, are discussed for covalent donor-bridge-acceptor (DBA) systems.

Keywords: Singlet excitation, triplet excitation, energy transfer, electron transfer, superexchange, bridged systems, through-bond coupling.

1. INTRODUCTION

The interaction between molecular subunits in supramolecular assemblies has been the topic for many research groups throughout the last 40 years. Knowledge of how to tune the electronic coupling between chromophores and thereby control the rates of electron transfer (ET), singlet excitation energy transfer (SEET), and triplet excitation energy transfer (TEET) is a necessary prerequisite to understand the natural photosynthetic machinery and also to develop man-made imitations of the natural systems (Gust *et al.*, 2001; Moore *et al.*, 2002; Sun *et al.*, 2001). Molecular scale electronics is still a dream, but recent developments hold much promise for the future in this field (Jortner and Ratner, 1997). If this revolution in nanoelectronics is going to come true, it is absolutely inevitable that a solid understanding of the components – *i.e.* the molecules and their mutual interaction – is in the hands of the next generation of electrical engineers. We are not yet at that point, and this review will try to summarize the current status of the field.

Although not Nature's primary choice for the formation of its complex donor-acceptor arrays with highly efficient and specific functions, and certainly not the proper choice for future construction of artificial analogues, the use of covalently linked donor-acceptor systems has so far proven superior in elucidating detailed understanding of the different transfer mechanisms. Therefore, in the limited space available, we have decided to concentrate on a number of illustrative examples of how the molecular bridge in covalent donor-bridge-acceptor (DBA) molecules controls the degree of electronic coupling. Thus, the intriguing and important results stemming from the considerable work devoted to understanding how, and to what extent, non-covalent interactions (Andersson *et al.*, 2000; Ward, 1997) and through-solvent interactions (Read *et al.*, 1999) contribute to the overall mediation of the transfer processes is not covered at all in this chapter. Neither is it comprehensive in the area of covalently linked DBA systems. To the many authors of all the fine work not included, we offer our humble apology.

2. THEORETICAL BACKGROUND

Many texts give a comprehensive account of the current theories for electron and energy transfer (e.g. Balzani, 2000) and this section is therefore limited to provide the most crucial mathematical relations, in most cases without formal derivation. The Fermi Golden Rule, Eq. 1, describes the rate of transfer between two adiabatic potential surfaces provided the electronic coupling is not too large:

$$k_{if} = (2\pi / \hbar)V_{if}^2 FCWD ,\tag{1}$$

where the electronic coupling, V_{if}, is defined as the effective electronic Hamilton matrix element that couples the initial (Ψ_i) and final (Ψ_f) states.

$$V_{if} = \langle \Psi_f | H' | \Psi_i \rangle .\tag{2}$$

The Franck-Condon weighted density of states (FCWD) describes the influence from the nuclear modes of the system and should be interpreted according to the transfer reactions studied. The electronic coupling between the initial and final diabatic states, Eq. 2, may be equally well derived from the adiabatic states as the energy splitting at the avoided crossing geometry. At this configuration the electronic coupling is given as half the energy splitting:

$$V_{if} = (E_i * - E_f *)/2 .$$

(3)

2.1. Excitation energy transfer

Electronic excitation energy is transferred between molecules by either a trivial radiative process (emission of a real photon followed by subsequent reabsorption) or in a non-radiative process. The non-radiative energy transfer process, in which a donor and an acceptor molecule with resonant vibronic states exchange excitation energy (a virtual photon), could be of significant magnitude at distances shorter than, say, 100 Å. The transfer of excitation energy either between different, freely diffusing, donor (D) and acceptor (A) molecules, or between D and A subunits of a supermolecule, could schematically be described by:

$$D * + A \rightarrow D + A *$$
$$D * BA \rightarrow DBA *$$

(4)

where B denotes a bridging subunit of the supermolecule. A necessary condition for the excitation energy transfer is that the relevant transitions, $D * \rightarrow D$ and $A \rightarrow A *$ of the donor and acceptor, respectively, are in resonance and that the states are coupled by suitable donor-acceptor interactions. The former requirement is fulfilled if the spectral overlap integral, Eq. 5, has a significant magnitude:

$$J = \int I_D(\widetilde{v}) \varepsilon_A(\widetilde{v}) d\widetilde{v}$$

(5)

$I_D(\widetilde{v})$ and $\varepsilon_A(\widetilde{v})$ are the normalized emission and absorption spectra of the donor and acceptor, respectively. The spectral overlap integral is proportional to the number of resonant transitions and thus proportional to the FCWD factor of Eq. 1.

The electronic coupling for excitation energy transfer (EET) can be divided into a Coulomb and an exchange part. The Coulomb term can be expanded into a multipole series that, for allowed donor and acceptor transitions, has a dominating dipole-dipole term. So doing, (Förster, 1946, 1948) derived the following expression for the energy transfer rate:

$$k_{EET} \propto \frac{f_D f_A}{R^6 \tilde{v}^2} J, \tag{6}$$

where f_D and f_A are the oscillator strengths of the donor and acceptor transitions, respectively, R is the (fixed) donor-acceptor distance, and J is the spectral overlap integral (Eq. 5). It is seen from Eq. 6 that both donor and acceptor transitions need to be allowed in order to have a large rate with the Coulomb mechanism. Förster was furthermore able to turn this equation into an operational expression for the rate of energy transfer expressed only in measurable quantities:

$$k_{EET} = (1/\tau_D)(R_0 / R)^6, \tag{7}$$

where τ_D is the excited lifetime of the donor in the absence of the acceptor and R_0 is the critical donor-acceptor distance, at which 50% of the donor decay is due to energy transfer. R_0 is given by:

$$R_0^6 = \frac{9000(\ln 10)}{128\pi^5 N_A} \frac{\kappa^2 \phi_D J'}{n^4}, \tag{8}$$

where ϕ_D is the emission quantum yield of the donor in absence of the acceptor, κ is an orientation factor, n is the solvent refractive index, and J' is an overlap integral similar to Eq. 5.

By taking a more general approach Dexter derived, in addition to the Coulomb term, an expression for the exchange term (Dexter, 1953). The exchange term depends on the overlap between the wavefunctions of the donor and acceptor and, since the molecular orbitals at sufficiently large separation have an exponentially decreasing overlap, the Dexter energy transfer is expressed as:

$$k_{EET} \propto J \cdot \exp(-2R / L), \tag{9}$$

where J is the spectral overlap integral (Eq. 5) and L is the effective orbital radius. The exchange interaction is by nature short-ranged (<10 Å) whereas the Coulomb interaction can extend over quite large distances (<100 Å).

2.2. Electron transfer

Starting with transition state theory and assuming simple quadratic potential surfaces with equal force constants (curvature) for the initial and final states, Marcus derived an expression for the rate of thermal electron transfer (Marcus, 1956):

$$k_{ET} = \sqrt{(\pi / \hbar^2 k_B T \lambda)} V^2 \exp(-(\Delta G^0 + \lambda)^2 / 4\lambda k_B T), \qquad (10)$$

where ΔG^0 is the standard free energy change for the electron transfer reaction. The reorganization energy, λ, is defined as the vertical excitation energy required to reach the potential energy surface of the final state from the initial state configuration, i.e. the energy required to reorganize the reactant and the solvent cage into the configuration of the product. Eq. 10 is valid for weak coupling of the initial and final states, at sufficiently high temperature, and with a classical description of the reaction coordinate. A so-called semi-classical treatment, starting with Eq. 1, leads to a slightly modified equation for the electron transfer rate (Jortner and Bixon 1988; Ulstrup and Jortner, 1975).

2.3 Superexchange

In a supermolecule where the donor and acceptor moieties are separated by a molecular bridge, the electronic coupling at a given distance is generally larger than through vacuum due to the mixing of the zeroth-order states, Ψ_D, Ψ_A, and Ψ_B. This gives rise to an increased electronic coupling that sometimes dominates over the direct coupling. By using perturbation theory, McConnell derived a compact expression for the electronic coupling through n identical intermediate virtual states each offset from the D/A energy by a common energy gap, Δ (McConnell, 1961):

$$\left| V_{if} \right| = \left| V_{DB_1} (v / \Delta)^{n-1} (V_{B_n A} / \Delta) \right|. \qquad (11)$$

In Eq. 11, v is the common electronic coupling between the bridge subunits and in order for the perturbative approach to hold it is assumed that the ratio of the different couplings and the energy gap is much smaller than unity. Both the direct DA coupling and the superexchange coupling are approximately decaying exponentially with distance, which predicts that the rate for electron transfer also decays exponentially:

$$k_{if} \propto V_{if}^2 \propto \exp(-\beta R) \; ; \tag{12}$$

in Eq. 12, R is the donor-acceptor (edge to edge) distance and β is a damping factor specific for each set of DBA systems.

3. DONOR-BRIDGE-ACCEPTOR SYSTEMS WITH π-BRIDGES

Conjugated organic systems have for several reasons been used as connectors between donor and acceptor molecules. In research directed towards assessment and development of transfer theories, their rigidity has been a major justification. In many of the early donor-connector-acceptor systems the connector was generally thought of as an inactive spectator during the transfer process, merely functioning as a scaffold holding the donor and acceptor in specific positions relative to each other. The well-defined geometries obtained for the donor-acceptor systems linked by such connectors made it possible to successfully match the experimental results with existing theories or, when there were discrepancies, prove the need for refinement of these theories. The view of the bridge as an inactive spectator gradually changed and today much effort is put into the search for connectors that effectively promote different types of transfer processes. Many examples of such connectors or molecular wires are large conjugated π-systems. Another reason is based on the fact that several conjugated organic systems show photochromism and thermochromism, both of which are explored as possible switching mechanisms in future molecular electronics.

3.1 Singlet Excitation Energy Transfer

Energy transfer according to the Coulomb mechanism is, in the Förster approximation (Förster, 1946, 1948), based on a dipole-dipole interaction that can be significant even at large donor-acceptor distances and that has no obvious bridge dependence. There is, however, a bulk solvent dependence on the refractive index (Eq. 8) that is assumed to take into account the properties of the local environment surrounding the donor and acceptor. In the presence of a bridge, the correctness of this assumption can be questioned and, in particular, how the bridge structure affects the local refractive index. The exchange mechanism, on the other hand, depends on the spatial overlap of the orbitals of the donor and the acceptor and can only be effective at short donor-acceptor distances. However, it is often pictured that a covalently linked bridge extends the donor and acceptor orbitals through its bonds, resulting in a decrease in effective donor-acceptor distance. This mechanism is called superexchange and is considered to show strong bridge dependence.

Several studies have been devoted to evaluation of the relative importance of the Coulombic and the exchange interaction in DBA systems. By preparation and studies of properly designed diporphyrin systems, together with a comparison against relevant data for similar systems found in the literature, Osuka and co-workers (Cho *et al.*, 2001) found that the relative importance of the Coulombic interaction increases with decreasing donor-acceptor distance. In the beginning this result may seem to be counterintuitive, however, one should remember that both the

Donor	Bridge	Acceptor
TPP, 1		
Ar = 3,5-di-*tert*-butylphenyl or phenyl	**a**	Ar = 3,5-di-*tert*-butylphenyl or phenyl
	b	
OEP, 2	**c**	

exchange and Coulombic interactions should increase as the bridge length decreases. The conclusion was based on time-resolved spectroscopy data obtained from studies on two closely related series of DBA systems. The two series differed, in the main, only in the ability of the donor and the acceptor to electronically couple to the bridge. The first of the two series was based on tetraphenylporphyrins (TPP, **1**) that have A_{2u} HOMO orbitals with substantial electron density at the meso carbons onto which the bridge is connected. The bridges in the second series were also connected to the meso carbons of porphyrins, although, this time 5,15-diaryl-β-octaalkylporphyrins and β-octaalkylporphyrins (OEP, **2**) that have A_{1u} HOMO orbitals with nodes at the meso carbons, were used. The marked effect of electron-density distribution in porphyrins on the attenuation of the electronic coupling was originally described by Lindsey and co-workers (Strachan *et al.*, 1997, Yang *et al.*, 1999) as the origin for the large observed differences in singlet exchange energy transfer (SEET) rates for a set of similar DBA systems. The poor condition precedent for electronic coupling between the donor or the acceptor and the bridge in the latter of the two series of compounds would suggest that the exchange contribution should be of minor importance and that the Coulombic transfer should compete successfully within this series. In fact, SEET rates in reasonable agreement with, although generally higher than, those calculated according to the Förster approximation were found for OEP systems bridged by long, linear π-conjugated bridges. In contrast, much higher transfer rates than predicted by the Förster approximation were observed for the TPP systems. The results can, thus, be summarized as follows: Coulombic interaction alone cannot account for the observed SEET efficiencies in general and in the TTP-type systems in particular. The pronounced dependence on orbital symmetry at the donor and acceptor indicates that the electronic interaction is very sensitive to the frontier orbital overlap between the donor or the acceptor and the bridge, thus suggesting that it is a superexchange interaction.

Both exchange and Coulomb interactions should increase as the bridge length decreases. Indeed, smaller bridges, such as 4,4'-biphenylene, 1,4-, and 1,3-phenylene (**1a-c, 2a-c**), gave larger SEET

Donor **Bridge** **Acceptor**

a

b

c

d

3

m = 2H

4

m = Fe(III)(Im)$_2$

5

m = Au(III)BF$_4$

rates but they also gave smaller rate differences between TPP-type and OEP-type diporphyrins of equal length and geometry, indicating an increased relative contribution of the Coulombic interaction in the overall electronic interaction. Cho *et al.* explain the large relative contribution of the short-range exchange interaction in the systems with longer bridges by the presence of low-lying unoccupied molecular orbitals at those bridges. This conclusion is strengthened by our studies on DBA systems (**3**) in which we varied the energetics of the bridge but kept other important structural parameters, in particular the donor-acceptor distance of the systems, the same regardless of the bridging chromophore. These studies also demonstrated that SEET is governed by a superexchange mechanism in parallel with the Förster mechanism (Kilså *et al.*, 1997, 1999). More importantly, it was shown that the rate of SEET correlated with the energy gap between the lowest singlet excited states of the bridge and the donor (Figure 1). The variation in energetics of the bridge was achieved by changing the central unit of the bridging chromophore; from bicyclo[2.2.2]octane *via* benzene and naphthalene to anthracene.

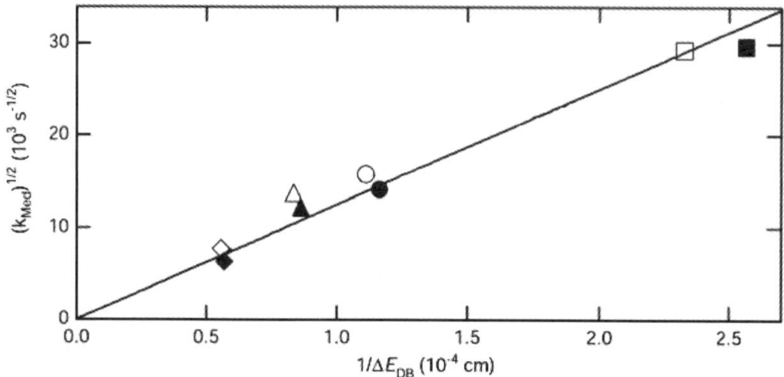

Fig. 1. SEET rate as a function of reciprocal energy separation (λE_{DB}) between the bridge and donor excited states.

In parallel with the numerous experimental investigations, computational chemistry can provide important additional information on the electronic coupling in DBA systems as well as guiding the development of new systems. Calculating the electronic coupling from Eq. 2 or 3 for simple model systems is, in principle, quite straightforward and several studies (see for examples the following reviews, books and thematic issues: Newton, 1991; Jortner *et al.*, 1997, 1999) have been performed to demonstrate the applicability of the different methods. The electronic coupling is equal to half the energy splitting between the interacting states (initial and final) at the avoided crossing point. For a real system, however, the position of the avoided crossing is unknown and usually very time-consuming to find. To circumvent these problems different approaches have been developed. A suitable symmetric model system might be used to represent the real DBA system and the interacting states are interpreted by the use of symmetry considerations. An alternative approach is to study the real unsymmetrical DBA molecule which is artificially brought to the avoided crossing by an external perturbation or by adjusting parameters in the quantum mechanical calculation. In addition, conformational flexibility and unknown structural parameters makes a quantitative comparison with experimental results difficult. Moreover, the outcome of an experimental study is usually the rate constant for electron or energy transfer whereas the electronic structure calculation yields primarily the electronic coupling. The electronic coupling could be determined experimentally by several methods, e.g. by studying the temperature dependence of the transfer reactions. In addition, the FCWD (Eq. 1) has in some cases been calculated from first principles in order to facilitate quantitative comparisons. Nevertheless, the number of studies found in the literature that allow direct quantitative comparison of calculated and experimentally determined electronic coupling values is quite limited.

Calculating the rate of SEET follows, by and large, the principles outlined in the preceding paragraph. In contrast to the electron transfer case, the FCWD is usually known for the SEET process from the spectral overlap integral (Eq. 5). In a series of papers Scholes, Ghiggino, Paddon-Row and co-workers theoretically investigated different mechanisms for singlet energy transfer; dipole-dipole, multipole-multipole, super-exchange, relayed Coulombic interactions etc. (Scholes *et al.*, 1996, 1994a, 1994b, 1995; Clayton *et al.*, 1996). They found that the different mechanisms for energy transfer act in parallel to form a total coupling between the acceptor and donor. For different donor-acceptor pairs and for different distances a particular mechanism may dominate.

A beautiful example of how information from calculations can aid in the understanding of energy transfer in complex multichromophoric systems is the work of Scholes and co-workers who modeled the energy transfer pathways in the bacterial photosynthetic reaction center (Scholes *et al.*, 2001, Jordanides *et al.*, 2001). This and other studies on the theoretical modeling of SEET have recently been reviewed (Scholes, 2003).

3.2 Triplet Excitation Energy Transfer

Harriman and co-workers found a remarkably small attenuation factor ($\beta = 0.11$ Å^{-1}) for triplet excitation energy transfer (TEET) in a series of binuclear metal complexes (**6**) linked by oligophenylacetylenes (Harriman *et al.*, 2000), *i.e.* the electronic coupling decreases only weakly with increasing donor-acceptor distance. Moreover, it was also found, by extrapolation, that the rate constant for electron exchange at donor-acceptor orbital contact ($A = 10^9$ s^{-1}) was unusually small. These

Donor	**Bridge**	**Acceptor**

6a–e; n = 1–5

results suggest that the first bridging unit introduces a significant barrier to electronic interaction but additional units have only modest effect on the barrier height. Calculation of the Förster-type dipole-dipole energy transfer indicates that this process is unlikely to contribute more than a few percent to the overall reaction. The TEET was, therefore, attributed to superexchange type interaction. Note, TEET by the Förster mechanism is a spin-forbidden process; however, the first excited states of these rather heavy metal complexes are not pure triplet states and the probability for energy transfer may be high compared to the slow intrinsic deactivation processes. The low value for the attenuation factor is explained as arising from exceptionally good blending of molecular orbitals on adjacent units in the bridge. The triplet state remains localized on either terminal without obvious population of the connector triplet.

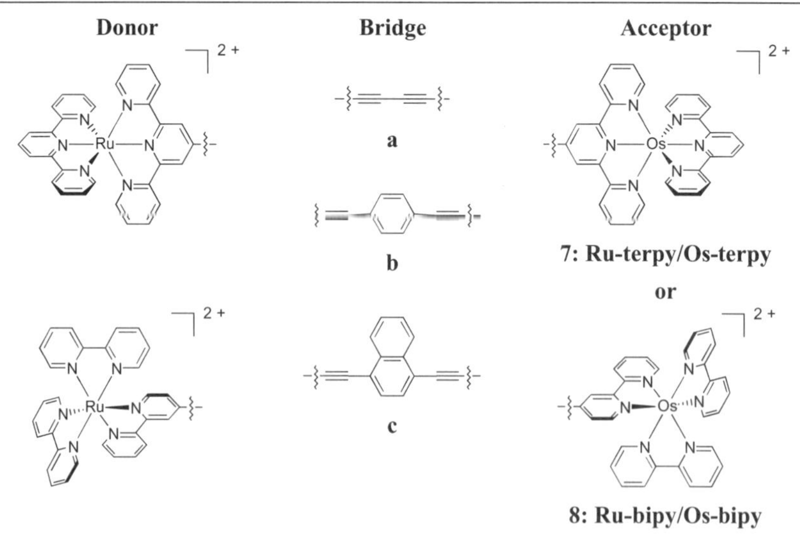

This is interpreted as that the latter triplet lies at too high an energy to be formed as a real intermediate, but it is able to mediate triplet-energy transfer by the superexchange interactions. That an aromatic unit incorporated in polyalkynes imposes a severe barrier for superexchange electronic interactions was also shown by Harriman and co-workers in their study of binuclear Ru-terpy/Os-terpy complexes (**7**) linked by diethynylarylenes (El-ghayoury *et al.*, 2000). The barrier arises, in part, from extended length, but it was also shown that there is a substantial difference between the coupling through different central aromatic units, i.e. phenylene and naphthylene. It is claimed that the severe drop in the degree of connectivity along the molecular axis must arise because of poor electronic coupling between the ethynylene and aromatic fragments The faster transfer rate and, consequently, higher electronic coupling V_{DA} for naphthylene relative to phenylene, can be ascribed to the relative energy differences between the triplet states of the donor and the bridges (ΔE_{DB} values).

The faster rate of energy transfer found for the bipyridyl-phenylene-based systems (**8b**) is argued to stem mainly from improved blending of orbitals along the connector. The energy transfer in the bipyridyl-naphthylene-based system proceeds by an intermediate population of the triplet localized on the connector. The two-step mechanism is, in this case, much faster than long-range transfer. This finding indicates the benefit of having an intermediary relay and provides the first opportunity to compare long-range and successive short-range triplet energy-transfer steps in structurally related molecular systems. (El-ghayoury *et al.*, 2000)

Triplet energy transfer is related to electron and hole transfer as has been both experimentally and theoretically shown (Closs *et al.*, 1988, 1989). Direct comparison between quantitative calculations and experiments on model systems are, however, quite rare. In a series of triads similar to the singlet energy transfer systems described above (**3**), Andréasson *et al.* (2002) were able to show that superexchange mediated TEET was responsible for the triplet state quenching of the donor porphyrin. In accordance with a companion electron transfer study (*vide infra*) only the fully conjugated bridges sustained strong enough electronic coupling to give measurable rates for TEET. Among the fully conjugated bridged triads the electronic coupling was proportional to the inverse energy gap between the lowest triplet states of the donor and bridge and, thus, varied according to the perturbation theory based superexchange mechanism. These experiments were supported by a series of time-dependent density functional theory (TDDFT/B3LYP) calculations on the same structures (Kyrychenko *et al.*, 2002). The electronic coupling was calculated to be close to zero for **3d** and varied systematically for the different fully conjugated bridge structures. In addition, the electronic coupling was shown to depend strongly on the bridge conformation in a way similar to the dependence shown in Fig. 2. Measurements of the rates at varying temperatures and solvent viscosities revealed a quite strong dependence indicating also experimentally the influence from bridge conformations on the electronic coupling for TEET (Andréasson *et al.*, 2002).

3.3 Electron Transfer

Osuka and co-workers (Osuka *et al.*, 1995) have compared polyynes (**9**) and polyenes (**10**) as mediators for electronic interaction between porphyrin electron donors and acceptors. Transient absorption spectra were recorded to confirm ET by the observation of the charge-separated state (ZnP$^{\bullet+}$). However, the charge-separated state was difficult to detect and fast charge recombination was discussed as one possible explanation. The ET is approximately two times more efficiently mediated by polyene-bridges than polyyne-bridges. The attenuation factor β (Eq. 12) was found to be 0.08 and 0.1 for the polyene and polyyne series, respectively. The conclusion drawn from the small attenuation factor is that the electronic coupling between the donor and acceptor is efficiently

Donor	Bridge	Acceptor

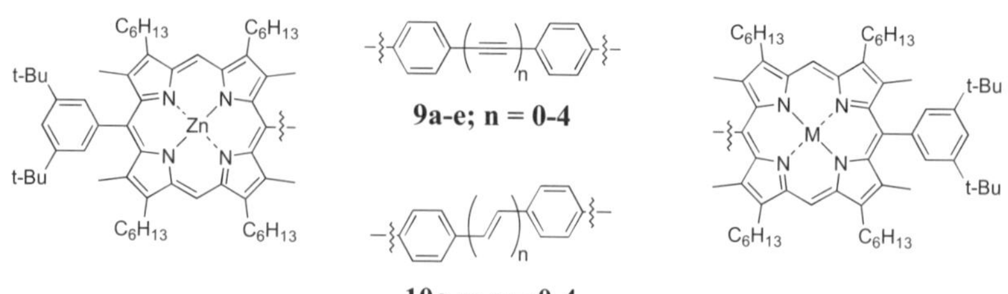

mediated by such conjugated bridges. The lower S_1-energies for the polyene spacers are advocated as the reason for the more efficient mediation by these spacers in accordance with the superexchange theory. The decline in the trough bond coupling upon elongation of unsaturated spacer may be offset partially by an extension of the spacer π-electronic system which leads to the decrease of the lowest excitation energy as well as a decrease of the one-electron oxidation and reduction potentials.

A particularly clear example of a bridged donor-acceptor system where the bridge energetics plays an instrumental role on the transfer mechanism was provided by Wasielewski and co-workers (Davis *et al.*, 2001). They investigated five DBA molecules in which the donor was tetracene, the acceptor was pyromellitimide, and the bridges were oligophenilenevinylenes (OPV) of different lengths (**11**). For the shorter members of the series electron transfer is governed by the superexchange mechanism and, thus, a fairly strong length dependence was observed. In contrast, for the longer members of the series a bridge-assisted electron hopping mechanism dominates the ET reaction, showing a much weaker length dependence, and the longer bridges were regarded as molecular wires (for this particular set of donors and acceptors). This clearly shows that, in addition to the length dependence of ET (exponential damping), the electronic properties of the bridge must also be considered when designing a DBA system.

Piotrowiak *et al.* (2003) have studied the electron injection to mesoporous nanocrystaline TiO_2 from Ru complexes *via* phenylethynyl bridges (**12**). The faster injection exhibited by the phenanthroline-based metal complex is explained by the larger electronic coupling between the phenanthroline fragment and the conjugated bis-phenylethynyl bridge,

Donor	Bridge	Acceptor

R = H or 2-ethylhexyl

11 a-e; n = 1-5

Donor	**Bridge**	**Acceptor**
 a or **b**	**12**	Mesoporous nanocrystalline TiO_2

compared to the corresponding coupling with the bipyridyl ligand. MO-calculations support this explanation. The pronounced sensitivity to the details of the electronic structure of the sensitizer suggests that, despite the fast injection rates, these systems may remain within the limits of weak coupling.

Electron transfer rates have been the primary target for theoreticians involved in modeling the electronic coupling between molecules. One of the earlier computational studies of electron transfer in π-bridged DBA systems is the pioneering work by S. Larsson (Larsson, 1981). The electronic coupling was estimated from extended Hückel calculations on bridged metal ions. This work was later extended to DBA systems with organic donors and acceptors (Larsson *et al.*, 1993). Although the work of Larsson in the early 1980s was purely theoretical it showed that computational chemistry could give electronic couplings for quite complicated systems with enough accuracy to allow meaningful comparison to experiments. The following few examples are calculations made with the aim of interpreting electron transfer reactions in specific DBA systems experimentally studied in parallel with the computations.

In an elegant series of combined theoretical and experimental work Heitele and Michel-Beyerle investigated the DBA systems **13** and **14** (Heitele *et al.*, 1985, 1987; Finckh *et al.*, 1988). The bridge comprised of

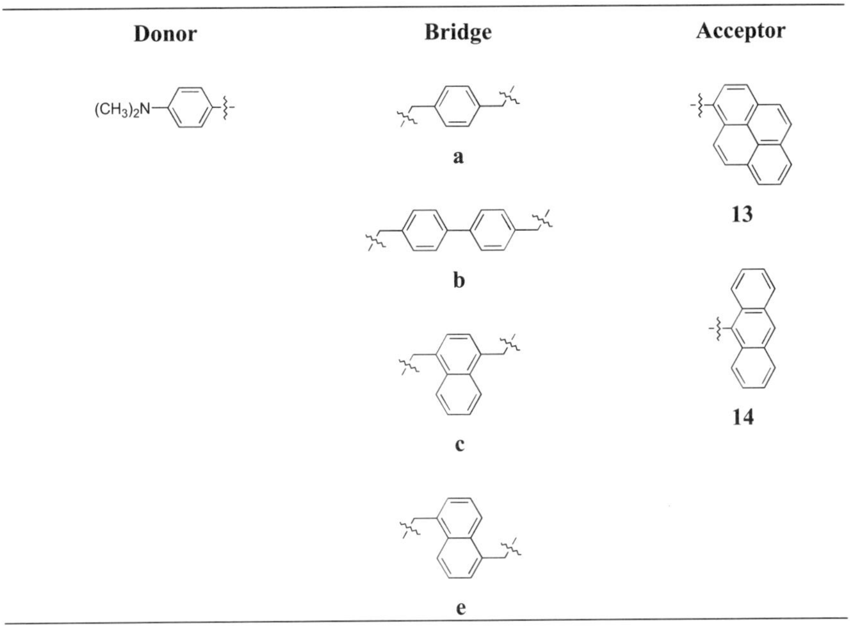

Donor	Bridge	Acceptor

oligomers of phenylene units (**a-b**) or differently connected naphthalenes (**c-d**), the electron donor was dimethylaniline and the acceptor either pyrene (**13**) or anthracene (**14**). Upon excitation of the acceptor the rate constants for electron transfer leading to charge separation were calculated from the measured fluorescence quenching. Varying the temperature and evaluating the ET rates by the Marcus equation (Eq. 10) yielded the electronic coupling from a least squares fitting procedure. The electronic structure of the bridge was shown to have a marked effect on the electron transfer rates and the results could qualitatively be interpreted by the superexchange model. The electronic coupling was also determined by quantum mechanical extended Hückel calculations. To simplify the calculations a model system comprised of the relevant bridging molecules, but with the donor and acceptor replaced by methylene ($-CH_2$) units, was studied. The p-orbital of the methylene unit represented the relevant molecular orbitals of the donor and acceptor. In order to get reasonable results the energy of the p-orbitals of the terminal CH_2 was arbitrarily adjusted. The electronic coupling was estimated from the splitting of the potential energy surfaces (Eq. 3). The computational results, which were performed at a specific optimal DBA conformation,

15

did not agree quantitatively with the experimental couplings (in general the calculated V was 4 times larger than the experimental V). However, the relative electronic couplings comparing the different bridges agreed very well between calculations and experiments.

A different theoretical approach was taken by Newton and co-workers in their study of the electronic coupling in oligophenylene ethynylene (OPE, **15**) bridged systems (Sachs *et al.*, 1997, Newton, 2000). The experiments in this case was interfacial electron transfer between a gold electrode and a ferrocene group covalently connected to the gold surface by thiolated OPEs. This system was also modeled by CH_2-terminated chains. The electronic coupling was calculated by summing pathways in the McConnell expression (Eq. 11). The parameters, V_{ij} and Δ, were estimated from the Generalized Mulliken-Hush method (Newton, 2000; Cave and Newton, 1996) using an INDO/S Hamiltonian. It was found that the electronic coupling strongly depends on the chain conformation in the OPEs. For a planar OPE-bridge the electronic coupling is large and depends only weakly on the chain lengths ($\beta = 0.4$ Å^{-1}, Eq. 12) whereas for perpendicular phenylene units the coupling is much smaller and the distance dependence is similar to that found for rigid alkane chains ($\beta = 1.0$ Å^{-1}). In reality the OPE chain adopts a mixture of conformations at room temperature since the barrier for rotation is of the order of the thermal energy (< 1 kcal/mol) and the observed coupling depends on the chain length with an intermediate attenuation factor ($\beta - 0.6$ Å^{-1}). This result is very important since it shows the interplay between bridge conformation and electronic coupling and since it also indicates that electron transfer could show a strong non-Marcus dependence on temperature.

By studying the electron transfer rates in model systems **4** and **5**, Kilså *et al.* were able to directly show how the energy of the relevant bridge states influence the superexchange mediated electronic coupling (Kilså *et al.*, 2001a, 2001b; Pettersson *et al.*, 2003). The inverse energy relation suggested in Eq. 11 was clearly demonstrated for the fully

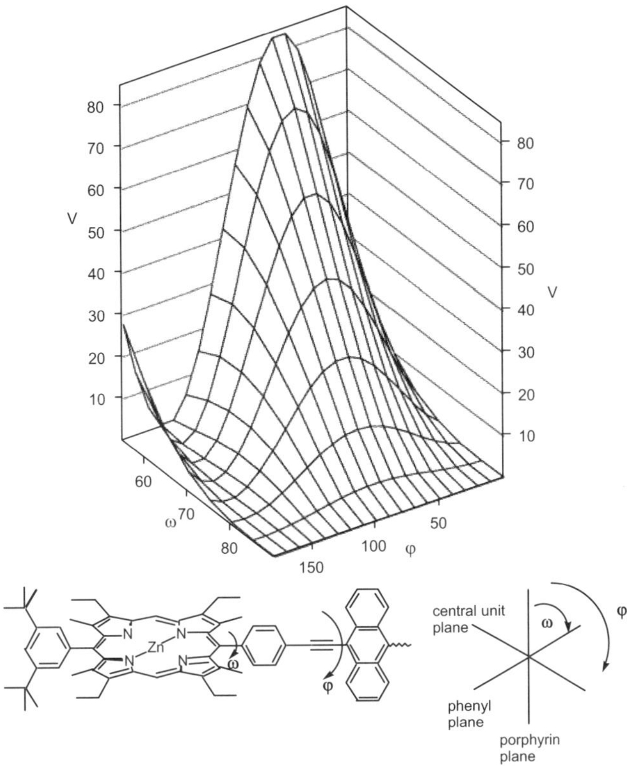

Fig. 2. Calculated electronic coupling (cm^{-1}) as a function of dihedral angles ω and φ for the ZnP-AB-H$_2$P system.

conjugated bridges (**4a-c** and **5a-c**) whereas no intramolecular electron transfer was detected for the bicyclo[2.2.2]octane bridged system (**4d** and **5d**). This trend was confirmed by calculations of the electronic coupling from the energy splitting between the LUMOs of the donor and acceptor with the CNDO/S method. The calculations were performed with the acceptor replaced by a free-base porphyrin (H$_2$P) and this system was brought to the avoided crossing by artificially perturbing the orbital energies by a charged species (H$_3$O$^+$). In addition, a strong dependence on the conformations of the bridge and the bridge to donor or acceptor linkage was observed. This dependence is shown in Figure 2, which again predicts a strong temperature dependence of the superexchange mediated coupling.

4. σ-BRIDGES AND SUPEREXCHANGE – THE THROUGH-BOND COUPLING MECHANISM.

Saturated organic systems, the obvious insulating bridge structure, show remarkable properties as mediators for both electron and energy transfer. Only by ingenious design and substantial synthetic effort in close combination with theoretical work has it been possible to reveal the most fundamental details of the characteristics of the transfer between donors and acceptors linked by saturated bridges. The most versatile and successful series of bridges in this context is undoubtedly the many rigid and symmetrical polybicycloalkyl systems, which affix the donor and acceptor at well-defined position relative to each other, developed by several different research groups. The research on the effect of σ-bridges on the different transfer processes is even more extensive than that for π-bridges, and has often preceded similar work on the latter.

4.1. Singlet Excitation Energy Transfer

The superexchange phenomenon, or the through-bond interaction as it is called when orbitals at different sites interact through their mutual mixing with σ-orbitals of the intervening framework, was suggested early on to efficiently promote SEET (Zimmerman *et al.*, 1980). Based on the observed distance dependence of the SEET, Zimmerman concluded that it was efficiently promoted by through-bond exchange interactions through the σ-bridges in the DBA systems **16a** and **16b**.

Efficient through-bond interactions were also observed by Verhoeven and co-workers (Oevering *et al.*, 1988, Verhoeven, 1990) when examining the SEET process in a series of bichromophoric

Donor	Bridge	Acceptor
	16a	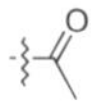
	**16b**	

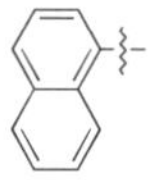

Donor	Bridge	Acceptor
(dimethoxynaphthalene)	**17a**	(EET)
	17b	or
	18a	(ET)
	18b	

molecules containing rigid polynorbornyl bridges, e.g. **17a** and **17b**, separating a dimethoxynaphthalene donor from a carbonyl acceptor by 4-10 σ-bonds. They found that in this series, where the two chromophores were connected *via* bridges with an *all-trans* conformation, the measured transfer rate constant, which was much larger than that calculated from the Förster model, depended exponentially on the number of σ-bonds separating the donor and acceptor.

In a bichromophoric system where the naphthalene donor and the anthracene acceptor were connected by a similar bisnorbornyl bicyclo[2.2.0]hexane bridge (**19**) SEET was prohibited to occur *via* a direct dipole-dipole coupling because the donor and acceptor transition dipole moments are orthogonal (Scholes *et al.*, 1993). Although vibronic coupling could make this process partially allowed, it was argued that it could not account for the measured high SEET rate. The results were instead explained in terms of a novel EET mechanism involving a combination of dipole-dipole interaction and exchange contribution, relayed through the bridge.

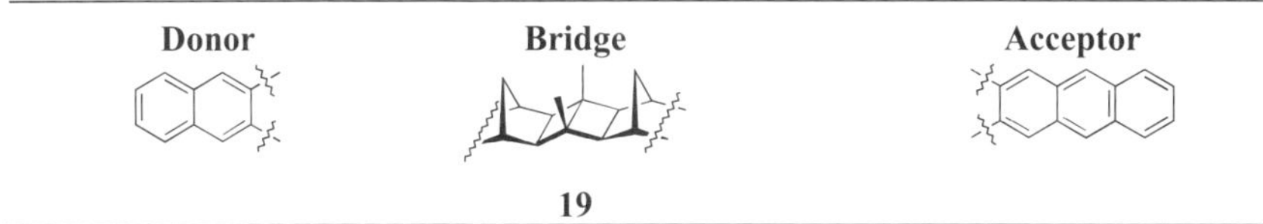

Donor	Bridge	Acceptor
	19	

Donor	Bridge	Acceptor
	2 • **20a**	
	2 • **20b**	

Paddon-Row and co-workers (Kroon *et al.,* 1990) demonstrated convincingly that SEET was hindered by adding one or more kinks to the *all-trans* bridge, by comparison of bichromophoric systems in which the two chromophores were covalently linked by rigid saturated bridges with an all-*trans* configuration, **17a** and **17b**, with those systems in which one or two *trans* arrangements of bonds in the bridge had been substituted for *cis* or *gauche* alignments, **18a** and **18b**. This change in configuration resulted in a much less efficient SEET process although it was shown by X-ray studies that such changeover had only a minor effect on the spatial separation of the chromophores, bringing these slightly closer together in the *gauche* arrangements. This behavior is usually referred to as the *all-trans* rule, which states that the coupling through a bridge is maximized for an *all-trans* (antiperiplanar) configuration of bridge bonds (Hoffmann, 1971; Paddon-Row, 1982).

The marked influence of bridge configuration might be further exemplified by the results obtained by Speiser and his group who investigated the excitation energy transfer in benzene-α-diketone bichromophoric molecules, **20**, (Toledano *et al.,* 1996). They could rationalize all of their observations on SEET in terms of a short-range transfer mechanism involving a Dexter-type exchange interaction, without considering any through-bond interactions (Levy *et al.,* 1992a, 1992b.). To test their interpretation the bichromophoric molecules were modified by methyl substitution at the carbonyl groups β-positions. It was argued that any through-bond interactions should be manifested as a difference between the SEET efficiencies obtained for the unsubstituted and the β,β'-dimethyl substituted systems, respectively, owing to the perturbation of the electronic structure of the bridge. The geometries of the unsubstituted and the substituted members in the pair, **20a** and **20b**,

Donor	Bridge	Acceptor
	21a	
or	**21b**	

were very similar, allowing the difference in SEET to be attributed to the electronic structure of the bridge. However, the SEET process observed for the bichromophoric molecules did not reflect any special features, which can be attributed to methyl substitution.

4.2. Triplet Excitation Energy Transfer

The first quantitative demonstration of efficient σ-bond spacer promoted long-distance TEET was given by Closs *et al.* (1988, 1989). They compared the rates of both TEET and ET in a series of compounds consisting of a 4-biphenyl or 4-benzophenoyl donor chromophores, a 2-naphtyl acceptor and cyclohexane, **21a**, or decaline, **21b**, spacers as interchromophore bridges.

For the triads **3c** and **3d,** that differ only in the replacement of a benzene unit for a bicyclo[2.2.2]octane unit, an enormous difference ($>10^6$ times) in TEET rates was observed (Andréasson *et al.*, 2000). This large difference in electronic coupling was also confirmed by TDDFT calculations (Kyrychenko *et al.*, 2002). Apparently, breaking the π-conjugation of the bridge in a DBA system can have a tremendous influence on the mediated electronic coupling.

4.3 Electron Transfer

The norbornylogous bridges **17-19** have also been extensively studied with respect to different electron transfer processes (Paddon-Row, 1994). As observed for the EET processes, an exponential decrease in transfer rates with increasing number of intervening σ-bonds is also found for the ET processes in accordance with through-bond coupling. For example, thermal intramolecular ET, studied in the radical anions of the system in which the dimethoxynaphthalene donor is linked by the **17**-bridges to a dicyanovinyl acceptor (Penfield *et al.*, 1987), shows a small distance dependence for the electronic coupling with a β value of ca. 0.35-0.49 (per bond). Photoinduced electron transfer from the electronically excited dimethoxynaphthalene to the acceptor dicyanovinyl across the same bridges has also been studied (Oevering *et al.*, 1987, Paddon-Row and Verhoeven, 1991). The rates were found to be fast, although not as fast as those deduced for the charge shift process in the corresponding radical anion systems. The Coulombic interaction that arises in the charge separation process was put forward as an explanation for the retardation. The distance dependence observed for the electron transfer rate is solvent dependent with attenuation coefficient values (per bond) 0.46 (THF), 0.50 (EtOAc), and 0.61 (MeCN). It is pointed out that this is the overall attenuation factor and, thus, includes not only the distance dependence of the electronic coupling term but also that of the FCWD term.

The fact that the singlet and triplet charge separated states are discrete from each other, and interconvert only slowly, if sufficient exchange interaction is mediated by the bridge separating the charges, has been used to create long-lived charged separated states (**22**) (Hviid *et al.*, 2001; Wegner *et al.*, 1999). The singlet-triplet energy gap has also been shown to be a useful indicator of the magnitude of the electronic coupling in ET and EET processes (Kobori *et al.*, 2000; Forbes *et al.*, 1996). For example, the calculated average damping factor for the distance dependence of the singlet-triplet energy gap for the series of charge separated states including **17a** and **17b** with the dicyanovinyl acceptor corresponds very closely to the average experimentally determined value for a variety of ET transfer processes occurring in the same series of molecules (Paddon-Row, 2001, Seischab *et al.*, 2000).

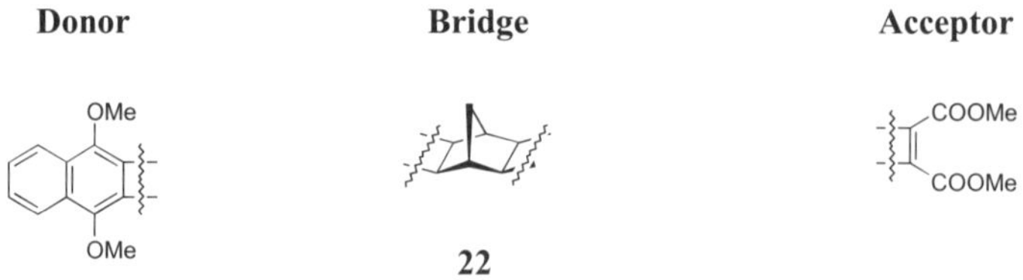

Donor	**Bridge**	**Acceptor**

22

This follows from the fact that both the ET rate and the singlet-triplet energy gap are proportional to the square of two similar electronic coupling elements that should follow the same trends upon variations in bridge parameters, such as length and shape. Comparison of systems in which the dimethoxynaphthalene donor is linked by either a bent bridge or an *all-trans* form, of **18** and **17** type, respectively, to a dicyanovinyl acceptor has confirmed the configuration dependence of through-bond orbital interactions. This has been done both theoretically, e.g. by calculating the singlet-triplet gap, and experimentally by studies of photoinduced ET (Olivier *et al.,* 1988; Paddon-Row and Shephard, 2002).

Donor	**Bridge**	**Acceptor**
	23a	
	23b	

Chow *et al.* (Chow *et al.,* 2003) have constructed a series of saturated hydrocarbon bridged donor-acceptor systems to gain insight into the factors controlling photoinduced long-distance electron transfer, in particular the effect of the reaction thermodynamics, but also the donor-acceptor orientation. They found a substantial difference in ET rates between the two systems **23a** and **23b** where the donor, benzene-1,2-dithioketal, and the acceptor, dicyanovinyl , were aligned at different orientations with their π-orbitals either coplanar or perpendicular to each other. Although the bridge is slightly longer in **23b** than that in **23a**, the key difference in transfer rate is believed to be in the electron-coupling matrix resulting from the orientation between the donor and acceptor moieties.

5. CONCLUSIONS

This review has attempted to outline the properties that govern electronic coupling between donor and acceptor molecules through a bridging structure. At present we can design bridging molecules that either promote electronic coupling or isolate the donor and acceptor. In general, conjugated π-systems are better "conductors" than σ-systems but quite subtle effects might change this generalization. It has been demonstrated, both theoretically and experimentally, that the prediction from the superexchange model, i.e. a Δ^{-1} energy gap dependence of the electronic coupling, holds for both π and σ-bridges. In addition, conformational effects, such as changed π-overlap or the configuration of a σ-system (trans-effect), has a large influence on the degree of coupling. These subtle effects might be the starting point for designing switching functions into molecular electronic components. Although the bridging structures can be designed to perform the desired transfer reactions, the donor-bridge and bridge-acceptor couplings are also crucial in order to get the net reaction. Naturally each donor-bridge or bridge-acceptor link has specific requirements to form a strongly coupled system. Here we know less and the theoretical models are not yet predictive enough to design a complete DBA system that will perform a specific task (e.g. have a minimum rate for electron transfer). This problem becomes even more obvious when interfacial (e.g. to/from a metal surface) or non-covalent couplings are concerned, and since it is anticipated that these motifs will be important in the development of molecular scale electronics, more research efforts are needed in this field.

References

1. Andersson M, Linke M, Chambron JC, Davidsson J, Heitz V, Sauvage JP and Hammarström L. Porphyrin-Containing 2-Rotaxanes: Metal Coordination Enhanced Superexchange Electron Transfer between Noncovalently Linked Chromophores. *J. Am. Chem. Soc.* 2000; **122**: 3526-3527.

2. Andréasson J, Kajanus J, Mårtensson J and Albinsson B. Triplet Energy Transfer in Porphyrin Dimers: Comparison Between π- and σ-Chromophore Bridged Systems. *J. Am. Chem. Soc.* 2000; **122**: 9844-9845.

3. Andréasson J, Kyrychenko A, Mårtensson J and Albinsson B. Temperature and Viscosity Dependence of the Triplet Energy Transfer Process in Porphyrin Dimers. *Photochem. Photobiol. Sci.* 2002; **1**: 111-119.

4. Balzani V., Ed., Electron Transfer in Chemistry, Wiley-VCH, Weinheim, 2001.

5. Cave RJ and Newton MD. Generalization of the Mulliken-Hush treatment for the calculation of electron transfer matrix elements *Chem. Phys. Lett.* 1996; **249**: 15-19.

6. Cho HS, Jeong DH, Yoon MC, Kim YH, Kim D, Jeoung SC, Kim SK, Aratani N, Shinmori H, Osuka A. Excited-State Energy Transfer Processes in Phenylene- and Biphenylene-Linked and Directly-Linked Zinc(II) and Free-Base Hybrid Diporphyrins. *J. Phys. Chem. A* 2001; **105**: 4200-4210.

7. Chow TJ, Chiu NR, Chen HC, Chen CY, Yu WS, Cheng YM, Cheng CC, Chang CP and Chou PT. Photoinduced Electron Transfer Reaction Tuned by Donor-Acceptor Pairs Via the Rigid, Linear Spacer Heptacyclo[6.6.0.0^{2,6}.0^{3,13}.0^{4,11}.0^{5,9}.0^{10,14}]tetradecane. *Tetrahedron* 2003; **59**: 5719-5730.

8. Clayton AHA, Scholes GD, Ghiggino KP, and Paddon-Row MN. Through-bond and through-space coupling in photoinduced electron and energy transfer: An ab initio and semiempirical study *J. Phys.* Chem 1996; 100: 10912-10918.

9. Closs GL, Johnsson MD, Miller JR, and Piotrowiak P. A connection between intramolecular long-range electron, hole, and triplet energy transfers. 1989; **111**: 3751-3753.

10. Closs GL, Piotrowiak P, Maccinis JM, and Fleming GR. Determination of long-distance intramolecular triplet energy transfer rates – a quantitative comparison with electron transfer. *J. Am. Chem. Soc.* 1988; **110**: 2652-2653.

11. Davis WB, Ratner MA, and Wasielewski MR Conformational gating of long distance electron transfer through wire-like bridges in donor-bridge-acceptor molecules. *J. Am. Chem. Soc.* 2001; **123**: 7877-7886.

12. Dexter DL. A theory of sensitized luminescence in solids. *J. Chem. Phys.* 1953; **21**: 836-850.

13. El-ghayoury A, Harriman A, Khatyr A, Ziessel R. Controlling Electronic Communication in Ethynylated-Polypyridine Metal Complexes. *Angew. Chem. Int. Ed.* 2000; **39**: 185-189

14. Finckh P, Heitele H, Volk M., and Michel-Beyerle ME Electron Donor/Acceptor Interaction and Reorganization Parameters from Temperature-Dependent Intramolecular Electron Transfer Rates. *J. Phys. Chem.* 1988; **92**: 6584-6590.

15. Forbes MDE, Ball JD and Avdievich NI. In Search of Through-Solvent Electronic Coupling in Flexible Biradicals. *J. Am. Chem. Soc.* 1996; **118**: 4707-4708.

16. Förster T. Energiewanderung und fluoreszenz. *Naturwiss.* 1946; **33**: 166-175.

17. Förster T. Zwischenmolekulare energiewanderung und fluoreszenz. *Ann. Phy.* 1948; **2**: 55-75.

18. Gust D, Moore TA, and Moore AL. Mimicking photosynthetic solar energy transduction. *Acc. Chem. Res.* 2001; **34**: 40-48.

19. Harriman A, Khatyr A, Ziessel R, and Benniston AC. An Unusually Shallow Distance-Dependence for Triplet-Energy Transfer. *Angew. Chem. Int. Ed.* 2000; **39**: 4287-4290.

20. Heitele H and Michel-Beyerle ME Electron-transfer through aromatic spacers in bridged electron-donor acceptor molecules. *J. Am. Chem. Soc.* 1985; **107**, 8286-8288.

21. Heitele H, Michel-Beyerle ME and Finckh P. Electron transfer through intramolecular bridges in donor/acceptor systems. *Chem. Phys. Lett.* 1987; **134**: 273-278.

22. Hoffmann R. Interaction of Orbitals through Space and through Bonds. *Acc. Chem. Res.* 1971; **4**: 1-9.

23. Hviid L, Brouwer AM, Paddon-Row MN and Verhoeven JW. Long-Lived Short-Distance Intramolecular Charge Separation Via Intermolecular Triplet Sensitization. *ChemPhysChem* 2001; **2**: 232-235.

24. Jordanides XJ, Scholes GD and Fleming GR. The mechanism of energy transfer in the bacterial photosynthetic reaction center *J. Phys. Chem.* B 2001; **105**: 1652-1669.

25. Jortner J and Bixon M. Intramolecular vibrational excitations accompanying solvent-controlled electron-transfer reactions. *J. Chem. Phys.* 1988; **88**: 167-170.

26. Jortner J and Bixon M., Eds., *Adv. Chem. Phys.* 1999; **106**: entire volume.

27. Jortner J. and Ratner MA, Eds., *Molecular Electronics*. Blackwell Science Ltd., Oxford, 1997.

28. Kilså Jensen K, van Berlekom S, Kajanus J, Mårtensson J, and Albinsson, B. Mediated Energy Transfer in Covalently Linked Porphyrin Dimers. *J. Phys. Chem. A* 1997; **101:** 2218-2220.

29. Kilså K, Kajanus J, Larsson S, Macpherson AN, Mårtensson J and Albinsson B Enhanced Intersystem Crossing in Donor/Acceptor Systems Based on Zinc/Iron or Free Base/Iron Porphyrins. *Chem. Eur. J.* 2001a; **7:** 2114-2121.

30. Kilså K, Kajanus J, Macpherson AN Mårtensson, J and Albinsson B Bridge-Dependent Electron Transfer in Porphyrin-Based Donor-Bridge-Acceptor Systems *J. Am. Chem. Soc.* 2001b; **123:** 3069-3080.

31. Kilså K, Kajanus J, Mårtensson J, and Albinsson B. Mediated Electronic Coupling: Energy Transfer in Porphyrin Dimers Enhanced by the Bridging Chromophore. *J. Phys. Chem. B.* 1999; **103:** 7329-7339.

32. Kobori Y, Akiyama K and Tero-Kubota S. Theoretical Analysis of Singlet-Triplet Energy Splitting Generated by Charge-Transfer Interaction in Electron Donor-Acceptor Radical Pair Systems. *J. Chem. Phys.* 2000; **113:** 465-468.

33. Kroon J, Oliver AM, Paddon-Row MN, and Verhoeven JW. Observation of a Remarkable Dependence of the Rate of Singlet Singlet Energy-Transfer on the Configuration of the Hydrocarbon Bridge in Bichromophoric Systems. *J. Am. Chem. Soc.* 1990; **112:** 4868-4873.

34. Kyrychenko A and Albinsson B. Conformer-Dependent Electronic Coupling for Long-Range Triplet Energy Transfer in Donor-Bridge-Acceptor Porphyrin Dimers. *Chem. Phys. Lett.* 2002; **366:** 291-299.

35. Larsson S and Braga M Transfer of Mobile Electrons in Organic Molecules. *Chem. Phys.* 1993; **176:** 367-375.

36. Larsson S. Electron-Transfer in Chemical and Biological Systems – Orbital Rules for Non-Adiabatic Transfer. *J. Am. Chem. Soc.* 1981; **103:** 4034-4040.

37. Levy ST and Speiser S. Calculation of the Exchange Integral for Short-Range Intramolecular Electronic-Energy Transfer in Bichromophoric Molecules. *J. Chem. Phys.* 1992a; **96:** 3585-3593.

38. Levy ST, Rubin MB and Speiser S. Photophysics of Cyclic Alpha-Diketone Aromatic Ring Bichromophoric Molecules - Structures, Spectra, and Intramolecular Electronic-Energy Transfer. *J. Am. Chem. Soc.* 1992b; **114:** 10747-10756.

39. Marcus RA. Theory of oxidation-reduction reactions involving electron transfer. 1. *J. Chem. Phys.* 1956; **24**: 966-978.

40. McConnell HM. Intramolecular charge transfer in aromatic free radicals. *J. Chem. Phys.* 1961; **35**: 508-515.

41. Moore TA, Moore AL, and Gust D. The design and synthesis of artificial photosynthetic antennas, reaction centers and membranes. *Philos. T. Roy. Soc. B* 2002; **357**: 1481-1498.

42. Newton MD. Quantum chemical probes of electron transfer kinetics – the nature of donor-acceptor interactions. *Chem. Rev.* 1991; **91**, 767-792

43. Newton, MD. Modeling donor/acceptor interactions: Combined roles of theory and computation. *Int. J. Quant. Chem.* 2000; **77**: 255-263.

44. Oevering H, Paddon-Row MN, Heppener M, Oliver AM, Cotsaris E, Verhoeven JW and Hush NS. Long-Range Photoinduced through-Bond Electron-Transfer and Radiative Recombination Via Rigid Nonconjugated Bridges - Distance and Solvent Dependence. *J. Am. Chem. Soc.* 1987; **109**: 3258-3269.

45. Oevering H, Verhoeven JW, Paddon-Row MN, Cotsaris E and Hush NS. Long-Range Exchange Contribution to Singlet-Singlet Energy-Transfer in a Series of Rigid Bichromophoric Molecules. *Chem. Phys. Lett.* 1988; **143**: 488-495.

46. Oliver AM, Craig DC, Paddon-Row MN, Kroon J and Verhoeven JW. Strong Effects of the Bridge Configuration on Photoinduced Charge Separation in Rigidly Linked Donor-Acceptor Systems. *Chem. Phys. Lett.* 1988; **150**: 366-373.

47. Osuka A, Tanabe N, Kawabata S, Yamazaki I, Nishimura Y. Synthesis and Intramolecular Electron- and Energy-Transfer Reactions of Polyyne- or Polyene-Bridged Diporphyrins. *J. Org. Chem.* 1995; **60**: 7177-7185.

48. Paddon-Row MN and Shephard MJ. A Time-Dependent Density Functional Study of the Singlet-Triplet Energy Gap in Charge-Separated States of Rigid Bichromophoric Molecules. *J. Phys. Chem. A* 2002, **106**. 2935-2944.

49. Paddon-Row MN and Verhoeven JW. Towards the Rational Design of Systems Exhibiting Rapid Long-Range Intramolecular Electron-Transfer Processes Caused by through-Bond Coupling. *New J. Chem.* 1991; **15**: 107-116.

50. Paddon-Row MN. In *Electron-Transfer in Chemistry*; Balzani, V., Ed.; Wiley-VCH: Weinheim, Germany, 2001; Vol. 3, Part 2, Chapter 1 p. 179.

51. Paddon-Row MN. Investigating Long-Range Electron-Transfer Processes with Rigid, Covalently-Linked Donor-(Norbornylogous Bridge)-Acceptor Systems. *Acc. Chem. Res.* 1994; **27:** 18-25.

52. Paddon-Row MN. Some Aspects of Orbital Interactions through Bonds - Physical and Chemical Consequences. *Acc. Chem. Res.* 1982; **15:** 245-251.

53. Penfield KW, Miller JR, Paddon-Row MN, Cotsaris E, Oliver AM and Hush NS. Optical and Thermal Electron-Transfer in Rigid Difunctional Molecules of Fixed Distance and Orientation. *J. Am. Chem. Soc.* 1987; **109:** 5061-5065.

54. Pettersson K, Kilså K, Mårtensson J and Albinsson B. Low versus high spin iron(III) porphyrins as electron acceptors in donor acceptor systems. Submitted manuscript 2003.

55. Piotrowiak P, Galoppini E, Wei Q, Meyer GJ, and Wiewior R. Subpicosecond photoinduced charge injection from "Molecular tripods" into mesoporous TiO2 over the distance of 24 angstroms. *J. Am. Chem. Soc.* 2003; **125:** 5278-5279.

56. Read I, Napper A, Kaplan R, Zimmt MB and Waldeck DH. Solvent-Mediated Electronic Coupling: The Role of Solvent Placement. *J. Am. Chem. Soc.* 1999; **121:** 10976-10986.

57. Sachs SB, Dudek SP, Hsung RP, Sita LR, Smalley JF, Newton MD, Feldberg SW, and Chidsey CED. Rates of Interfacial Electron Transfer through π-conjugated Spacers. *J. Am. Chem. Soc.* 1997; **119:** 10563-10564.

58. Scholes GD and Ghiggino KP. Electronic interactions and interchromophore excitation transfer. *J. Phys. Chem.* 1994a; **98:** 4580-4590.

59. Scholes GD and Ghiggino KP. Rate expressions for excitation transfer 1. Radiationless transition theory perspective. *J. Chem. Phys.* 1994b; 101: 1251-1261.

60. Scholes GD and Ghiggino KP. Rate expressions for excitation transfer 4. Energy migration and superexchange phenomena. *J. Chem. Phys.*1995; **103:** 8873-8883.

61. Scholes GD and Harcourt RD Configuration interaction and the theory of electronic factors in energy transfer and molecular exciton interactions. *J Chem. Phys.* 1996; **104:** 5054-5061.

62. Scholes GD Long-range resonance energy transfer in molecular systems. *Ann. Rev. Phys. Chem.* 2003; **54:** 57-87.

63. Scholes GD, Ghiggino KP, Oliver AM and Paddon-Row MN. Intramolecular Electronic-Energy Transfer between Rigidly Linked

Naphthalene and Anthracene Chromophores. *J. Phys. Chem.* 1993; **97:** 11871-11876.

64. Scholes GD, Jordanides XJ, and Fleming GR. Adapting the Förster theory of energy transfer for modeling dynamics in aggregated molecular assemblies. *J. Phys. Chem. B* 2001; **105:** 1640-1651.

65. Seischab M, Lodenkemper T, Stockmann A, Schneider S, Koeberg M, Roest MR, Verhoeven JW, Lawson JM, and Paddon-Row MN. Investigation of the Primary Electron Transfer Processes in Some Rigidly Bridged Dyads and Triads by (Sub)Picosecond Pump-Probe Experiments. *PhysChemChemPhys.* 2000; **2:** 1889-1897.

66. Strachan JP, Gentemann S, Seth J, Kalsbeck WA, Lindsey JS, Holten D, and Bocian DF. Effects of Orbital Ordering on Electronic Communication in Multiporphyrin Arrays. *J. Am. Chem. Soc.* 1997; **119:** 11191-11201.

67. Sun LC, Hammarström L, Åkermark B, and Styring S. Towards artificial photosynthesis: ruthenium-manganese chemistry for energy production *Chem. Soc. Rev.* 2001; **30:** 36-49.

68. Toledano E, Rubin MB, and Shammai Speiser S. Dependence of the intramolecular electronic energy transfer in bichromophoric molecules on the interchromophore bridge, *Angew. Chem. Int. Ed.* 1996; **94:** 93-100 .

69. Ulstrup J and Jortner J. Effect of intramolecular quantum modes on free-energy relationships for electron-transfer reactions. *J. Chem. Phys.* 1975; **63:** 4358-4368.

70. Verhoeven JW. Electron-Transport via Saturated-Hydrocarbon Bridges - Exciplex Emission from Flexible, Rigid and Semiflexible Bichromophores. *Pure Appl. Chem.* 1990; **62:** 1585-1596.

71. Ward MD. Photo-Induced Electron and Energy Transfer in Non-Covalently Bonded Supramolecular Assemblies. *Chem. Soc. Rev.* 1997; **26:** 365-375.

72. Wegner M, Fischer H, Koeberg M, Verhoeven JW, Oliver AM and Paddon-Row MN. Time-Resolved CIDNP from Photochemically Generated Radical Ion Pairs in Rigid Bichromophoric Systems. *Chem. Phys.* 1999; **242:** 227-234.

73. Yang SI, Seth J, Balasubramanian T, Kim D, Lindsey JS, Holten D, and Bocian DF. Interplay of Orbital Tuning and Linker Location in Controlling Electronic Communication in Porphyrin Arrays. *J. Am. Chem. Soc.* 1999; **121:** 4008-4018.

74. Zimmerman HE, Goldman TD, Hirzel TK and Schmidt SP. Photochemical Series .124. Rod-Like Organic-Molecules - Energy-Transfer Studies Using Single-Photon Counting. *J. Org. Chem.* 1980; **45:** 3933-3951.

Chapter 7

ENERGY TRANSFER AND TRAPPING IN ENGINEERED MACROMOLECULES

Kenneth P. Ghiggino and Trevor A. Smith

The dynamics and efficiency of electronic excitation energy transfer (EET), migration and trapping are investigated in a series of novel macromolecular systems. These systems have been engineered specifically with the aim of controlling the efficiency of the energy migration though the structure of the macromolecule. Three systems are discussed. Firstly, time-resolved fluorescence measurements of poly(acenaphthylene) homopolymer, P(ACE), are compared with similar measurements on an acenaphthylene/maleic anhydride copolymer, P(ACE/MA), in which the MA groups act as spacers between the ACE chromophores. The rate of energy migration is shown to be extremely rapid in P(ACE) but restricted in the copolymer. Secondly, we illustrate the structural control now achievable through living free radical polymerization methods, which permit the design and synthesis of polymers with structures optimized for energy migration and trapping. Finally, we discuss the results of time-resolved fluorescence anisotropy measurements in the study of EET in several generations of porphyrin-based dendrimers.

Keywords: electronic energy migration, energy transfer, synthetic polymers, macromolecules.

1. INTRODUCTION

1.1. Electronic Energy Transfer, Migration and Trapping

The fate of excitation energy residing in multichromophoric macromolecules such as vinyl aromatic polymers, following irradiation with light, has been an active research area over a number of years. (Guillet, 1985; Guillet *et al.*, 1987; Hoyle *et al.*, 1987; Phillips, 1985; Webber, 1990). Due to the high effective local concentration, and hence close proximity of the chromophores within the macromolecule (along the polymer backbone), there are possibilities for energy migration and final trapping of the excitation at sites far removed from the initial light-absorbing chromophore. The redistribution of energy within the macromolecule can also have important consequences in determining the final site of photodegradation or in activating specific chromophores for further photochemical processes.

Electronic energy migration (EEM) is a process by which energy absorbed initially at one particular site or chromophore in a system, can be relocated nonradiatively from the initially excited molecular chromophore to other chromophores of the same type elsewhere in the system. This phenomenon is believed to occur in at least some synthetic polymers and other macromolecules, and is fundamentally important in such diverse areas as the excitation of rare-earth lasers, (Macfarlane *et al.*, 1987) organic photochemistry, (Birks, 1970) photon collection in photosynthetic systems, (Cogdell *et al.*, 1999; Gnanakaran *et al.*, 1999; Hu *et al.*, 1998; Hu *et al.*, 1997; Pullerits *et al.*, 1996; Sundström *et al.*, 1999; van Grondelle *et al.*, 1999; van Grondelle *et al.*, 1997) polymer photophysics, (Allen *et al.*, 1985; Ghiggino *et al.*, 1993; Guillet, 1985; Phillips, 1985; Webber, 1990) and structural and functional studies on proteins (Van der Meer *et al.*, 1994).

The mechanisms by which excitation energy might migrate within a macromolecule, such as a synthetic aromatic polymer, have been discussed in detail elsewhere (Ghiggino *et al.*, 1993; Webber, 1990). In systems in which the coupling between individual molecules or chromophores is very weak, such as photosynthetic systems and synthetic aromatic polymers (especially in solution), it is reasonable to regard the electronic excitation as being temporarily completely localized, and to interpret the energy migration process as a series of sequential energy transfer steps or hops between local chromophore sites on the molecular lattice or chain. In this case, the energy can be thought of as migrating as an incoherent exciton (Pearlstein, 1972). The energy migration process in polymers is often thought (Fox *et al.*, 1972; Harrah,

1972) to involve segmental and chromophore rotation/diffusion in combination with the series of electronic energy transfer (EET) hops between chromophores of the same type. This exciton hopping continues until either a site suitable for excimer (excited-state dimer) formation (or another form of energy sink) is encountered, at which point the energy is trapped, or one of the chromophores excited in the migration process deactivates spontaneously, either radiatively or non-radiatively.

Whether EEM actually occurs within a particular multichromophoric array, and if so with what rate and efficiency, and by what mechanism, are important factors still to be determined if this process is to be fully understood and harnessed. Through this knowledge, and the design and synthesis of optimized macromolecular structures, EEM efficiency may be maximized for use in applications in photomolecular devices exploiting the 'photon-harvesting' apparatus (Fox, 1993).

The structural configuration of a particular macromolecule can influence strongly the dynamic behavior of its excited state, particularly through the processes of excimer formation and EEM. In this chapter we discuss the dynamics and efficiency of electronic energy transfer, migration and trapping in a series of novel macromolecular systems. The three molecular systems under discussion have been engineered specifically with the aim of controlling the efficiency of the energy migration through the structure of the macromolecule. Firstly, time-resolved fluorescence measurements of poly(acenaphthylene) homopolymer, P(ACE), are compared with similar measurements of an acenaphthylene/maleic anhydride copolymer, P(ACE/MA), in which the MA groups act as spacers between the ACE chromophores. The rate of energy migration is shown to be extremely rapid in P(ACE) and restricted in the copolymer. Secondly, we illustrate the structural control now achievable through living polymerization methods, which permit the design and synthesis of polymers with structures optimized for energy migration and trapping. We discuss this with reference to a series of acenaphthyl(ACE)-containing polymers synthesized by the method of reversible addition-fragmentation chain transfer polymerization (RAFT) (Chiefari *et al.*, 1998; Chong *et al.*, 1999). Finally, we discuss EET in several generations of porphyrin-based, hyperbranched, three-dimensional molecular structures known as dendrimers. The controlled synthesis of these dendrimers results in macromolecules with well-defined and regular molecular size and architecture, making them ideal for investigations of the redistribution of excitation energy.

2. EXPERIMENTAL DETAILS

2.1. Polymer Synthesis and Sample Preparation

2.1.1   ACE-Containing Homo- and Co-polymers

Acenaphthylene was recrystallized from redistilled methanol three times. Insoluble solids were removed by hot filtration (m.p.= 92 - 93.5 °C). The bright yellow solution was vacuum filtered and dried at the pump prior to being dried thoroughly in a desiccator *in vacuo*. Maleic anhydride (MA) was recrystallized from UV spectroscopic grade dichloromethane. The solid was vacuum filtered and dried at the pump prior to further drying in a dessicator *in vacuo* (m.p.= 53.2 - 54 °C). Both the purified ACE and MA were stored in the dark to avoid photodegradation.

Several P(ACE) and P(ACE/MA) polymers were prepared by free radical polymerization (using AIBN as initiator) using a range of co-monomer and initiator feedstock concentrations. In this chapter only results from the copolymer made from a 1:1 mole ratio and 6% initiator are reported. Repeated freeze-pump-thaw cycles were used to degas the feedstocks which were then reacted in an oil bath for 19 hours (P(ACE/MA)) and 49 hours (P(ACE)) at 76°C. The polymers were purified by multiple reprecipitations from the minimum amount of dichloromethane into approximately 400 ml of distilled methanol. The methanol/polymer solutions were filtered using Millipore filter paper (0.45 μm pores). Molecular weights of the P(ACE) and P(ACE/MA) polymers were determined by gel permeation chromatography to be ~5400 and 4000 respectively. The copolymerization behavior of ACE with MA, including the reactivity ratios of $r_{ACE} = 0.50$ and $r_{MA} = 0.05$, have been reported elsewhere (Ballesteros *et al.*, 1977). Since maleic anhydride does not undergo any homopolymerization it will react only with the ACE co-monomer, and so copolymers of ACE groups separated by a single unit of MA should be produced, although ACE-ACE units are still likely.

Polymer solutions were prepared in 2-methyl tetrahydrofuran (MTHF, Aldrich, 99%) which was first thoroughly dried and purified by repeated refluxing and distillation until no impurity fluorescence was detectable at the relevant excitation and emission wavelengths. This solvent forms a clear glass at ~90 K and when freshly purified, is transparent at the excitation wavelengths required to excite the molecules under investigation (Brocklehurst *et al.*, 1994; Murov *et al.*, 1993). Low temperature measurements were carried out in a temperature-controlled liquid nitrogen cryostat (Oxford Instruments). All samples were degassed thoroughly using multiple freeze-pump-thaw cycles (to 10^{-7} Torr) in specially constructed quartz fluorescence cuvettes with long glass necks to facilitate insertion into the cryostat.

2.1.2 RAFT Polymerization

The living radical polymerization method of reversible addition fragmentation chain transfer (RAFT) polymerization was used to synthesize a series of polymers containing varying numbers of adjacent acenapthyl groups, with each polymer containing a terminal anthryl energy acceptor group. Acenaphthyl and 9-anthryl chromophores were selected as the energy donor and acceptor respectively as they fulfill the spectral overlap requirements for radiationless energy transfer, and acenaphthyl can be preferentially excited with minimal direct excitation of the anthracene from 290 to 320 nm (Fox, 1992; Ghiggino *et al.*, 1993; Webber, 1990). The anthryl energy acceptor group was introduced into the RAFT agent to produce RAFT-AN by the coupling of RAFT-acid and 9-anthracenemethanol in the presence of dicyclohexylcarbodiimide (DCC) as coupling agent and 4-dimethylaminopyridine (DMAP) as catalyst in dichloromethane (Scheme 1, overleaf).

Polymers were synthesized by RAFT methods described previously (Chiefari *et al.*, 1998; Chong *et al.*, 1999) using RAFT-AN (see Scheme 2) and precipitated twice in MeOH to remove unreacted monomers.

The dithiobenzoyl groups have been shown to also quench the fluorescence emission of acenaphthyl chromophores (Chen *et al.*, 2002).

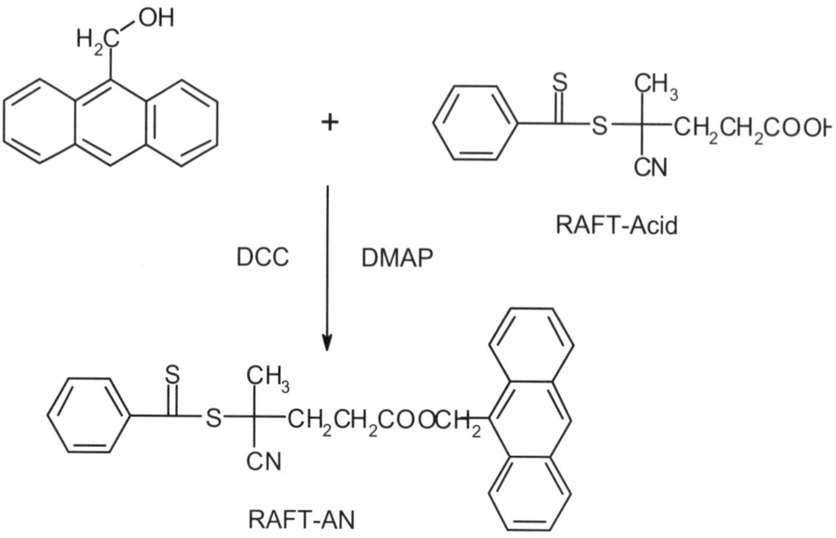

Scheme 1. Synthesis of RAFT-AN

To avoid this effect of the dithiobenzoyl groups on the energy transfer efficiency from acenaphthyl to anthryl groups, two strategies were used (Scheme 3): (i) treating the polymers with amine to convert the dithiobenzoyl group to a thiol, SH-P(ACE)-AN, and (ii) inserting a photoinactive 'spacer' between the dithiobenzoyl group and acenaphthyl sequences by block copolymerization with methyl acrylate, P(MA)-*b*-P(ACE)-AN.

The polydispersity (M_W/M_n) of the P(ACE)-AN sample before precipitation was determined to be 1.08 by gel permeation chromatography (GPC). Matrix-assisted laser desorption/ionization time-of-flight mass spectrometry (MALDI-TOF/MS) was used to determine the actual molecular weights of the polymers. The polymer mass peaks observed are consistent with the predicted structure. This also serves to confirm the mechanism of RAFT polymerization.

ACE-containing oligomers were also made by a stepwise radical addition-fragmentation process in our laboratories (Chen *et al.*).

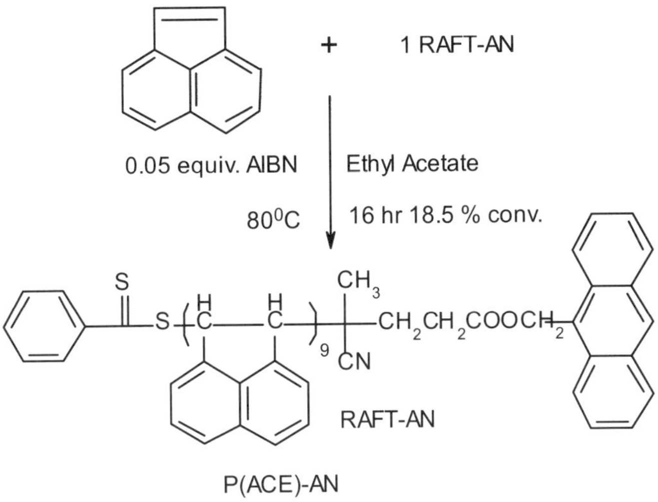

Scheme 2. Synthesis of P(ACE)-AN

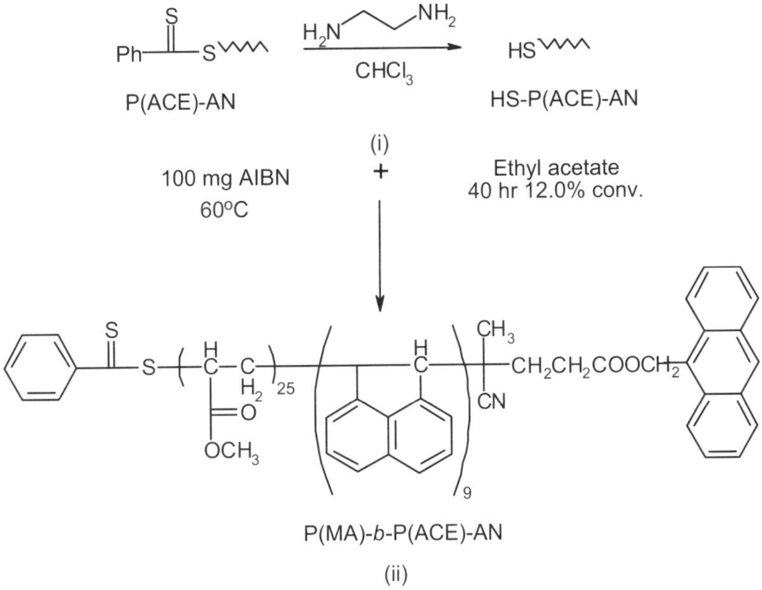

Scheme 3. Removal of dithiobenzoyl groups from RAFT polymers (i) conversion of dithiobenzoyl group to a thiol, and (ii) insertion of photoinactive 'spacer' group.

2.1.3 Porphyrin Dendrimers

Porphyrin-functionalized dendrimers and the mono-porphyrin model compound G0P1 were synthesized and characterized in the research groups of M. Crossley (School of Chemistry, University of Sydney), and B. Meijer (Laboratory of Macromolecular and Organic Chemistry, Eindhoven University of Technology) (Yeow *et al.*, 2000). Solutions of all compounds were prepared in diethyl ether-isopentane-ethanol (EPA) solvent to optical densities of less than 0.2 at 595 nm for both steady-state and time-resolved fluorescence measurements. For time-resolved measurements the excitation wavelengths used were 595 nm and 610 nm, and emission was monitored at 655 nm. EPA is an organic solvent mixture that forms a clear glass at 77 K with a low cracking frequency. It is composed of ethanol (Ajax; spectroscopy grade), isopentane (Sigma-Aldrich; HPLC grade) and diethyl ether (Ajax; analytical reagent grade) in a ratio of 2:5:5.

2.2. Spectral and Temporal Measurements

Absorption measurements were carried out using Hitachi 150-20 or Varian Cary Bio50 absorption spectrometers. Steady-state fluorescence measurements were recorded using Hitachi 4010, Varian Eclipse or Spex Fluorolog-2 spectrofluorimeters. Steady-state fluorescence polarization spectra were recorded using a Hitachi 4500 spectrofluorimeter equipped with a fluorescence polarization accessory, and using excitation and emission bandpass values of 3 nm.

Time-resolved fluorescence measurements were carried out using the technique of time-correlated single photon counting (TCPSC) (O'Connor et al., 1983) using as an excitation source the frequency doubled output of a cavity dumped rhodamine 6G dye laser (Spectra Physics 3500) pumped synchronously by a mode-locked argon ion laser (Spectra Physics 2030). The emission was collected at right angles relative to the excitation beam using an *f*1 air-spaced fused silica condenser lens (Melles Griot 01CMP119) and focused onto the entrance slit of a monochromator (Jobin-Yvon, H-20). The fluorescence was detected using a microchannel plate photomultiplier (Hamamatsu R-1564-U01). Time-resolved fluorescence anisotropy measurements (TRAMS) were carried out by collecting decays with the emission polarization analyzer (Polaroid HNP'B) set parallel and then perpendicular to the direction of polarization (vertical) of the input pulses. The emission polarization analyzer was switched between the parallel and perpendicular positions for preset periods (usually 30 second dwell times) and the two decays collected in separate channels of a multichannel analyzer (Tracor Northern ECON II) over total collection periods exceeding an hour in order to enhance the signal to noise ratio of the anisotropy function.

The time-resolved fluorescence anisotropy function, $r(t)$, is calculated using Equation 1 in which $I_\parallel(t)$ and $I_\perp(t)$ are the individual decays collected with the polarization analyzer set parallel and perpendicular to the vertically polarized excitation light. The G factor is included to account for any polarization bias of the detection system. The influence of this term was minimized by arranging the polarization analyzer to be the first element in the detection system and using a polarization pseudo-scrambler (Oriel 28115) immediately prior to the emission monochromator slit.

$$r(t) = \frac{I_\parallel(t) - GI_\perp(t)}{I_\parallel(t) + 2GI_\perp(t)} \tag{1}$$

Time-resolved fluorescence anisotropy measurements were carried out in low temperature, transparent glasses in order to eliminate any fluorescence depolarization caused by molecular rotation. The samples were thoroughly degassed by repeated freeze-pump-thaw cycles to remove any dissolved gases before being cooled to a glassy state and maintained at 77 K using a variable temperature liquid nitrogen cryostat (Oxford Instruments Optistat DN).

3. ENGINEERED POLYMER SYSTEMS

3.1. Acenaphthylene-Containing Polymers

The acenaphthyl group (ACE, Figure 1) is a particularly interesting aromatic chromophore for studies of energy migration and trapping that can be incorporated as pendant moieties on a polymer backbone.

Polymers containing the acenaphthyl chromophore have been investigated previously (Ghiggino *et al.*, 1995) with the P(ACE) homopolymer (Figure 1c) being particularly well characterized (Agbaje *et al.*, 1986; Agbaje *et al.*, 1987; Muller, 1968). The ACE chromophore differs from many of the other commonly studied polymer chromophores such as naphthyl-, pyrenyl- and carbazolyl-containing polymers, in that the aromatic group is bound to the polymer backbone by two covalent bonds, rather than just one. This restricts the rotational motion of the chromophore with respect to the polymer backbone. Since every alternate C-C bond of the polymer main chain is bridged by a bulky naphthyl substituent, the P(ACE) chain is expected to exhibit considerable inflexibility and an expanded coil, spiral (Tsvetkov *et al.*, 1971) configuration in solution, although the rigidity of the chain has been the subject of some debate, (Barrales-Rienda *et al.*, 1967; Chu, 1988; Küçükyavuz *et al.*, 1993; Kulkarni *et al.*, 1991; Moacanin *et al.*, 1967) and theories relating to flexible coils have been shown to be applicable in determining the dilute solution properties of P(ACE) (Agbaje *et al.*, 1986; Agbaje *et al.*, 1987). This restricted motion inhibits, but does not completely eradicate, excimer formation in the P(ACE) homopolymer in solution, an energy trapping process that may decrease the efficiency and range of energy migration in polymers. It has been shown through molecular modeling (Mendicuti *et al.*, 1990) and excited-stated semi-empirical calculations, (Scholes *et al.*, 1992) that despite the expected rigidity of the P(ACE) structure, nearest-neighbor excimer formation in ACE-containing systems cannot necessarily be ignored.

The rigidity of the ACE chromophore may also greatly influence the efficiency of energy migration in ACE-containing polymers compared to other common synthetic aromatic polymers. In the light of recent studies of efficient energy transfer in linked bichromophores (Ghiggino *et al.*, 1996) and the role of rigidity of the bridge in the enhancement of through bond energy transfer, (Smith *et al.*, 2002) the ACE chromophore may provide a useful link between the studies on bichromophores and synthetic aromatic polymers.

Evidence for energy migration in synthetic aromatic polymers is often circumstantial, commonly being provided by observations of enhanced efficiency of emission from energy trap sites. Steady-state and, more particularly, time-resolved emission polarization studies can provide more compelling evidence for EET, but this relies on the premise that there is substantial loss of the degree of polarization with each energy hopping step. We have previously reported results of P(ACE) end-capped with an anthracene group which suggested the existence of efficient energy migration occurring in this polymer through emission from the anthracene moiety being enhanced compared to the efficiency expected from calculations based on a single, Förster-type energy transfer step from ACE to the anthracene group (Ghiggino *et al.*, 1993). Here we report on energy migration in P(ACE) through direct measurement of the fluorescence depolarization.

Time-resolved fluorescence anisotropy measurements (TRAMs) can provide compelling evidence for the occurrence of EEM within polymers. Fluorescence anisotropy can, under the conditions used here, vary between −0.2 and 0.4, with a value of zero indicating complete randomization of the emission polarization. In the absence of chromophoric, segmental or polymer coil rotation, and assuming that the solution is dilute enough to negate the possibility of *inter*chain interactions, a time-independent anisotropy value proportional to the intrinsic anisotropy of the fluorophore (related to the angle between the absorption and emission dipoles) should be observed. Any time-dependent decay of $r(t)$ can then be attributed to EEM, assuming that each step of the energy migration process results in a loss of some degree of the initial emission polarization of the system. An exception to this was reported for the model compound, acenapthene, Figure 1(a), in which we observed complex excitation and emission wavelength-dependent fluorescence polarization behavior. An unusual, time-dependent emission polarization behavior was observed that was attributed to the excitation of, emission from, and interaction between close-lying excited states. (Smith *et al.*, 2000) The rate of this process was found to occur over tens of nanoseconds.

TRAMs carried out on dilute solutions of P(ACE) in MTHF glasses are shown in Figure 2a. The anisotropy decays to zero within the time resolution of our current TCSPC instrumentation (<100 ps), indicating extremely rapid and efficient EEM. The additional emission depolarizing

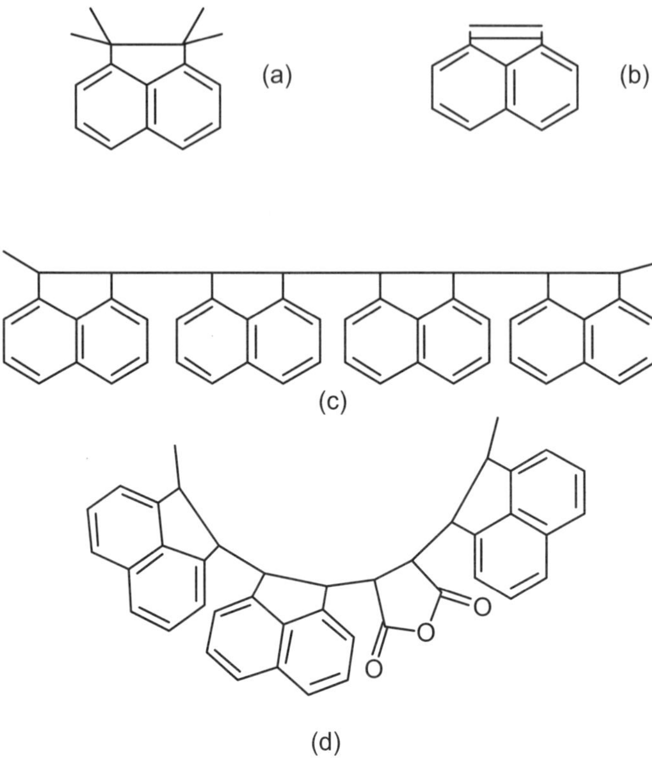

Fig. 1. Structures of ACE-compounds: (a) acenaphthene, (b) acenaphthylene monomer, (c) P(ACE) and (d) P(ACE/MA).

mechanism mentioned above for acenaphthene is too slow to compete with the energy migration process, and so the observed depolarization is dominated by EEM.

We have previously attributed the high efficiency and rate of EEM in P(ACE) to the possible involvement of superexchange mechanisms arising from the rigidity of the linking polymer backbone (Ghiggino *et al.*, 1996). However, excimer formation is also reduced in this polymer compared to other naphthyl-containing polymers in which the chromophore is bonded to the backbone by a single sigma bond. The reduced excimer formation may also contribute to the high efficiency of energy migration in P(ACE).

As discussed above, maleic anhydride is known to polymerize only with other co-monomers (Ballesteros *et al.*, 1977), so the P(ACE/MA)

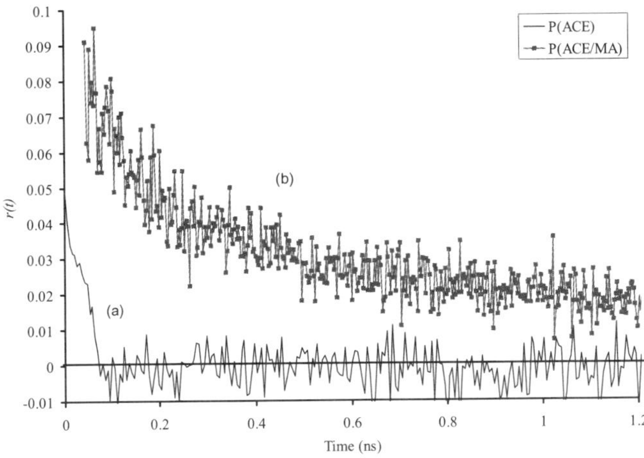

Fig. 2. Low temperature time-resolved fluorescence anisotropy decays from (a) P(ACE) and (b) P(ACE/MA). λ_{ex}= 295 nm, λ_{em}= 330 nm

copolymer (Figure 1(d)) prepared as described should not contain any MA-MA segments, but there will be a proportion of ACE-ACE segments. The purpose of the MA is therefore to space the ACE groups out and, since energy transfer/migration is strongly chromophore distance dependent, reduce the rate of intramolecular electronic energy transfer, k_{eet}.

The inclusion of the maleic anhydride units into the polymer backbone results in a significantly reduced rate of emission depolarization, as shown in Figure 2b. While this may be related to the time-dependent decay observed from the model ACE compound referred to above, the observed fluorescence depolarization occurs on a faster timescale and does not show the same excitation and emission wavelength dependence. Therefore, in the case of the P(ACE/MA) copolymer, we interpret the $r(t)$ decay as evidence for EEM, but with the structural changes induced by the inclusion of the MA units being effective in inhibiting the rate of EEM between ACE units. The decreased EEM rate can be attributed to deceases in electronic coupling between chromophores arising from large interchromophore separations and, possibly, changes in polymer chain conformation and flexibility.

Thus through carefully engineering the structure of such polymers, the rate (and efficiency) can clearly be controlled. In the following section we discuss ACE-containing polymers that have been engineered with far more structural control than has been achieved in the past.

3.2. RAFT Polymerization of ACE-Containing Polymers

We have shown previously, in non-engineered ACE-containing polymers with terminal anthryl groups, that the efficiency of energy transfer to the anthryl trap site was well above that predicted by Förster long-range Coulombic theory (Ghiggino *et al.*, 1993). We have also shown, in the previous Section and elsewhere, that the ACE chromophore is capable of efficient EEM and that the efficiency of this process can be altered to some extent through structural variations. However, control over polymeric structure and molecular weight distribution through conventional free radical polymerization is limited and somewhat random.

Polymers can be prepared with a much higher degree of structural control through reversible addition-fragmentation chain transfer (RAFT) polymerization methods, than is achievable through conventional free radical polymerization methods. In this Section we report on the use of RAFT synthesis methods to engineer ACE-containing polymers with well-defined sequences of nine ACE ($n = 9$) chromophores with energy-accepting anthryl terminal groups (Chen *et al.*, 2002). The efficiency of energy transfer to the anthryl group, following excitation of the ACE groups, has been determined with a view to optimizing the design and construction of systems for enhanced 'light harvesting' or 'antenna' effects.

A comparison of the fluorescence spectra of the different polymers is presented in Figure 3. Photoexcitation of P(ACE)-AN, SH-P(ACE)-AN and P(MA)-*b*-P(ACE)-AN (see Schemes 2 and 3 for structures) at 295 nm, where absorption is almost exclusively due to acenaphthyl groups, results in fluorescence being predominantly emitted from the 9-anthryl end groups (Figure 3). Furthermore, the fluorescence excitation spectrum of the 9-anthryl emission contained a contribution attributable to acenaphthyl absorption.

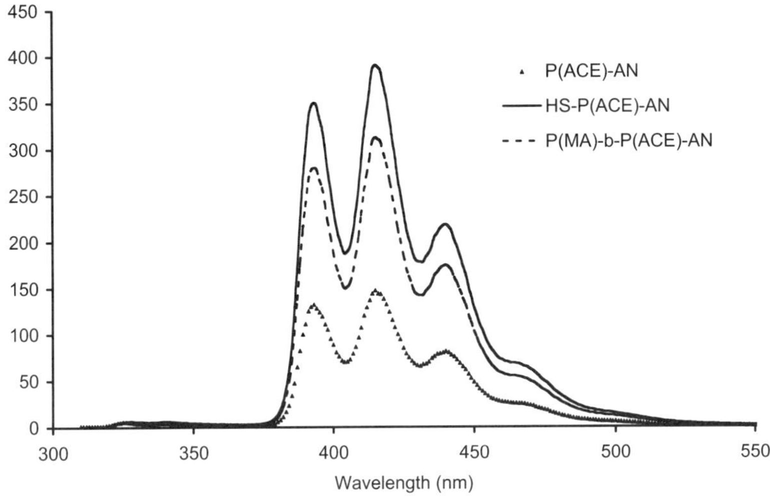

Fig. 3. Fluorescence spectra of acenaphthylene polymers in dilute, degassed dichloromethane solution excited at 295 nm. All solutions have the same absorbance at 295 nm.

These observations confirm that energy transfer from the ACE groups to the anthryl end-group is occurring. At the solution concentrations used (polymer concentrations < 10^{-5} M) only intra-chain energy transfer processes are possible during the excited state lifetime. The efficiency of excitation energy transfer in the polymers can be determined by comparison of the fluorescence excitation spectrum and the absorption spectrum as described previously (Bai *et al.*, 1986; Ghiggino *et al.*, 1995). The efficiency of acenaphthyl to anthryl energy transfer in P(ACE)-AN is 15%. This low value can be attributed to the presence of a competing non-fluorescent energy trap (i.e. the dithiobenzoyl group remaining after RAFT polymerization). After removing the dithiobenzoyl group, the energy transfer efficiency increased to 70% in SH-P(ACE)-AN, confirming this hypothesis. Introducing 'spacer' monomer units between the dithiobenzoyl group and the acenaphthyl donors as in P(MA)-*b*-P(ACE)-AN also increased the energy transfer efficiency to 56%. The smaller effect on energy transfer efficiency by synthesizing the block copolymer is most likely due to the ability of the dithiobenzoyl group to fold back and quench the excited acenaphthyl groups, due to the flexibility of the poly(methyl acrylate) chain.

The mechanism of excimer formation within the P(ACE) homopolymer has been the subject of some controversy in the literature. The observation of significant excimer emission from P(ACE) suggests a breakdown of Hirayama's Rule which states that only parallel aromatic rings separated by three carbon atoms can form excimers. In P(ACE), the restricted rotation imposed by the dual linkages of the aromatic chromophore to the polymer backbone on the ACE units, with respect to one another and the backbone, means that nearest-neighbor monomer moieties cannot assume the required orientation and separation to form a stable excimer configuration. David *et al.*, (1972) concluded therefore that excimer formation must be due, in P(ACE), to the interaction of next-to-nearest neighbors, which was supported by later experimental work (Phillips *et al.*, 1980; Wang *et al.*, 1975). On the other hand, theoretical modeling on acenaphthylene diads (Mendicuti *et al.*, 1990; Scholes *et al.*, 1992) suggested the possibility that nearest-neighbor interactions should not be ignored in describing P(ACE) excimer formation.

The RAFT technique discussed above can also be used to produce smaller, yet still structurally well-defined, ACE-containing oligomers and polymers. We have recently synthesized the ACE-ACE dimer, the trimer, as well as higher P(ACE) oligomers and the homopolymer (Chen *et al.*). Such systems should help elucidate the mechanism of excimer formation in ACE-containing polymers for the first time.

The emission spectra of the ACE oligomers and the high molecular weight homopolymer are shown in Figure 4. No excimer emission is evident from the ACE dimer or trimer. Evidence of excimer emission is observed in the emission spectra of P(ACE)$_7$ (an oligomer of 7 ACE repeat units) and the homopolymer. This indicates that excimer formation between nearest-neighbor ACE groups does not occur, and that even next-to-nearest neighbor interactions are not a significant excimer forming pathway in P(ACE). The excimer emission from P(ACE)$_7$ and P(ACE) suggests that sufficient chain length (n>3) is required to provide the conformational flexibility to bring two well-separated ACE chromophores together into the required co-facial arrangement.

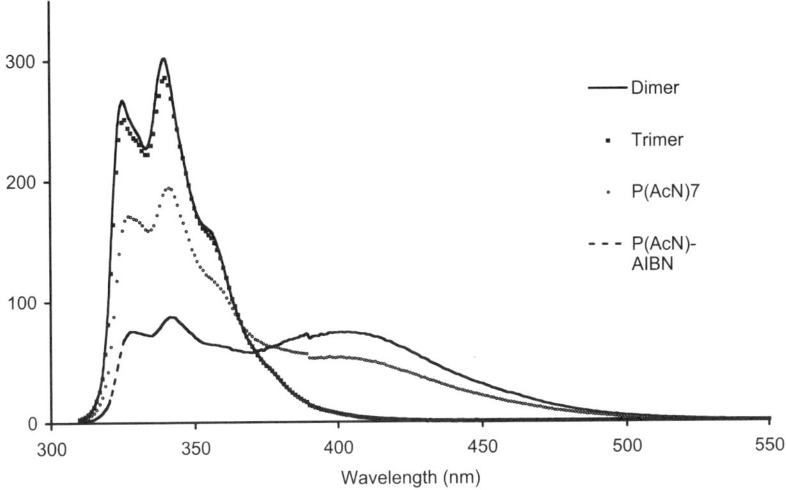

Fig. 4. Emission spectra of the ACE dimer, trimer, P(ACE)$_7$ and P(ACE) homopolymer (average molecular weight 15,900)

3.3. Porphyrin-Containing Dendrimers

Dendrimers are now recognized to have important existing and potential roles in many fields. Recently, a number of studies have been focused on using dendrimers to mimic the photosynthetic light-harvesting antennae system (Bar-Haim *et al.*, 1997; Kopelman *et al.*, 1997; Shortreed *et al.*, 1997; Swallen *et al.*, 1998). One important feature of dendrimers, especially those of high generations, is the large overall cross-section offered by the array of terminal chromophores for excitation energy collection. This is reminiscent of the primary events in photosynthesis, where photons are harvested in arrays of chlorophyll molecules found in the light harvesting complexes.

We have investigated (Yeow et al., 2000) the dynamics of electronic energy transfer (EET) for the 1st (G1P4), 3rd (G3P16) and 5th (G5P64) generation multi-porphyrin functionalized dendrimers, G*n*P*m*, where *n* is the generation number and *m* is the number of porphyrin end-groups (see Figure 5), using time-resolved fluorescence anisotropy measurements (TRAMS).

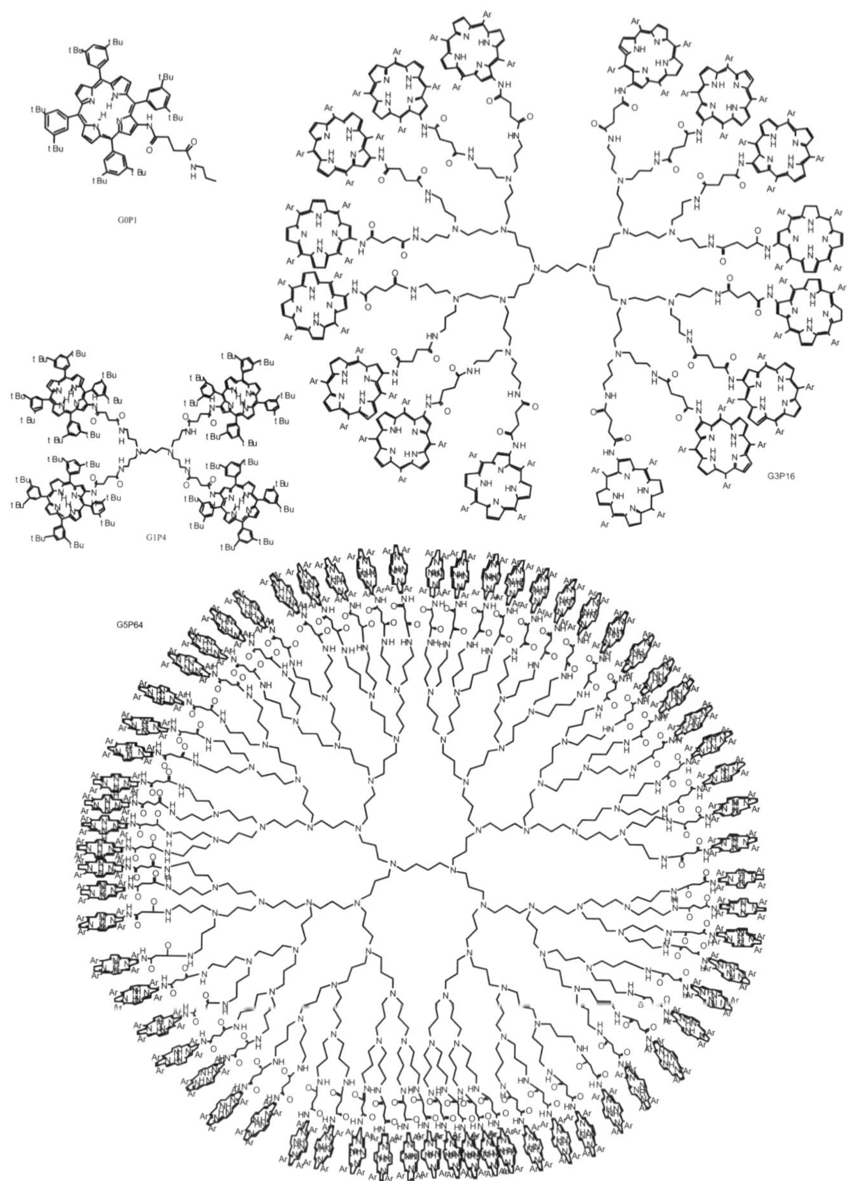

Fig. 5. Structures of the porphyrin dendrimers studied.

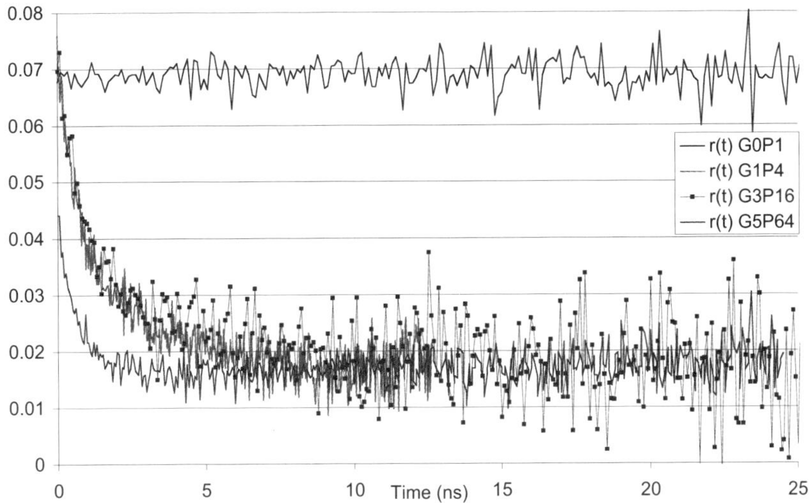

Fig. 6. Time-dependent fluorescence anisotropy decays for G0P1, G1P4, G3P16 and G5P64 measured at an emission wavelength of 655 nm when excited at 595 nm in an EPA glass

The time-resolved fluorescence anisotropy decay profiles from G0P1, G1P4, G3P16 and G5P64, measured at an emission wavelength of 655 nm when excited at 595 nm in an EPA glass, are shown in Figure 6. For G0P1, the fluorescence anisotropy remains reasonably constant ($r_0 \approx$ 0.07) throughout the observed time range. The observation that the initial anisotropy is not 0.4 indicates that mixed polarization transitions are involved in fluorescence from porphyrins. Gouterman *et al.* (1962) have shown that the intensity of the $Q_x(1 \leftarrow 0)$ band arises from vibrational mixing with the intense Soret bands, which introduces both x- and y-polarized intensity to the Q-band. The absence of any decay in the fluorescence anisotropy for the model compound is expected since no time-dependent depolarization process should occur in the rigid glass.

In contrast to G0P1, the fluorescence anisotropy from G1P4 (see Figure 6) is observed to decay over a time scale of a few nanoseconds before leveling off to a residual anisotropy value of $r_\infty = 0.0175$. This indicates that excitation energy is transferred between the porphyrin molecules and eventually becomes equally distributed among the four chromophores (*i.e.*, $r_\infty \approx r_0/4$). This anisotropy decay has been successfully modeled based on the Förster energy transfer mechanism and employing the Pauli Master Equation (Yeow *et al.*, 2000) assuming that the chromophores are arranged along the perimeter of a circular disk of radius 12 Å such that the chromophores are positioned at the corners of an enclosed rectangle where the nearest separation between the porphyrins is 15 Å and the furthest displacement is equivalent to the diameter of the circular disk (*i.e.*, 24 Å).

The time-resolved fluorescence anisotropy profile for G3P16, is also shown in Figure 6. The limiting anisotropy value is not 0.0044 ($r_\infty = r_0/16$) expected if an equilibrium distribution of excitation on each of the randomly oriented sixteen chromophores is achieved. Instead, $r_\infty = 0.0175$, which is essentially identical to the limiting anisotropy observed for G1P4. This suggests that EET is essentially confined to only the four porphyrins contained within a single dendron. The rather long and flexible branching units would enable each dendron and associated porphyrins to be randomly arranged on a quarter of the dendrimer's spherical surface. Monte Carlo methods were used (Yeow *et al.*, 2000) to assign the coordinates of the chromophores and predict theoretical anisotropy decays and thereby to determine the maximum rate of EET within a dendron to be 0.81 ns^{-1}.

Complex fluorescence anisotropy decay behavior is observed for the 5th generation dendrimer, G5P64 (see Figure 6). A very rapid initial anisotropy decay occurs to a value of $r(t) \approx 0.015$ (at $t = 4$ ns) before $r(t)$ gradually regains a little intensity and levels off at $r_\infty \approx 0.018$. These features allude to complex energy transfer dynamics, which cannot be easily explained using the models proposed for the lower generation dendrimers. This unusual 'dip' and 'rise' behavior in the anisotropy curve suggests that there are more than one species in the system whose contributions to the effective $r(t)$ vary with time (Smith *et al.*, 1998). In G5P64, it has been proposed that there are two species contributing to the effective anisotropy. The first species (core porphyrins) are packed together closely enough such that strong couplings are experienced, which ultimately give rise to the rapid initial anisotropy decay. Their total contributions to $r(t)$ are important at short times but diminish rapidly when $r(t)$ regains intensity at longer times due to the second contributing species, porphyrins with the long lifetime component (i.e. adventitious porphyrins). The limiting anisotropy clearly suggests that EET occurs only between four adventitious porphyrins, which are too distant from the core chromophores to allow any effective energy transfer. It was stressed that the rapid EET dynamics between the core porphyrins may be more complex than suggested and not revealed fully in the experimental data.

The novel multi-porphyrin functionalized dendrimers are able to absorb light and efficiently delocalize the excitation energy over the chromophore arrays with minimal loss during the energy migration process. The dendrimers are thus acting as excellent light-harvesting antennae.

4. CONCLUSIONS

We have demonstrated that macromolecular systems can be specifically engineered, through careful design and novel synthesis pathways now available, for the study of electronic energy transfer, migration and trapping mechanisms. Through these methods, oligomers, polymers and dendrimers can be developed which may ultimately allow the mechanisms and dynamics of EET, EEM and trapping in macromolecules to be investigated and harnessed.

ACKNOWLEDGMENTS

The authors gratefully acknowledge the Australian Research Council for funding. The many co-workers and students who have contributed to this work are also acknowledged, in particular Ming Chen, Edwin Yeow, Greg Scholes and David Haines.

References

1. Agbaje H and Springer J. Production and properties of polyacenaphthylene - V Dilute solution properties. *Eur. Polym. J.* 1986; **22**: 943-948.

2. Agbaje H and Springer J. Production and properties of polyacenaphthylene - VII Unperturbed dimensions and chain conformational studies of polyacenaphthylene. *Eur. Polym. J.* 1987; **23**: 283-286.
3. Allen NS and Rabek JF (Ed): *New Trends in Photochemistry of Polymers* London: Elsevier; 1985.
4. Bai F, Chang CH, and Webber SE. Photon-harvesting polymers: Singlet energy transfer in anthracene-loaded alternating and random copolymers of 2-vinylnaphthalene and methacrylic acid. *Macromolecules* 1986; **19**: 2484-2494.
5. Ballesteros J, Howard GJ, and Teasdale L. Acenaphthylene copolymers I. Copolymerization behaviour. *J. Macromol. Sci. - Chem.* 1977; **A11**: 29-37.
6. Bar-Haim A, Klafter J, and Kopelman R. Dendrimers as controlled artificial energy antennae. *J. Am. Chem. Soc.* 1997; **119**: 6197-6198.
7. Barrales-Rienda JM and Pepper DC. The dilute solution properties of polyacenaphthylene I - Light scattering, osmotic pressure and viscosity measurements. *Polymer* 1967; **8**: 337-360.
8. Birks JB: *Photophysics of Aromatic Molecules*: Wiley, Chichester; 1970.
9. Brocklehurst B and Young RN. Fluorescence anisotropy decays and viscous behaviour of 2-methyltetrahydrofuran. *J. Chem. Soc. Faraday Trans.* 1994; **90**: 271-278.
10. Chen M, Ghiggino KP, Mau AWH, Rizzardo E, Thang SH, and Wilson GJ. Synthesis of light harvesting polymers by RAFT methods. *Chem. Comm.* 2002: 2276-2277.
11. Chen M, Ghiggino KP, Smith TA, Thang SH, and Wilson GJ. Mechanisms of excimer formation in poly(acenaphthylene). *Aust. J. Chem.* (submitted).
12. Chiefari J, Chong YK, Ercole F, Krstina J, Jeffery J, T. P. T. Le R, Mayadunne TA, Meijs GF, Moad CL, Moad G, Rizzardo E, and Thang SH. Living free-radical polymerization by reversible addition-fragmentation chain transfer: The RAFT process. *Macromolecules* 1998; **31**: 5559-5562.
13. Chong YK, Le PT, G. Moad, Rizzardo E, and Thang SH. A more versatile route to block copolymers and other polymers of complex architecture by living radical polymerization: The RAFT process. *Macromolecules* 1999; **32**: 2071-2074.
14. Chu EF. Local stiffness vs global flexibility in polyacenaphthylene. *Polym. Mat. Sci. Eng.* 1988: 1041-1043.
15. Cogdell RJ, Isaacs NW, Howard TD, McLuskey K, Fraser NJ, and Prince SM. How photosynthetic bacteria harvest solar energy. *J. Bacteriol.* 1999; **181**: 3869-3879.
16. David C, Lempereur M, and Geuskens G. Energy transfer in polymers - V. Singlet and triplet energy transfer in polyacenaphthylene. *Eur. Polym. J.* 1972; **8**: 417-427.

17. Fox MA. Polymeric and supramolecular arrays for directional energy and electron transport over macroscopic distances. *Acc. Chem. Res.* 1992; **25**: 569-574.

18. Fox MA. Light-harvesting polymer systems. *Chem. & Eng. News* 1993: March 15, 38-48.

19. Fox RB, Price TR, Cozzens RF, and McDonald JR. Photophysical processes in polymers. IV. Excimer formation in vinylaromatic polymers and copolymers. *J. Chem. Phys.* 1972; **57**: 534-541.

20. Ghiggino KP, Haines DJ, Smith TA, and Wilson GJ. Energy migration and transfer in styrene-containing polymers with inhibited excimer formation. *Can. J. Chem.* 1995; **73**: 2015-2020.

21. Ghiggino KP and Smith TA. Dynamics of energy migration and trapping in photoirradiated polymers. *Prog. React. Kin.* 1993; **18**: 375-436.

22. Ghiggino KP, Yeow EKL, Haines DJ, Scholes GD, and Smith TA. Mechanisms of excitation energy transport in macromolecules. *J. Photochem. Photobiol A: Chem.* 1996; **102**: 81-86.

23. Gnanakaran S, Haran G, Kumble R, and Hochstrasser RM: Energy transfer and localization: Applications to photosynthetic systems. In *Resonance Energy Transfer*. Edited by Andrews DL, Demidov AA: John Wiley, Chichester; 1999: 308-365.

24. Gouterman M and Stryer L. Fluorescence polarization of some porphyrins. *J. Chem. Phys.* 1962; **37**: 2260-2266.

25. Guillet J: *Polymer Photophysics and Photochemistry. An Introduction to the Study of Photoprocesses in Macromolecules*. University Press, Cambridge; 1985.

26. Guillet JE, Takahashi Y, and Gu L: Polymer models for photosynthesis. In *Photophysics of Polymers*. Edited by Hoyle CE, Torkelson JM: American Chemical Society; 1987: 412-421.

27. Harrah LA. Excimer formation in vinyl polymers. I. Temperature dependence in fluid solution. *J. Chem. Phys.* 1972; **56**: 385-389.

28. Hoyle CE and Torkelson JM (Ed): *Photophysics of Polymers*. American Chemical Society, Washington; 1987.

29. Hu X, Damjanovic A, Ritz T, and Schulten K. Architecture and mechanism of the light-harvesting apparatus of purple bacteria. *Proc. Natl. Acad. Sci. USA* 1998; **95**: 5935-5941.

30. Hu X and Schulten K. How nature harvests sunlight. *Physics Today* 1997; **50** (8): 28-34.

31. Kopelman R, Shortreed M, Shi Z-Y, Tan W, Bar-Haim A, and Klafter J. Spectroscopic evidence for excitonic localization in fractal antenna supermolecules. *Phys. Rev. Lett.* 1997; **78**: 1239-1242.

32. Küçükyavuz S, Göksel C, and Küçükyavuz Z. Solution properties and chain stiffness of poly(acenaphthylene). *J. Macromol. Sci. - Pure Appl. Chem.* 1993; **A30**: 907-917.

33. Kulkarni R, McIntyre D, and Rinaldi P. The local stiffness and global entanglements of polyacenaphthylene. *Polym. Prep.* 1991; **32**: 128-129.

34. Macfarlane RM and Shelby RM. Homogeneous line broadening of optical transitions of ions and molecules in glasses. *J. Lumin.* 1987; **36**: 179-207.

35. Mendicuti F, Kulkarni R, Patel B, and Mattice W. Identification of the short-range intramolecular excimers in polyacenaphthalene. *Macromolecules* 1990; **23**: 2560-2566.

36. Moacanin J, Rembaum A, Laudenslager RK, and Adler R. Dilute solution properties and conformation of polyacenaphthylene. *J. Macromol. Sci. - Chem.* 1967; **A1**: 1497-1518.

37. Muller J-C. Polymerisation radiochimique de L'acenaphthylene. *J. de Chimie Physique* 1968; **65**: 567-577.

38. Murov SL, Carmichael I, and Hug GL (Ed): *Handbook of Photochemistry* Marcel Dekker, New York; 1993.

39. O'Connor DV and Phillips D: *Time-Correlated Single Photon Counting.* Academic Press, New York; 1983.

40. Pearlstein RM. Impurity quenching of molecular excitons. I. Kinetic comparison of Förster—Dexter and slowly quenched Frenkel excitons in linear chains. *J. Chem. Phys.* 1972; **56**: 2431-2442.

41. Phillips D (Ed): *Polymer Photophysics. Luminescence, Energy Migration and Molecular Motion in Synthetic Polymers.* Chapman and Hall, New York; 1985.

42. Phillips D, Roberts AJ, and Soutar I. A time-resolved fluorescence spectroscopic study of excimer formation in polyacenaphthylene and an alternating acenaphthylene/maleic anhydride copolymer. J. Polym. Sci. Polym. Lett. 1980; **18**: 123-129.

43. Pullerits T and Sundström V. Photosynthetic light-harvesting pigment-protein complexes: Toward understanding how and why. *Acc. Chem. Res.* 1996; **29**: 381-389.

44. Scholes GD and Ghiggino KP. Excimer formation in poly(acenaphthalene) diads. *Chem. Phys. Lett.* 1992; **188**: 140-144.

45. Shortreed MR, Swallen SF, Shi Z-Y, Tan W, Xu Z, Devadoss C, Moore JS, and Kopelman R. Directed energy transfer funnels in dendrimeric antenna supermolecules. *J. Phys. Chem. B* 1997; **101**: 6318-6322.

46. Smith TA, Haines DJ, and Ghiggino KP. Fluorescence polarization studies of acenaphthene. A case for dual emission. *J. Fluorescence* 2000; **10**: 365-373.

47. Smith TA, Irwanto M, Haines DJ, Ghiggino KP, and Millar DP. Time-resolved fluorescence anisotropy measurements of the adsorption of

rhodamine-B and a labelled polyelectrolyte onto colloidal silica. *Coll. Polym. Sci.* 1998; **276**: 1032-1037.

48. Smith TA, Lokan N, Cabral N, Davies SR, Paddon-Row MN, and Ghiggino KP. Photophysics of novel donor-{saturated rigid hydrocarbon bridge}-acceptor systems exhibiting efficient excitation energy transfer. *J. Photochem. Photobiol. A: Chem.* 2002; **149**: 55-69.

49. Sundström V, Pullerits T, and van Grondelle R. Photosynthetic light harvesting: Reconciling dynamics and structure of purple bacterial LH2 reveals function of photosynthetic unit. *J. Phys. Chem. B* 1999; **103**: 2327-2346.

50. Swallen SF, Shi Z-Y, Tan WX, Z., Moore JS, and Kopelman R. Exciton localization hierarchy and directed energy transfer in conjugated linear aromatic chains and dendrimeric supermolecules. *J. Lumin.* 1998; **76&77**: 193-196.

51. Tsvetkov VN, Vitovskaya MG, Lavrenko PN, Zakharova EN, Gavrilenko IF, and Stefanovskaya NN. The conformational properties and the rigidity of polyacenaphthylene macromolecules in solutions. *Polym. Sci., USSR* 1971; **13**: 2845-2858.

52. Van der Meer BW, Coker GI, and Chen S-Y: *Resonance Energy Transfer.* VCH, New York; 1994.

53. van Grondelle R and Somsen OJG: Excitation energy transfer in photosynthesis. In *Resonance energy transfer*. Edited by Andrews DL, Demidov AA: John Wiley, Chichester; 1999: 366-398.

54. van Grondelle R and Valkunas L: (Organizers) Light-harvesting, Physics Workshop. Bristonas, Lithuania, September 14-17, 1996: *J. Phys. Chem. B*: 1997: **101**: 7197-7359.

55. Wang Y-C and Morawetz H. Excimer fluorescence of polymers with aromatic substituents. 1. *Die Makromolekulare Chemie*, Suppl 1975; **1**: 283-295.

56. Webber SE. Photo-harvesting polymers. *Chem. Rev.* 1990; **90**: 1469-1482.

57. Yeow EKL, Ghiggino KP, Reek JNH, Crossley MJ, Bosman AW, Schenning APHJ, and Meijer EW. The dynamics of electronic energy transfer in novel multiporphyrin functionalized dendrimers: A time-resolved fluorescence anisotropy study. *J. Phys. Chem. B* 2000; **104**: 2596-2606.

Chapter 8

DENDRIMER-BASED DEVICES: ANTENNAE AND AMPLIFIERS

Ophir Flomenbom, Joseph Klafter, Roey J. Amir and Doron Shabat

In this chapter we focus on two applications of dendrimers that stem from their unique branched structures: dendrimers as light harvesting antennae, and dendrimers as novel platforms for drug delivery (molecular amplifiers). Both applications are formulated theoretically within a framework based on the master equation. The quantities of interest are the first passage time probability density function and its moments. We examine how this function depends on the geometric and energetic characteristics of the dendrimeric system. In particular, we investigate the dependence of the first passage time properties on the number of generations (dendrimer size), and on the system bias. We present analytical expressions for the first passage times of very efficient dendrimeric antennae and of dendrimeric amplifiers. For these cases the mean first passage time scales linearly with the system length, and fluctuations around the mean are negligible for large systems. Light harvesting processes under different system-bias conditions are discussed.

Keywords: Light harvesting, dendrimeric antennae, drug platform, dendrimeric amplifiers, first passage time.

1. INTRODUCTION

Dendrimers are highly defined artificial macromolecules, which are characterized by a combination of a high number of functional groups, and a compact molecular structure.[1] The macromolecule constituents are organized in a branching form from a central core, creating a sphere of chemical end-groups at the periphery that can be tailored according to the requirements,[2] see Figure 1. The concept of repetitive-growth with branching can create unique spherical mono-dispersed dendrimer formations, which are defined by their generation number.[3] As dendrimers are built from AB_z-type monomers, each layer or generation (G) of branching unit doubles or triples, for $z = 2$ or $z = 3$, the number of peripheral groups. For example, for $z = 2$, a first generation dendrimer, which is denoted *G1*, will have one branching unit, and a second-generation dendrimer (*G2*) will have an additional two branching unit, *etc.* In addition the core branching C, can be chosen independently from z.

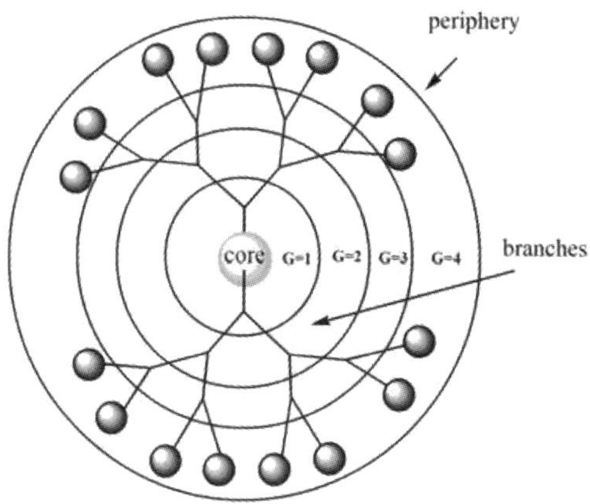

Fig. 1 Schematic representation of a dendrimer. Shown are the generation indices, the dendrimer core and end groups. Here $z = 2$.

Although dendrimers are large molecules, they can be synthesized and characterized with a precision similar to that possible with smaller organic molecules. They do not suffer from the problem of 'poly-dispersity' that troubles linear macromolecules: that is, constituents of a given set of dendrimers can have exactly the same molecular weight, rather than being a mixture of chains with a distribution of molecular weights. The large number of identical chemical units in the branching sub-units as well as those at the periphery confers greater versatility; see Figures 2 and 3 for examples.

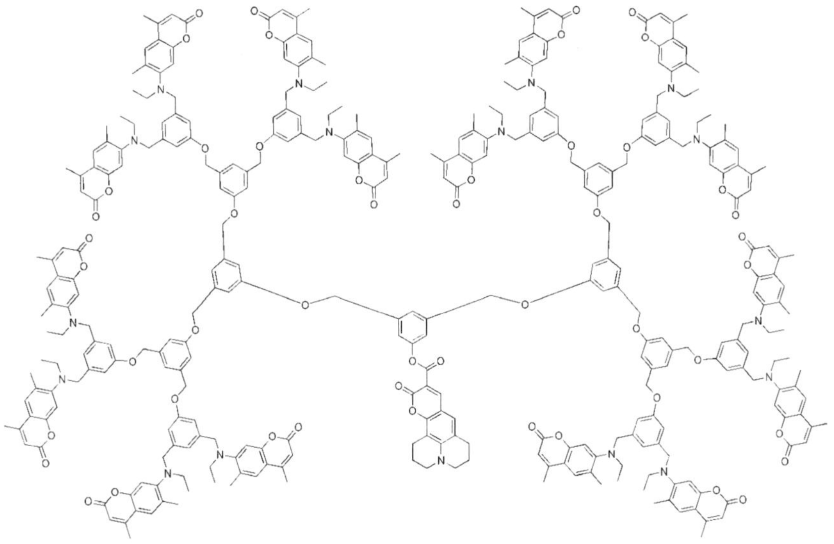

Fig. 2 A four-generation dendrimer, built *via* meta positions.[4]

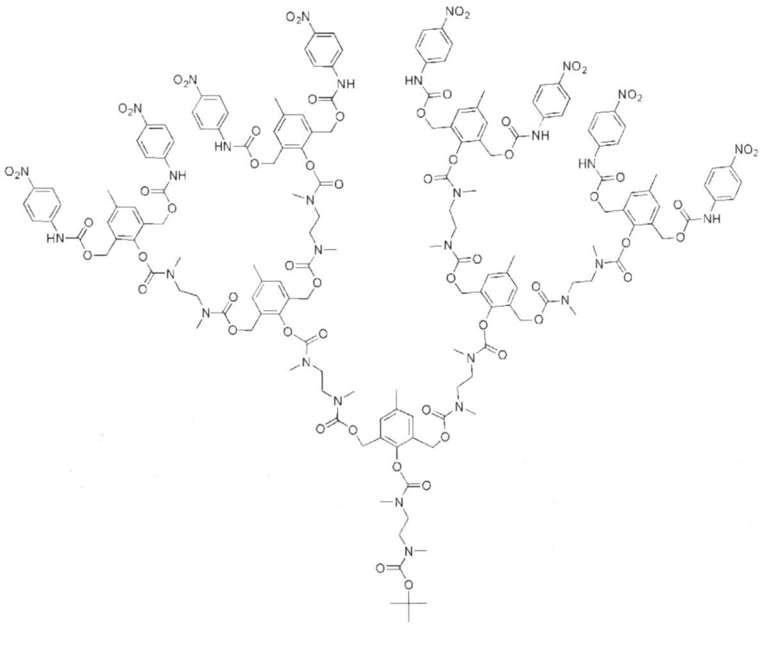

Fig. 3 A three-generation dendrimer, designed to release its end groups upon trigger activation (Boc deprotection).[5]

The special features of dendrimers make them promising candidates for a large number of applications. For example, one can vary the type of the end-groups, such that the specifically designed macromolecule can be used for sensing, catalysis or biochemical activity.[6] Most of the applications of dendrimers have been based mainly on the high number of functional groups and not on the unique structure of the molecule. Although the enhanced effect that stems from lots of identical end-groups being present at the same time and place is of great importance, the combination of utilizing the multivalency and the precise architecture as an active framework able to function more than merely a scaffold, opens the way for novel and exciting dendrimeric devices.

In this work, we present two related concepts of dendrimer use which have evolved in recent years. Both conceal the multivalency and the active framework features, although in so called 'opposite directions'. The first concept, which is introduced in section 2, is the idea to use of dendrimers as light harvesting antennae.[7] Here, the active framework serves as an energy funnel, which directs excitation energy from donors placed at the periphery of the dendrimer to its core, where an acceptor molecule is placed. We describe the process as a random walk of excitations on the dendrimeric framework, and present character-ization of the efficiency of such possible antennae (see Figure 4).

The second concept is a recently reported application of dendrimers as molecular amplifiers.[5,8,9] Here the dendrimeric skeleton is constructed in such a way that it can disintegrate into its subunits, releasing all of its functional end-groups as a result of a signal initiated at the core of the

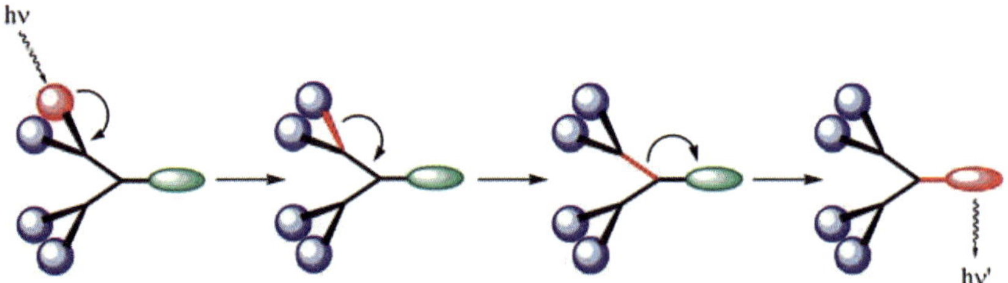

Fig. 4. Schematic illustration of a dendrimer as antennae. The excitation energy *hv* (shown in red) migrates along the dendrimeric framework until it reaches the core where the energy is used for some purpose such as a chemical reaction.

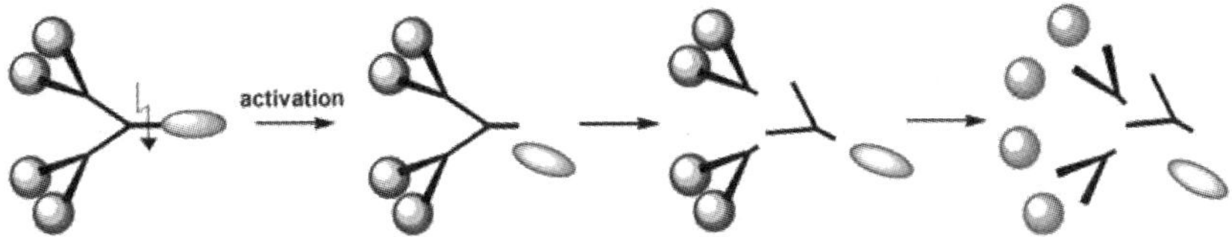

Fig. 5. Dendrimer as an amplifier. Upon trigger activation the dendrimer framework undergoes a spontaneous disassembly to release its end-groups.

dendrimer. The amplification is established due to the exponential increase of the number of subunits along the dendrimer generations (see Figure 5). It has been suggested that when drug molecules are attached as end groups at the periphery, the dendrimer can be used as an efficient drug delivery platform. We address this concept in section 3.

2. DENDRIMERS AS LIGHT HARVESTING ANTENNAE

In recent years efforts have been devoted to create and manipulate molecular scale systems that can be used as efficient light harvesting antennae.[7] The main motivation for studying these systems can be attributed to the desire to mimic similar biological systems.[10] Dendrimers, due their special architecture and features, have been proposed as candidates for serving as antennae. Yet, an issue for debate is whether the energy from the periphery, where chromophores are placed, is being transferred through space directly to the dendrimeric core, or alternatively, through the bonds of the dendrimeric framework, thus using the unique structure of the molecule, as shown in Figure 4. The main argument for direct energy transfer relies on the fact that most dendrimers are synthesized through *meta*-position branching that prevents resonative conjugation among the benzene rings, and leads to localization of the π-electron excitations.[11] However, it has been demonstrated that by designing dendrimers with varying generation lengths, one can create an energetic funnel that is directed towards the dendrimeric core.[11] Such dendrimers are termed *extended*, in contrast to compact dendrimers for which the length between generations is fixed, see Figure 6. In particular, for the extended dendrimers the length between generations decreases for distant core generations, see Figure 7. Building dendrimers in such a way that utilizes the dendritic framework, may result in creating an energetic funnel from that framework; namely the energy of generation G depends on its distance from the core.[11] More recently dendrimers that are branched through *meta*-position and *para*-

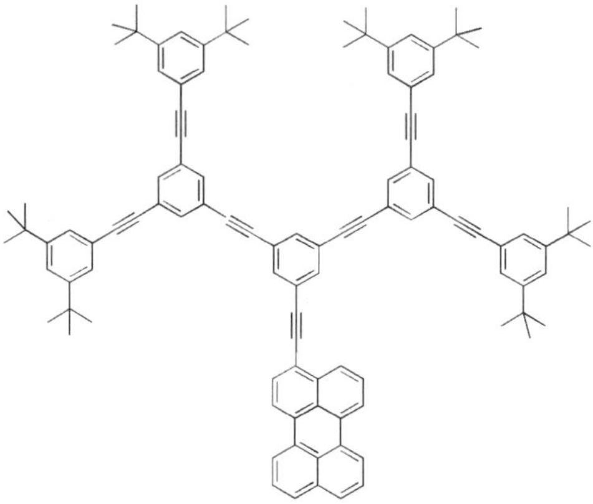

Fig. 6. Compact dendrimer, see Ref. 11 and references therein. The length between generations is fixed.

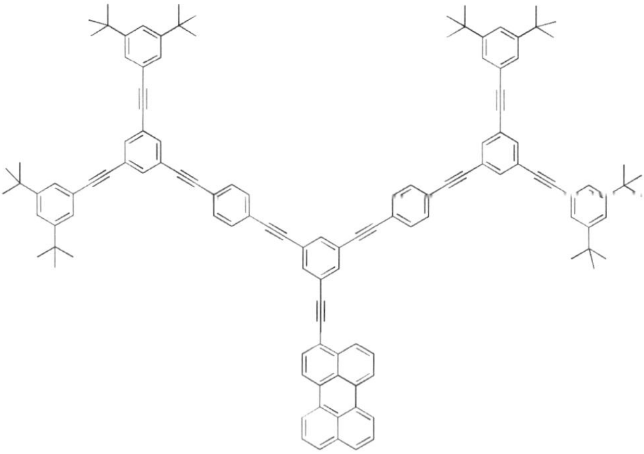

Fig. 7. Extended dendrimer, see Ref. 11 and references therein. The length between the generations changes as a function of the generation index.

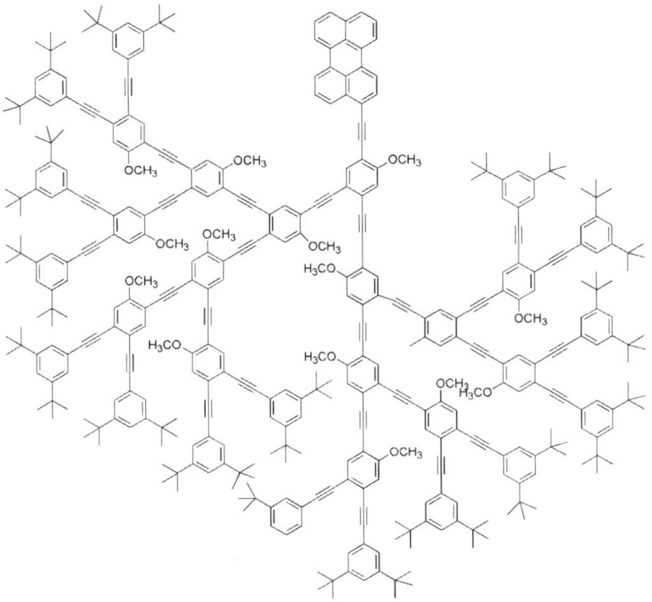

Fig. 8. *Para* dendrimer, see Ref. 12 for details.

position have been synthesized, see Figure 8,[12] thus opening another possibility for producing antenna systems.

In this section we present a theory that models a dendrimeric antenna assuming that the excitation energy migrates along the molecular bonds. In the first subsection we formulate the model, which is then studied in details in subsections 2.2 through 2.5. The presented model is general and can be applied to other systems as well, see examples in Refs. 13, 14. Nevertheless throughout this section relationships to dendrimers are emphasized, and in particular subsection 2.6 is dedicated to the thermodynamic aspects of dendrimeric antennae.

2.1 Formulation of the Model

The dynamics of an excitation (signal) spreading over a dendrimer can be described by mapping the problem on a one-dimensional system, as we are interested only in the distance of the excitation from the core, and not its exact position within a generation, since there are equivalent locations at each generation. According to this mapping, the time evolution of the signal can be modeled by a set of coupled kinetic equations with an absorbing and reflecting boundary conditions:

$$\frac{\partial}{\partial t} P_0(t) = a_1^f P_1(t) \text{ , absorbing site} \tag{1a}$$

$$\frac{\partial}{\partial t} P_1(t) = -(a_1^f + a_1^b) \cdot P_1(t) + a_2^f \cdot P_2(t) \text{ ,} \tag{1b}$$

$$\frac{\partial}{\partial t} P_m(t) = a_{m-1}^b P_{m-1}(t) - (a_m^f + a_m^b) \cdot P_m(t) + a_{m+1}^f \cdot P_{m+1}(t) \text{ ;}$$

$$2 \leq m \leq n-1 \text{ ,} \tag{1c}$$

$$\frac{\partial}{\partial t} P_n(t) = a_{n-1}^b \cdot P_{n-1}(t) - a_n^f \cdot P_n(t) \text{ , reflecting site.} \tag{1d}$$

The meaning of these boundary conditions is that, due to the reflecting boundary at the periphery (site $j = n$), the signal stays in the system as long as it does not reach the absorbing site, $j = 0$. Once the signal reaches the absorbing site it is captured there. The other symbols appearing in Eqs. (1a)-(1d) are defined as follows: $P_j(t)$ is the probability density function (PDF) that the signal occupies site j at time t, and a_j^f and a_j^b are the transition rates from site j to site j-1 and j+1, respectively. Figure 9 presents a schematic illustration of the system. The coupled equations (1a)-(1d) are also referred to as the system's master equation.

The site index j represents the jth generation of the dendrimer. The dependence of the transition rates on j should reflect both the exponential branching growth of the dendrimer end-groups with an increase in the generation number, which can be viewed as an entropic bias towards the periphery,[7] and a possible energetic funnel towards the core. We elaborate on this issue in subsection 2.6.

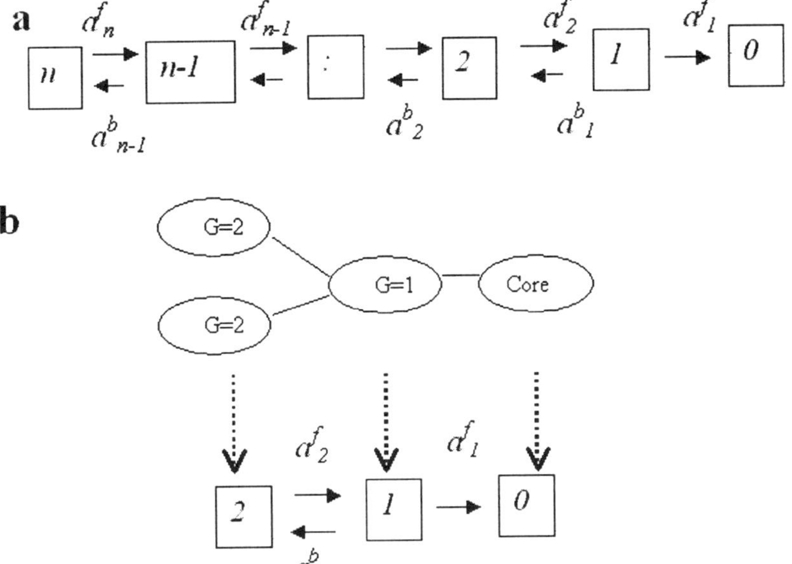

Fig. 9. (*a*) Schematic illustration of the escape process from a system with *n* sites. Once the signal reaches site $j = 0$ it is absorbed, namely site $j = 0$ is a trap. (*b*)The mapping of a two-generation dendrimer onto a one-dimensional system. See text for discussion.

It is convenient to write the master equation expressed in equations (1a)-(1d) in a matrix representation:

$$\frac{\partial}{\partial t}\vec{P}(t) = \mathbf{A}\vec{P}(t), \tag{2}$$

which has the formal solution;

$$\vec{P}(t) = \exp(\mathbf{A}t)\vec{P}_0. \tag{3}$$

Here,

$$\vec{P}(t) = \begin{pmatrix} P_1(t) \\ P_2(t) \\ \vdots \\ P_n(t) \end{pmatrix},$$

and,

$$\mathbf{A} = \begin{pmatrix} -a_1^f - a_1^b & a_2^f & 0 & \vdots & 0 \\ a_1^b & -a_2^f - a_2^b & a_3^f & \vdots & 0 \\ \vdots & \vdots & \vdots & \vdots & \vdots \\ 0 & \vdots & a_{n-2}^b & -a_{n-1}^f - a_{n-1}^b & a_n^f \\ 0 & \vdots & 0 & a_{n-1}^b & -a_n^f \end{pmatrix},$$

where the jth element of the initial condition vector is given by $\left(\vec{P}_0\right)_j = \delta_{j,x}$, namely the process starts at site (generation) x with probability one. The propagation matrix $\mathbf{A}$ is a tridiagonal n-dimensional square matrix (the absorbing site is not included in the propagation matrix) that contains information about the time evolution of the signal, in term of transition rates.

Due to the normalization condition we have

$$P_0(t) + \sum_{j=1}^{n} P_j(t) = 1 ,$$

where $P_0(t)$ is the PDF to occupy the trap as a function of the time. We further define

$$S(t) = \sum_{j=1}^{n} P_j(t) ,$$

as the survival probability, which is the PDF that the signal has not been absorbed, *i.e.* the signal has survived within the system. Note that $P_0(t)$ increases monotonically with time, while $S(t)$ decreases monotonically with time. Both functions depend on the initial condition x. Accordingly, from this point on, we add the initial condition x as an additional argument of the survival probability, $S(t,x)$, and functions that are derived from it. A function that plays a central role in the theory of random walks in finite and semi-infinite intervals, and is of great importance for possible applications, is the first passage time PDF $\Phi(t,x)$,[15] defined by

$$\Phi(t,x) = \frac{\partial}{\partial t}[1 - S(t,x)] , \tag{4}$$

which is equivalent to Eq. (1a); namely, the first passage time PDF is the rate of change of the trap occupation probability. From Eq. (3) and from the definition of the survival probability we realize that, in order to obtain the survival probability, one should sum up the elements of the vector that solves the master equation:

$$S(t,x) = \vec{U}_n e^{\mathbf{A}t} \vec{P}_0 , \tag{5}$$

where $\vec{U}_n = (1,1,:,1)$ is the summation row vector on n dimensions. By diagonalizing matrix $\mathbf{A}$ we rewrite Eq. (5) as:

$$S(t,x) = \sum_{i=1}^{n} b_{i,x} e^{\lambda_i t} , \tag{6}$$

where the $b_{i,x}$'s are given by: $b_{i,x} = c_{i,x}^{-1} \sum_{j=1}^{n} c_{j,i}$. Here, $c_{i,j}$ and $c_{i,j}^{-1}$ are the ith row, j^{th} column elements of the matrix $\mathbf{C}$ and its inverse $\mathbf{C}^{-1}$, respectively. These matrices are the eigenvector matrix and its inverse of matrix $\mathbf{A}$, by which the eigenvalue matrix λ,

$$\lambda = \begin{pmatrix} \lambda_1 & 0 & 0 & : & 0 \\ 0 & \lambda_2 & 0 & : & 0 \\ : & : & : & : & : \\ 0 & : & 0 & \lambda_{n-1} & 0 \\ 0 & : & 0 & 0 & \lambda_n \end{pmatrix} ,$$

is obtained through the similarity transformation, $\lambda = \mathbf{C}^{-1}\mathbf{A}\mathbf{C}$.

Substituting Eq. (6) into Eq. (4) we get:

$$\Phi(t,x) = -\sum_{j=1}^{n} \lambda_i b_{i,x} e^{\lambda_i t} , \tag{7}$$

from which numerical results can be obtained for any choice of the transition rates. Thus, by comparing numerical results to experimental results, one can extract the system transition rates that appear in Eqs. (1a) - (1d). We note that in a recent work,[16] the first passage times PDF was studied and related to dendrimeric antennae although by a different approach for which matrix $\mathbf{A}$ is specified. The importance of computing the PDF of the first passage times and not just its moments is emphasized, for example, by single molecule experiments where the PDF is measured directly. Such experiments involving individual dendrimers were reported recently.[17]

2.2 The First Passage Times PDF

Below we show numerical solutions of Eq. (7) for several invariant systems that are of interest for dendrimers. By an invariant system we mean that the transition rates are taken to be independent of the site (generation) index j, namely, $a_j^b = k_-$ and $a_j^f = k_+$ for all j. For these systems the ratio between the forward and the backward transition rates, $Q \equiv k_- / k_+$, is the important parameter. We distinguish between three cases, $Q \ll 1$, $Q = 1$ and $Q \gg 1$:

- The first case, $Q \ll 1$, represents a system that displays a bias towards the absorbing site. Such a system can describe a dendrimer that has a large energetic bias towards its core,[7,18] or a dendrimer that irreversibly dissociates as the signal propagate across the molecule,[5,8,9] to be discussed in the next section.
- The second case, $Q = 1$, describes bias-free dynamics, which means in the dendrimeric antennae case that the entropic bias is canceled by the energetic bias.
- The third case, $Q \gg 1$, represents an escape process against a constant force, which for our purposes is translated into a dendrimer that experiences an energetic bias towards the periphery.

Figure 10a shows $\Phi(\tau, n)$ for an invariant system and for which $Q = 0.01$, as a function of the dimensionless time $\tau \equiv tk_+$ for several system sizes, $n = 2, 3, 4, 6$, and $x = n$, namely, the process starts at the reflecting site (the periphery). $\Phi(\tau, n)$ is mono-peaked for all n values and decays exponentially at large times, which is easily seen from Eq. (7). Figure 10b shows three characteristics of a density function, here for $\Phi(\tau, 6)$. These are the mean time $\langle \tau \rangle$, the standard deviation

$\sigma = \sqrt{\langle \tau^2 \rangle - \langle \tau \rangle^2}$, and their ratio $R = \sigma / \langle \tau \rangle$, also known as the relative error of a PDF. The relative error is an important characteristic that gives a normalized measure for a spread of a density function, and is discussed further in subsection 2.4. In Figs. 10c - 10d we show the dependence of these characteristic behaviors on n. Note the linear scaling of $< \tau(n) >$ with n, and the $\sqrt{1/n}$ scaling of $R(n)$, for $Q \ll 1$.

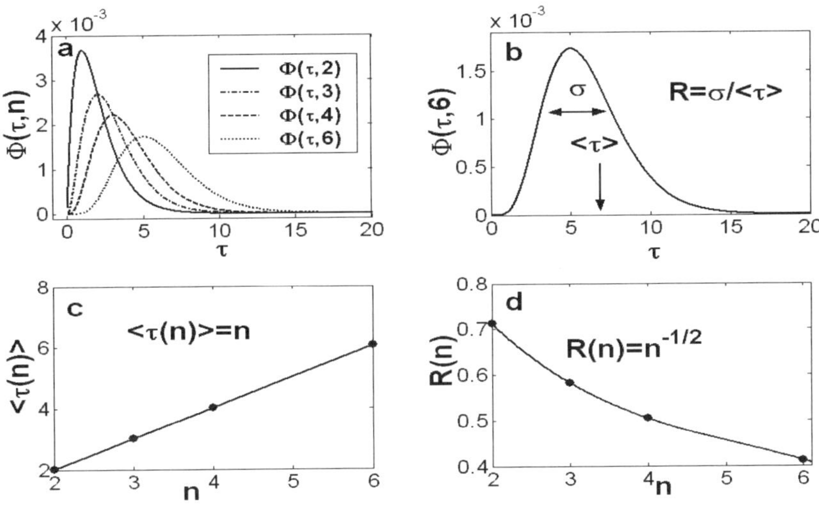

Fig. 10. (*a*) $\Phi(\tau,n)$ for $Q = 0.01$, and values of $n = 2$, 3, 4, and 6 corresponding to the full, dotted-dashed, dashed, and dotted lines, respectively. (*b*) The parameters R, σ, and τ for $\Phi(\tau,6)$. (*c*) The mean of the density functions displays linear scaling with n. (*d*) The relative error R decays as $1/\sqrt{n}$.

In Figure 11a we compare $\Phi(\tau,4)$ of an invariant *symmetric* system, namely $Q = 1$, to a trap-oriented ($Q \ll 1$) system with $Q = 0.01$. Although the location of the peak of the PDFs is similar for both cases, $\Phi(\tau,4)$ for the symmetric case is broader. This is also reflected in the slower saturation of the cumulative probability function of the first passage times PDF $G(\tau,n) = \int_0^\tau \Phi(s,n)ds$, shown in the inset of Figure

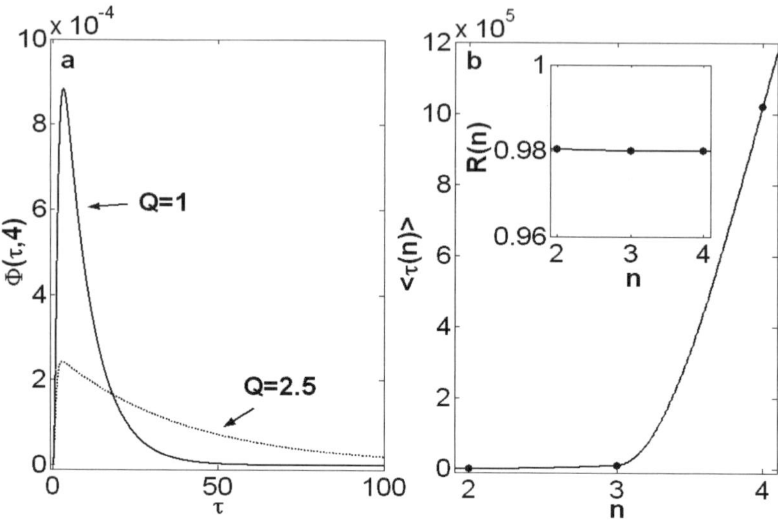

Fig. 11. (*a*) $\Phi(\tau,4)$ and $G(\tau,4)$ (inset) for two values of Q; $Q = 0.01$ (full curves), and $Q = 1$ (dashed curves). (*b*) The mean for the symmetric case scales as n^2, and R reaches its asymptotic value [$= \sqrt{2/3}$, see Eq. (19)] for considerably small systems (inset).

11a for $n = 4$. For some applications, $G(\tau,n)$ is the direct information one gets from experiments,[19] and can be used to obtain the mean of $\Phi(\tau,n)$ by performing the time integral of $1 - G(\tau,n)$, see Eq. (9) in the next subsection. For the symmetric case the mean of the PDF scales as n^2 (Figure 11b), while the relative error R is independent of n for large systems (inset of Figure 11b), in contrast to the trap-biased system.

Finally, as shown in Figure 12a for an invariant system with a bias towards the periphery, and set $Q = 2.5$, $\Phi(\tau,4)$ decays slowly relative to its symmetric invariant counterpart. Note that as Q increase the smallest absolute eigenvalue, $|\lambda_{\min}|$, dominates the PDF behavior, which leads to the relation $\langle \tau \rangle \approx 1/|\lambda_{\min}|$. Figure 12b shows that for this case the mean of the PDF grows exponentially with the system size, $\langle \tau \rangle \propto e^{n\log Q}$, and the relative error is approximately unity (inset).

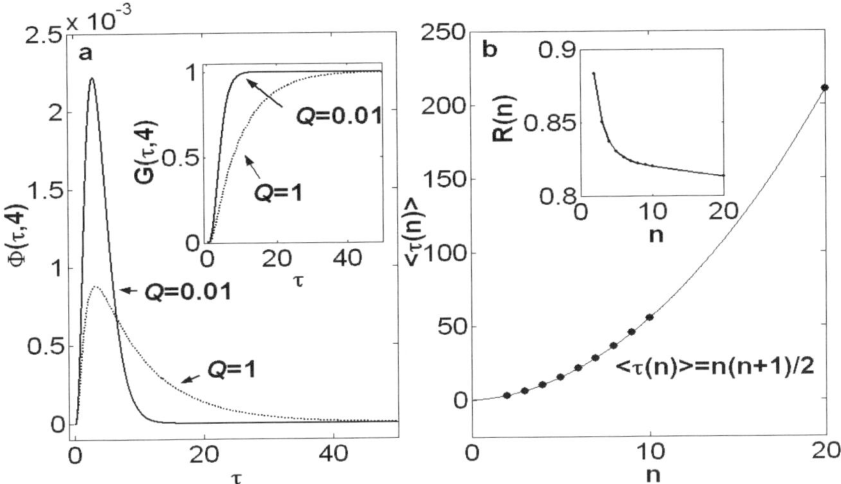

Fig. 12. (*a*) The slowly decaying $\Phi(\tau,4)$ for $Q = 2.5$ in comparison to the symmetric case. (*b*) The characteristics of the PDF for $Q = 50$ are $<\tau(n)> \approx e^{n\log Q}$, and $R \approx 1$ (inset).

In the next subsections we derive analytical expressions for the first two moments of invariant systems. We show that the moments' behavior presented here for a given system length holds for any system size.

2.3 The Mean First Passage Time

The characteristics of a PDF are its moments and their interrelations. The $m > 0$ moment of the first passage times PDF is defined by:

$$\left\langle t^m(x) \right\rangle = \int_0^\infty t^m \Phi(t,x)dt = m\int_0^\infty t^{m-1} S(t,x)dt \; , \tag{8}$$

where the second equality is obtained by integration by parts, using Eq. (4), and noticing that the boundary term vanishes due to the fact that the survival probability is zero at infinite time. Substituting $m = 1$ in Eq. (8), the mean first passage time (MFPT) is obtained:

$$\left\langle t(x) \right\rangle = \int_0^\infty S(t,x)dt = -\vec{U}_n \mathbf{A}^{-1} \vec{P}_0 \; , \tag{9}$$

where the second equality is obtained by using Eq. (5). Thus, one needs to invert the matrix $\mathbf{A}$ to calculate the MFPT. Note that as long as all the forward transition rates are finite, matrix $\mathbf{A}^{-1}$ exists. Equation (9) can be written as

$$\langle t(x) \rangle = -\sum_{j=1}^{n} \mathbf{A}^{-1}_{j,x} \; ,$$

where the element $-\mathbf{A}^{-1}_{j,x}$ defines the mean residence time (MRT) on site j when starting at site x, before trapping occurs.[20] We shall denote $-\mathbf{A}^{-1}_{j,x} \equiv t_{j,x}$. By straightforward calculations the MRTs are found to be:[20]

$$t_{1,x} = \frac{1}{a_1^f} \qquad\qquad j = 1 \quad (10a)$$

$$t_{j,x} = t_{j-1,x}\frac{a_{j-1}^b}{a_j^f} + \frac{1}{a_j^f} \qquad\qquad 1 < j \leq x \quad (10b)$$

$$t_{j,x} = t_{j-1,x}\frac{a_{j-1}^b}{a_j^f} \qquad\qquad x < j \; . \, (10c)$$

Equation (10a) shows that the MRT of the first site (generation), *i.e.* the site (generation) which is the closest to the absorbing site, is independent of the initial state x. Equation (10b) implies that for $x \geq j$ the MRT $t_{j,x}$ is the same as the MRT $t_{j,x+y}$ for all y fulfilling $x + y < n$. It then follows that the MFPT for starting on the last site, *i. e.* at the reflecting site (periphery), is given by the trace of $-\mathbf{A}^{-1}$, $<t(n)> = -Tr(\mathbf{A}^{-1})$. From Eqs. (10a) - (10c) a general expression for the MFPT is valid. As a function of the initial condition this expression should contain two terms that reflect the recursion relations given by Eq. (10b) and Eq. (10c). After several manipulations the expression for the MFPT is;

$$\langle t(x) \rangle = \sum_{j=1}^{n} \omega_j + \sum_{j=1}^{x-1} \omega_j^{-1} (a_j^b)^{-1} \sum_{s=j+1}^{n} \omega_s , \qquad (11)$$

where $\omega_j = \dfrac{1}{a_j^f} \prod_{i=1}^{j-1} \rho_i$, $\rho_i = \dfrac{a_i^b}{a_i^f}$, and $\omega_1 = 1$. The first term in Eq. (11)

is the MFPT for starting at the first site, as is easily seen from Eq. (10c).

From Eq. (11) the MFPT for several special cases can be obtained, thus providing a clearer understanding regarding the dependence of the MFPT on the system size, and the transition rates. As discussed in the previous subsection, when computing the first passage times PDF, the cases which are of interest to dendrimers require invariant system, namely $a_j^b = k_-$ and $a_j^f = k_+$ for all generations j. For these cases, the MFPT reads:

$$\langle t(x) \rangle = \frac{1}{\Delta k}\left[x + \frac{Q^{n+1}}{1-Q}(1 - Q^{-x}) \right] \ \text{ for } \ Q \equiv \frac{k_-}{k_+} \neq 1 \quad (12\text{a})$$

$$\langle t(x) \rangle = \frac{x(2n+1-x)}{2k} \qquad \text{ for } k_- = k_+ \equiv k \quad (12\text{b})$$

where $\Delta k = k_+ - k_-$.

For large biases, $Q \ll 1$ and $Q \gg 1$, Eq. (12a) reduces to

$$< t(x) >= \begin{cases} x/\Delta k & ; & Q \ll 1 \\ e^{(n+1)\ln Q}(1 - Q^{-x})/[(1-Q)\Delta k]; & Q \gg 1 \end{cases} \quad (12\text{c})$$

Equations (12b) - (12c) demonstrate the dependence of the MFPT on the system size. For a system that displays a bias towards the center of the dendrimer, the average time to be trapped scales linearly with the initial site of the process, x. For a system which is biased towards the periphery, an exponential dependence of the MFPT with the system size is exposed, regardless of the initial site of the process. For a system with no bias at all, namely $k_- = k_+$, the scaling of the MFPT with the size of the system depends on the initial site; when $x = 1$ a linear scaling with the system size is evident, while a square scaling with the system size is obtained for $x = n$.

For the special case of an invariant system and $k_- \to 0$ (we term this system a "death" system) a general expression for the mth moment is valid. For starting at the reflecting site $x = n$, we have:

$$< t^m (n) >= \frac{1}{k^m} \frac{(n+m-1)!}{(n-1)!} \ . \tag{13}$$

Equation (13) gives a full characterization of first passage time PDF of a "death" system. This system suits to describe a dendritic amplifier discussed in section 3, where the signal spreads irreversibly and causes dissociation of the dendrimer from the core to the periphery to release reporter molecules from the end-groups.

We note that the results in this subsection give a measure for the efficiency of dendrimeric antennae when assuming that the excitation energy transfers through bonds and that multiple excitations don't play a role. However for multi-excitation system the MFPT of the first excitation to reach the core is shorter than the MFPT presented here, and depends on the number of excitation the process starts with. [21]

2.4 The Second Moment of the First Passage Times PDF

The second moment of a PDF provides information about its spread. For our model, the expression for the second moment $\langle t^2 (x) \rangle$ reads;

$$\langle t^2 (x) \rangle = 2 \vec{U}_n \mathbf{A}^{-1} \mathbf{A}^{-1} \vec{P}_0 \ , \tag{14}$$

which is obtained by using Eqs. (5) and (8). Rewriting Eq. (14) as

$$\langle t^2 (x) \rangle = 2 \sum_{j=1}^{n} \langle t(j) \rangle t_{j,x} \ , \tag{15}$$

the second moment can be calculated for an arbitrary choice of the transition rates by using Eqs. (10a) - (10c), and Eq. (11). As mentioned, the cases of interest for dendrimeric antennae suggest invariant systems. By straightforward calculations we obtain the second moment for an invariant system and for $Q \neq 1$,

$$\left\langle t^2(n) \right\rangle = \left(\frac{\sqrt{2}}{\Delta a}\right)^2 \left[\frac{n(n+1)}{2} + \frac{Q^{n+1}(2n+1) - Q^{n+2}(3n-1) + Q^{2n+2} - 2Q}{(1-Q)^2}\right],$$

(16)

which asymptotically reduces to,

$$\left\langle t^2(n) \right\rangle = \begin{cases} (n+1)n/2 & ; \qquad Q \ll 1 \\ 2Q^{2n+2}/[\Delta a(1-Q)]^2 & ; \qquad Q \gg 1 \end{cases}$$

(17)

For the symmetric invariant system $Q=1$, we get

$$< t^2(n) >= \frac{n(n+1)}{12k^2}(5n^2 - 5n + 2) \ .$$

(18)

The relative error, introduced in subsection 2.1, is an essential parameter in statistics. The relative error is the ratio between the standard deviation, σ, and the average of a density function, $R(x) = \sigma(x)/ < t(x) >$, where $\sigma(x) = \sqrt{< t^2(x) > - < t(x) >^2}$, here depending on the initial site. This ratio gives a measure for the spread of a density function regardless of its argument values. For a sufficiently large system, we get for invariant systems:

$$R(n) = \begin{cases} \sqrt{1/n} & ; \qquad Q \ll 1 \\ \sqrt{2/3} & ; \qquad Q = 1 \\ 1 & ; \qquad Q \gg 1. \end{cases}$$

(19)

Equation (19) can be compared with Figures (8) - (10). Note that only a trap-oriented system exhibits the desirable behavior of $R(n)$, which vanishes (as $1/\sqrt{n}$) for large systems.

2.5 Equilibrium Distribution

Important characteristics of physical systems are derived from equilibrium distributions. Although for the system shown in Figure (7) no equilibrium exists, because at infinite time the occupation probability at each site (but the trap) is zero, for a system for which $a_1^f = 0$, a nontrivial equilibrium distribution exists. To find the equilibrium distribution for such a system, one should solve the corresponding equilibrium equation,

$$\vec{0} = (\mathbf{A} + \mathbf{T})\vec{P}_{eq} \,, \tag{20}$$

where $\vec{0}$ is the null vector of n dimensions, $\mathbf{A}$ is given earlier under Eq. (3), and

$$\mathbf{T} = \begin{pmatrix} a_1^f & : & 0 \\ : & : & 0 \\ 0 & 0 & 0 \end{pmatrix}.$$

The first column of $\mathbf{A}^{-1}$ solves Eq. (20).[7] To see this we first use $\mathbf{A}\mathbf{A}^{-1} = \mathbf{I}$, where $\mathbf{I}$ is the unit matrix, and then $(\mathbf{T}\vec{v})_j = \delta_{j,1} a_1^f v_1$, for any arbitrary vector $\vec{v}$ of the appropriate dimensions. Using Eq. (12a) [with a minus sign due to Eq. (9)] completes the proof. To retain the meaning of probability we need to preserve normalization, thus obtaining a general expression for the occupation probability of site j at equilibrium $P_{eq,j}$;

$$(\vec{P}_{eq})_j \equiv P_{eq,j} = t_{j,1} / <t(1)> \,. \tag{21}$$

Using Eqs. (10a), (10c), (12a) - (12c), we find from Eq. (21) that for invariant systems $P_{eq,j}$ obeys

$$P_{eq,j} = \begin{cases} (1-Q)Q^{j-1} & ; & Q \ll 1 \\ 1/n & ; & Q = 1 \\ (Q-1)Q^{j-1-n} & ; & Q \gg 1 \end{cases} \tag{22}$$

Note that for a symmetric system the equilibrium distribution is uniform, whereas for biased systems the equilibrium distribution tends to accumulate in the corresponding ends of the interval, *i.e.* at the reflecting site for $Q \gg 1$, and at the first site for $Q \ll 1$. Furthermore, for both biased cases the equilibrium distribution decays exponentially with the distance from the corresponding end sites.

2.6 Application to Dendrimers

In this subsection we start by describing thermodynamically a dendrimeric antenna,[22] which leads to explicit expressions of the transition rates that appear in Figure 9, up to one undetermined rate. We show in Figure 13 a schematic illustration of a dendrimer, which has a $C = 2$ core branching, and the branching parameter $z = 2$.

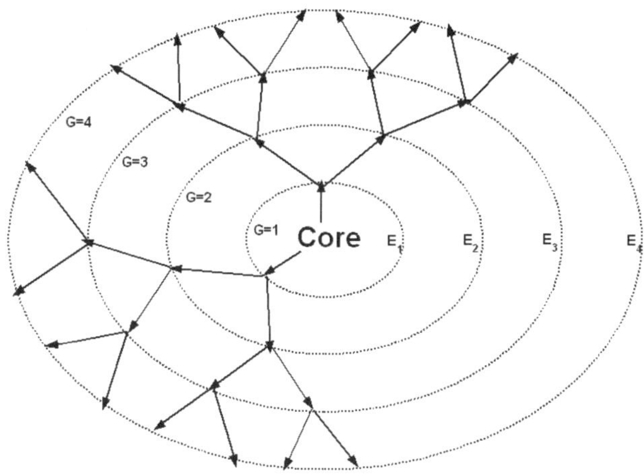

Fig. 13. Schematic illustration of a four-generation dendrimer with $z = 2$ and $C = 2$. Also shown are the energies of each generation, $E_4 > E_3 > E_2 > E_1$.

We consider an excitation that migrates on a dendrimer by nearest neighbor jumps. The core is assumed to capture the energy for some time, depending on its release rate. Namely, the core functions as a reversible trap. Energy levels are assigned for each generation, such that a funnel towards the core is created

$$
\begin{aligned}
E_0 &= \varepsilon_0 \\
E_G &= \varepsilon_0 + \varepsilon_1 + (G-1)U \; ; \; 1 < G \le n,
\end{aligned}
\tag{23}
$$

where ε_0 is the core excitation energy, ε_1 is the excitation energy difference between the core excitation energy and the first generation excitation energy, and U is the excitation energy difference between each nearest neighbors generations. Figure 14 shows the excitation energy as a function of the generation index. The excitation energy levels descend from the periphery to the core, which creates an energetic funnel.

The structure of the dendrimers that has z branching towards the periphery, and a core branching C, leads to the degeneracy

$$f_G = Cz^{G-1} . \tag{24}$$

Having assigned excitation energies and degeneracy to each of the generations, the partition function of the dendrimer is

$$Z = e^{-\varepsilon\beta} + \sum_{G=1}^{n} f_G e^{-E_G\beta} = e^{-\varepsilon\beta} + Ce^{-(\varepsilon+\varepsilon_1)\beta} \sum_{G=1}^{n} [ze^{-U\beta}]^{G-1} , \tag{25}$$

where $\beta^{-1} = k_B T$, k_B is the Boltzmann constant and T is the temperature. The equilibrium occupation probabilities of the various generations are

$$\begin{aligned} P_{0,eq} &= e^{-\varepsilon\beta} / Z \\ P_{G,eq} &= P_{0,eq} Ce^{-\varepsilon_1\beta} [ze^{-U\beta}]^{G-1} \end{aligned} \tag{26}$$

from which the free energy of each generation follows,

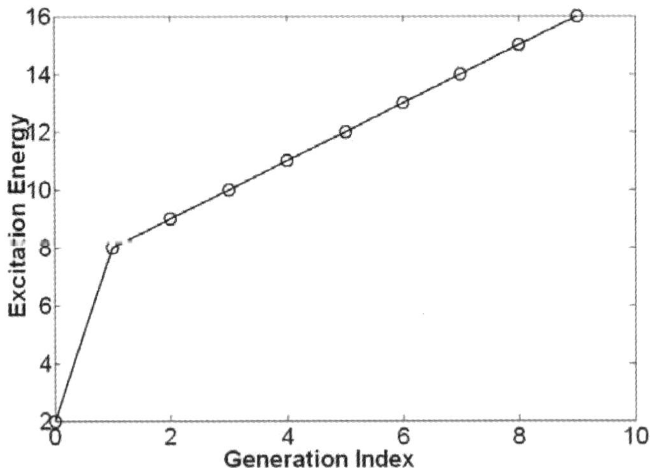

Fig. 14 The excitation energy as a function of the generation index. Here $\varepsilon = 2$, $\varepsilon_1 = 6$, $U = 1$, and $n = 9$. The first point corresponds to the core (trap).

$$F_G = -\beta^{-1}\ln(P_{G,eq}Q) = E_G - \beta^{-1}\ln(f_G) \ . \tag{27}$$

Figure 15 shows the free energy as a function of the generation index for three values of the parameter $\Delta = \ln(z)/(U\beta)$, $\Delta = 0,1,2$. For $\Delta = 0$ an energetic funnel towards the trap, similar to that shown in Figure 14, is created. For $\Delta = 1$ there is no energetic preference to be at a specific generation (excluding the trap), whereas for $\Delta = 2$ an energetic preference towards the periphery exists. Namely at high temperatures the energetic funnel becomes less efficient relative to the geometric one.

Now, we wish to translate the energetic picture into the dynamical model by obtaining expressions for the transition rates from the thermodynamic picture. To do so, we used the equilibrium condition;[22,23]

$$P_{eq,G}a_G^b = P_{eq,G+1}a_{G+1}^f \ . \tag{28}$$

From Eqs. (26) and (28) it follows that the transition rates are given by

$$\begin{aligned} k_- &= k_+ z e^{-U\beta} \\ a_0^b &= k_+ C e^{-\varepsilon_1\beta} \end{aligned} \tag{29}$$

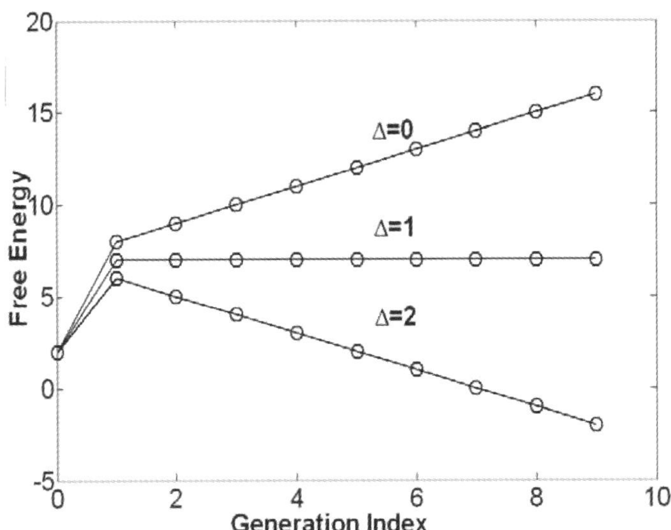

Fig 15 The free energy as a function of the generation number, for three different values of Δ, $\Delta = 0, 1, 2$. Here $z = 2$ and $C = 2$.

and k_+ is arbitrary. Note that k_- has two components; the component z of geometric origin, whereas $e^{-U\beta}$ emerges from energetic considerations. Taking the limit $\varepsilon_1 \beta \gg 1$ in Eq. (29) leads to $a_0^b = 0$, we thus recover the description of an irreversible trap shown in Figure 7. Accordingly, the expression for the moments of the first passage times PDF, Eqs. (11) - (13) and (16) - (18), can be written now in terms of the energetic model described in this subsection.

3. DENDRIMERIC AMPLIFIERS

In recent years much effort has been devoted to the design of novel drug delivery platforms.[24] The need to increase the drug localization at desired tissues has made dendrimers a preferred platform for drug delivery.[25,26] These macromolecules, due to their unique structure and properties, were used as multi-drug carriers, allowing the delivery of a large number drug molecules to target cells.[27] The drug molecules are attached to the dendrimer periphery or are encapsulated inside the cavities in the structure. Usually, each bond cleavage generates the release of a single drug molecule.[28] This strategy is limited in use when the activating agent, an enzyme for example, is present in small quantities. The release of active drug units from the dendrimer is slow and insufficient.

Dendrimers that are able to release all of their end-groups upon a single cleavage event could have a clear advantage over the existing dendrimeric drug delivery systems. Recently the design and synthesis of novel molecular amplifiers, termed self-immolative dendrimers (SIDs) was reported, almost simultaneously by three independent groups.[5,8,9] SIDs are a novel class of dendrimers capable of amplifying an external signal. The new dendrimeric structure is based on the idea of a molecular amplifier, which serves both as a core and as a branching unit. The molecular amplifier is constructed of three functional groups, one input port that is attached to a trigger and two output ports capable of generating bond cleavage. An input signal could be either a chemical or biological reaction. Activation of the trigger results in a spontaneous disassembly of the dendrimer framework, thus releasing all the tail groups as an output signal (Figure 16). The process can be formulated as a particular case (Q $\ll$ 1) of the master equation discussed in section 2.

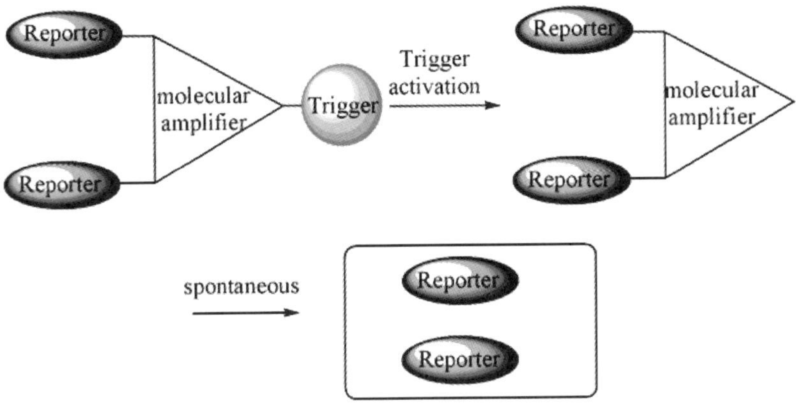

Fig. 16. General structure of G1 self-immolative dendrimer.

SIDs attract growing attention in the scientific community,[29,30,31] most likely, since they can serve as novel drug delivery platforms. They can be tailored for specific requirements, depending on the activating agents and the pharmaceutical activity, by incorporating the desired trigger and drugs. A G2-SID can be prepared by attaching the input port of two G1 units to the output ports of an additional G1 unit with a trigger. In this case the trigger cleavage initiates a cascade of spontaneous fragmentation that result in the release of four reporter units (Figure 17).

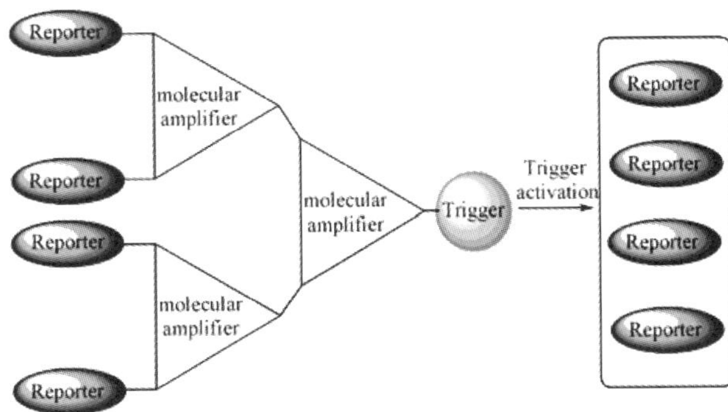

Fig. 17. General structure of G2 self-immolative dendrimer

Three related structures were used by Amir *et al.*,[5] de Groot *et al.*,[8] and Szalai *et al.*[9] All of them utilize a series of either a double 1,8,[8] double 1,4[5] or a combination[9] of a 1,4 and a 1,6 quinone methide type eliminations in order to achieve the cleavage of two bonds upon a single triggering event. The amplification mechanism, which describes the disassembly of our SIDs to release the reporter groups, is illustrated in Figure 18.

3.1 The Amplification Mechanism

The amplifier subunit of our dendrimer is based on 2,6-bis (hydroxymethyl)-*p*-cresol, a commercially available compound, which has three functional groups (Figure 18). The two hydroxybenzyl groups are attached through a carbamate linkage to reporter molecules and the phenol functionality is linked to a trigger through a short spacer (N,N'-dimethylethylenediamine) to form a G1 dendrimer. Cleavage of the trigger initiates a sequence of self-immolative reactions of the generated free amine intermediate, starting with spontaneous cyclization to form an N,N'-dimethylurea derivative. The generated phenol then undergoes a 1,4-quinone methide rearrangement, followed by spontaneous decarboxylation to liberate one of the reporter molecules. The quinone methide species is rapidly trapped by a water molecule (from the reaction solvent) to reform a phenol, which again undergoes a 1,4-quinone methide rearrangement to liberate the second reporter molecule. The generated quinone methide species is then trapped by another water molecule to form 2,6-bis (hydroxymethyl)-*p*-cresol.

Higher generations can be synthesized by attaching two G1 subunits to the hydroxybenzyl functionalities of the first molecular amplifier, thus a dendritic structure that holds four reporter molecules is obtained. The attachment is performed through a double carbamate linkage using N,N'-dimethylethylenediamine as a spacer (Figure 19). The carbamate linkage between the units is highly stable until the trigger is cleaved and the self-immolative reaction sequence is initiated. G1 and G2 SIDs (Figure 20) were synthesized, employing 2-nitro 4,5-dimethoxy benzyl alcohol as a photolabile trigger and aminomethylpyrene as reporter unit. The SIDs were activated by UV radiation, ($\lambda = 360$ nm), and the release of

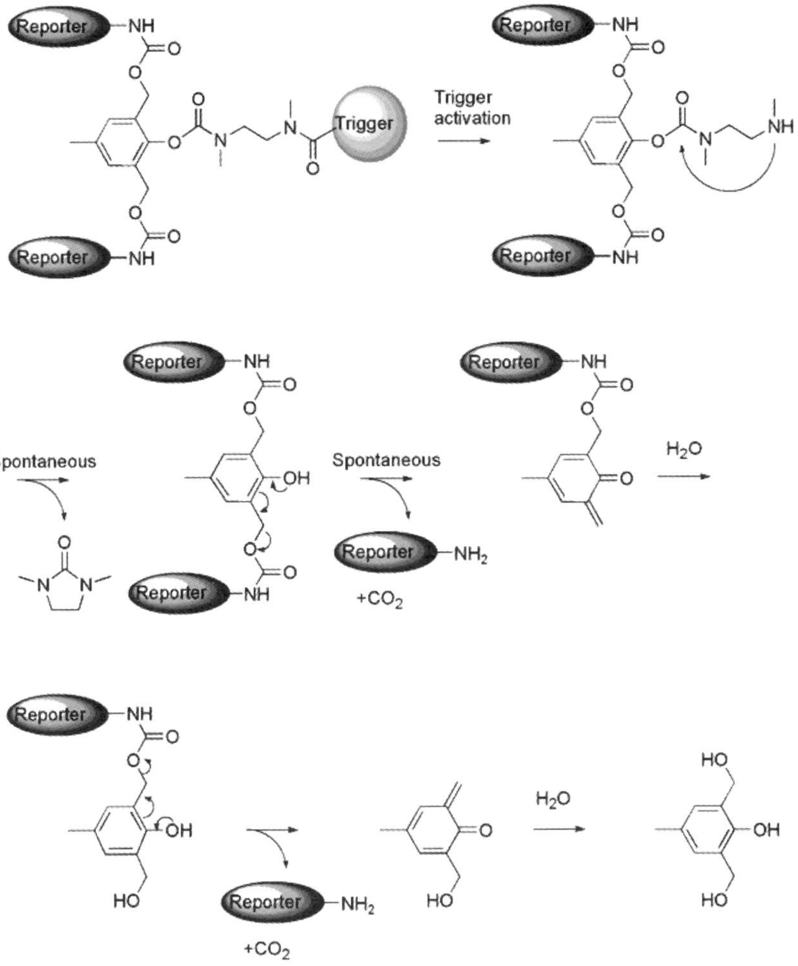

Fig. 18. Mechanism of reporter release from G1 self-immolative dendrimer.

aminomethylpyrene (R) was monitored by HPLC. The dendrimers' platform is constructed of building blocks which, upon receiving a photonic input signal, have the capability to fragmentize through a process of self-immolative chain reactions, based on cyclization (N,N'-dimethylethylenediamine) and elimination (1,4-quinone methide rearrangement) reactions. The only intermediates that were observed in

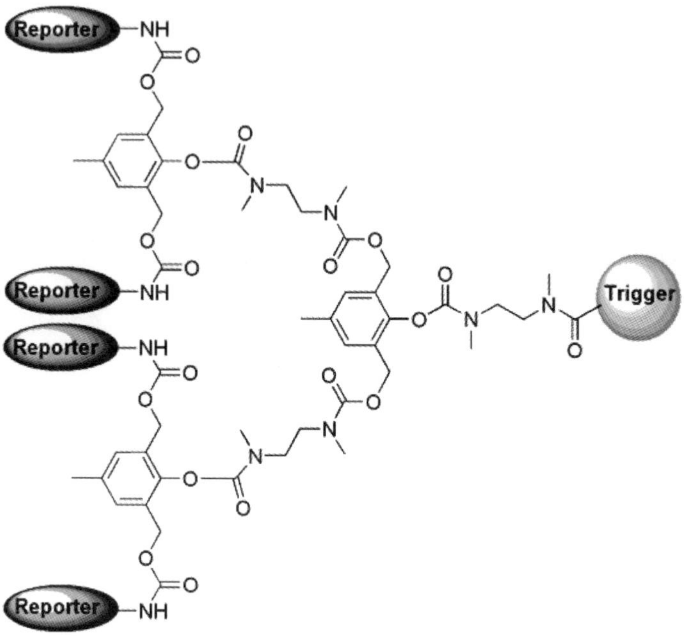

Fig. 19. General chemical structure of a G2 self-immolative dendrimer.

the HPLC chromatograms of the G1 and G2 SIDs were the amine intermediates AI1 and AI2 that are formed after the trigger activation of G1 and G2 SIDs respectively. Therefore, we have assumed that the rate limiting step of both disassembly processes of the G1 and G2 dendrimers is the cyclization of the diamino-spacer in AI1 and AI2, to form a cyclic urea derivative and a free phenol function. The latter then further undergoes a double 1,4 elimination, which serves as the amplifying mechanism of the adaptor subunit.

3.2 Kinetic Calculations

Additional support to the suggested release mechanism of the SIDs, is obtained from kinetic calculations for the G1 and G2 self-immolative dendrimers. Assuming that the self-immolative fragmentation of the G1

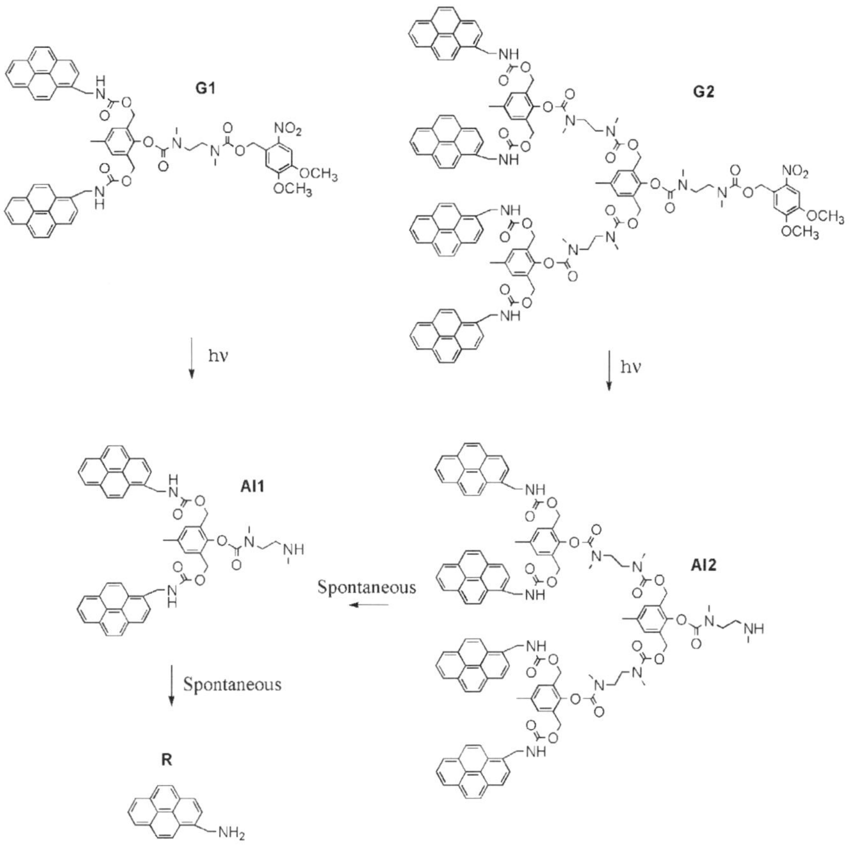

Fig. 20. Photochemical activation of G1, and G2 SIDs to release aminomethylpyrene as reporter units (*R*).

amine-intermediate AI1 follows a first order reaction, we have

$$\frac{\partial}{\partial t} P(\text{AI1}) = -k_+ P(\text{AI1}) \ . \tag{30}$$

Equation (30) is written in terms of a PDF, $P(AI1) = [AI1(t)]/[AI1(t=0)]$ {where $P[AI1(t)] \equiv P(AI1)$}, which is a special case of the master equation presented in section 2, for $n = 1$. The solution of Eq. (30) is thus given by Eq. (3) for $\mathbf{A} \to -k_+$ and $\vec{P}_0 = 1$;

$$P(AI1) = e^{-k_+ t} \tag{31a}$$

$$P(R) = 1 - P(AI1) \; . \tag{31b}$$

Due to the exponential increase in the number of monomers in a given generation as the generation is closer to the periphery, a relation between the PDF of occupying generation (site) j, $P_j(t)$ (as is denoted in section 2), and the actual measured concentration of that generation $[C_j(t)]$ should contain the amplification factor 2^{n-j}.

Here $[C_j(t)]$, for $j > 0$, is the concentration of an intermediate dendrimer of j generations, *i.e.* for G1 SID $[C_1(t)] = [AI1(t)]$. For $j = 0$, $[C_0(t)] = [R(t)]$, which is the concentration of the reporter units R (amino-methylpyrene) and is equivalent to the trap, site $j = 0$, shown in Figure 9a. Note, that for the amplifier case, the mapping of the dendrimer structure onto a one-dimensional model demands the redefinition of the site index, so that now $n \rightarrow n-1$, because the core, as defined in section 2, is merely the trigger for this process, and $j \rightarrow n-j$, as the signal propagates from the core to the periphery, which is opposite to the antennae case. This illustrated in Figure 21.

The relation between the calculated $P_j(t)$'s and the measured $[C_j(t)]$'s is given by;

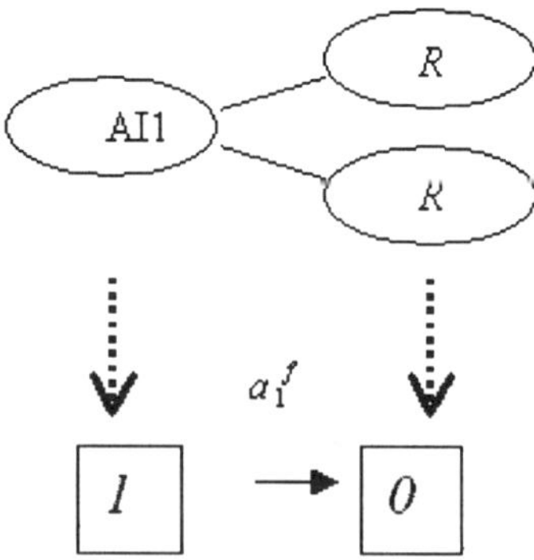

Fig. 21. The mapping of a two-generation dendrimer onto a one-dimensional system for the amplifier case. The reporter units are the analogue of the trap in the antenna case.

$$[C_j(t)] = P_j(t)[C_n(t=0)]2^{n-j} , \tag{32}$$

where the process starts with the dendrimer population of n generations. Accordingly, by solving the master equation, the experimental concentration $[C_j(t)]$ can be recovered by multiplying the corresponding PDF by the factor $[C_n(t=0)]2^{n-i}$. For example, for large enough reaction time, where $P(R)=1$, the reporter concentration, $[R]$, is given by, $[C_0(t \to \infty)] \equiv [R(t \to \infty)] = [AI1(t=0)] \cdot 2$, when using Eq. (32) with $n = 1$, *i.e.* for the G1 SID.

$P(AI1)$ and $P(R)$ as obtained from the experiment are shown in Figure 22a. From the slope of $\log[P(AI1)]$, when plotted versus time, the dissociation rate k_+, which appears in Eq. (30), can be extracted from the data. As shown in Figure 22b k_+ is found to be 2.2×10^{-3} min^{-1}.

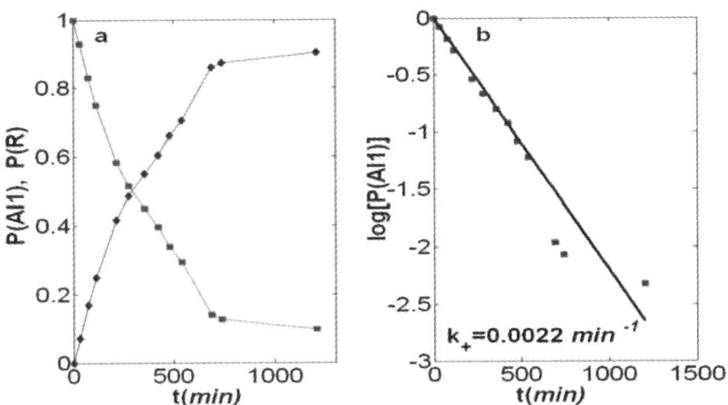

Fig. 22. (*a*) Temporal behavior of the experimental PDFs of the intermediate AI1, $P(AI1)$ (■), and the reporter, $P(R)$ (♦). (*b*) The logarithm of $P(AI1)$ as a function of time shows a linear dependence with a slope that corresponds to $k_+ = 2.2 \times 10^{-3}$ min^{-1}.

The kinetic equations for the dissociation of AI2 dendrimer to the intermediate AI1, of the self-immolation of the intermediate AI1 from G2 dendrimer, and of the reporter from intermediate AI1, are given (in terms of PDFs) by;

$$\frac{\partial}{\partial t} P(AI2) = -k_+ P(AI2) \tag{33a}$$

$$\frac{\partial}{\partial t}P(\mathrm{AI1}) = k_+ P(\mathrm{AI2}) - k_+ P(\mathrm{AI1}), \tag{33b}$$

$$\frac{\partial}{\partial t}P(R) = k_+ P(\mathrm{AI1}), \tag{33c}$$

where we take the dissociation rate of AI1 to have the same value as the dissociation rate of AI2, and assuming that no recombination of the AI1 fragments to form the AI2 fragments occurs.

Equations (33a - 33c) are equivalent to an invariant master equation in section 2, for $n = 2$ and $Q \rightarrow 0$. The solutions for Eqs. (33a - 33c) are given by;

$$P(\mathrm{AI2}) = e^{-k_+ t}, \tag{34a}$$

$$P(\mathrm{AI1}) = k_+ \cdot t \cdot e^{-k_+ t}, \tag{34b}$$

$$P(R) = (1 - e^{-k_+ t} - k_+ \cdot t \cdot e^{-k_+ t}), \tag{34c}$$

and are obtained by substituting the solution of Eq. (33a), which is an exponential decay with a rate k_+ and an amplitude 1, into Eq. (33b) and solving the resulted equation. The same procedure is followed for Eq. (34c) after obtaining the solution of Eq.(34b). For a G2 SID, the reporter concentration for large time increases by a factor of 4 relative to the initial concentration of the dendrimer, thus emphasizing the amplifying nature of this system.

Performing a similar analysis as for the G1 amine-intermediate AI1, the HPLC results obtained from the G2 self-immolative dendrimer are shown in Fig 23a, in terms of PDFs. The self-immolative mechanism of the G2 dendrimer to release two AI1 fragments shows a similar kinetic pattern to the one observed for the G1 SID; namely, an exponential temporal decay for AI2 concentration. As we expected the value of the rate constant for AI2 degradation was found to be equal to the one calculated for AI1 (Figure 23b). This can be rationalized when one considers that in both G1 and G2 fragmentation reactions, the rate-limiting step is the cyclization of the amine intermediate. An excellent

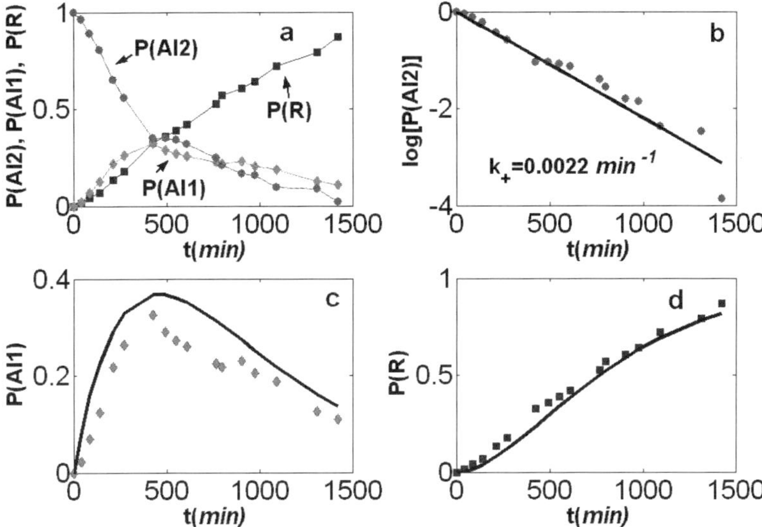

Fig. 23. (*a*) *P(AI2)* (●) *P(AI1)* (♦) and *P(R)* (■)plotted functions of time. (*b*) log[AI2(t)] as a function of time shows a linear dependence, with $k_+ = 0.022$ min^{-1}. (*c*) and (*d*) *P(AI1)* and *P(R)* fit very well to the calculated functions Eqs. (34b-c). Note that all the fitting curves presented in (*c*) and (*d*) are obtained using the same parameter, which is the dissociation rate k_+.

accord is found between the calculated and experimental PDFs $P(AI2)$, $P(AI1)$, and $P(R)$, shown in Figs. 23b-d respectively, supporting the assumption of one dissociation rate for describing the dissociation process.

In general, a new class of dendrimeric molecules that were termed Self-Immolative-Dendrimers was designed and synthesized. These structurally unique dendrimers can release all of their tail units, through a self-immolative chain fragmentation, which is initiated by a single cleavage at the dendrimer's core. The structural branching unit of the SIDs can be viewed as an amplifier of a chemical signal. The input of the unit is a single chemical bond cleavage, which is amplified to an output of two signals in the form of a double cleavage. The linking of additional two structural units to the output of the first, will consequently amplify one signal into four, *etc*. The kinetic analysis of G1 and G2 SIDs may serve as a powerful tool for characterizing other SID systems, and evaluating the rate constants of the disassociation process. Of particular interest is the dependence of the rate constants on possible different substituents attached to the core ring, and on the modification of the linker component. SIDs might be applied as a general platform for pro-drugs or sensor molecules for enzymatic activity. Bioactivation of dendrimeric pro-drugs with catalytic antibody has been successfully accomplished recently.[32]

Dendrimers have been shown to play an interesting role as nano-devices, making use of their special architecture and internal geometric bias. Both periphery-to-core and core-to-periphery processes were shown to lead to possible applications exemplified here as antennae and amplifiers.

References

1. Tomalia DA, and Fréchet JMJ. *J. Polym. Sci.* 2002; Part A, **40**: 2719.
2. Patri AK, Majoros IJ, and Baker JR. *Curr. Opin. Chem. Biol.* 2002; **6**: 466 (2002). Stiriba S-E, Frey H, and Haag R. *Angew. Chem.* 2002; **114**: 1385; *Angew. Chem. Int. Ed.* 2002; **41**: 1329.
3. Tomalia DA, Naylor AM, and Goddard III WA. *Angew. Chem.* 1990; **102**: 119; *Angew. Chem. Int. Ed. Engl.* 1990; **29**: 138.
4. Gilat SL, Adronov A, and Fréchet JMJ. *Angew. Chem. Int. Ed. Engl.* 1999; **38**: 1422.
5. Amir RJ, Pessah N, Shamis M, and Shabat D. *Angew. Chem.* 2003; **115**: 4632; *Angew. Chem. Int. Ed.* 2003; **42**: 4494.
6. Grinstaff MW. *Chem. Eur. J.* 2002; **8**: 2839; Seebach D, Herrmann GF, Lengweiler UD, Bachmann BM, and Amrein W. *Angew. Chem.* 1996; **108**: 2969; *Angew. Chem. Int. Ed. Engl.* 1996; **35**: 2795; Kim Y, and Zimmerman SC. *Curr. Opin. Chem. Biol.* 1998; **2**: 733; Padilla De Jesús OL, Ihre HR, Gagne L, Fréchet JMJ, and Szoka, Jr. FC. *Bioconjugate Chem.* 2002; **13**: 453.
7. Bar-Haim A, Klafter J, and Kopelman R. *J. Am. Chem. Soc.* 1997; **119**: 6197; Bar-Haim A, Eizenberg N, and Klafter J, *Organic Mesoscopic Chemistry*: Oxford: Blackwell Science; 1999.
8. de Groot FMH, Albrecht C, Koekkoek R, Beusker PH, and Scheeren HW. *Angew. Chem. International Edition.* 2003; **42**: 4490.

9. Szalai ML, Kevwitch RM, and McGrath DV. *J. Am. Chem. Soc.* 2003; **125**: 15688.

10. Alberts B, Roberts K, Bray D, Lewis J, Raff M, and Watson JD, *Molecular Biology of The Cell*. New York and London: Garland Publishing; 1994.

11. Kopelman R, Shortreed M, Shi ZY, Tan WH, Xu ZF, Moore JS, BarHaim A, and Klafter J. *Phys. Rev. Lett.* 1997; **78**: 1239.

12. Pan Y, Lu M, Peng Z, and Melinger JS. *J. Org. Chem.* 2003; **68**: 6952.

13. Zwanzig R, *Nonequilibrium Statistical Mechanics*. New York: Oxford University Press; 2001.

14. Flomenbom O, and Klafter J. *Phys. Rev. E* 2003; **68**: 041910.

15. Redner S, *A Guide to First-Passage Process*. Cambridge: University Press; 2001.

16. Bentz JL, Hosseini FN, and Kozak JJ. *Chem. Phys. Lett.* 2003; **370**: 319.

17. Gronheid R, Hofkens J, Köhn F, Weil T, Reuther E, Müllen K, and De Schryver FC. *J. Am. Chem. Soc.* 2002; **124**: 2418.

18. Bar-Haim A, and Klafter J. *J. Lumin.* 1998; **76&77**: 197.

19. Bates M, Burns M, and Meller A. *Biophys. J.* 2003; **84**: 2366.

20. Bar-Haim A, and Klafter J. *J. Chem. Phys.* 1998; **109**: 5187.

21. Drager J, and Klafter J. *Phys. Rev. E* 1999; **60**: 6503.

22. Bar-Haim A, and Klafter J. *J. Phys. Chem. B* 1998; **102**: 1662.

23. Risken H, *The Fokker-Planck Equation*. Berlin: Springer Verlag; 1989.

24. Gillies ER, and Fréchet JMJ. *Chem. Comm.* 2003; **14**: 1640.

25. Esfand R, and Tomalia DA. *Drug Discovery Today.* 2001; **6**: 427 *and references therein*.

26. Patri AK, Majoros IJ, and Baker JR. *Curr. Opin. Chem. Biol.* 2002; **6**: 466 *and references therein*.

27. Malik N, Duncan R, Tomalia DA, and Esfand R. *U.S. Pat. Appl. Publ.* 2003, cont.-in-part of U.S. Ser. No. 881,126.

28. Cordova A, and Janda KD, *J. Am. Chem. Soc.* 2001; **123**: 8248.

29. Borman S. *Chem. & Eng. News.* 2003; **81** (39): 4.

30. Meijer WE, and van Genderen MHP, *Nature* 2003; **426:** 128.

31. Borman S, *Chem. & Eng. News.* 2003; **81** (51): 39.

32. Shamis M, Lode HH, and Shabat D, *J. Am. Chem. Soc.* 2004: *in press*.

Chapter 9

ENERGY HARVESTING IN SYNTHETIC DENDRITIC MATERIALS

Gemma D. D'Ambruoso and Dominic V. McGrath

In the past two decades dendrimers have emerged as a distinctive branch of macromolecular chemistry. Tailoring of dendrimer structure yields precise placement of chromophores that can serve as energy harvesters, mimicking photosynthesis. The unique architecture afforded by dendrimers allows for multiple energy harvesters which can transfer their energy to a single core. These materials can have impact in optoelectronic applications such as organic light emitting diodes (OLEDs). This chapter reviews artificial energy-harvesting systems that employ a dendritic architecture, emphasizing the energy transfer characteristics that these dendrimers provide rather then their synthesis.

Keywords: Dendrimers, energy harvesting, light harvesting, antenna effect

1. INTRODUCTION

Dendrimers are an intriguing scaffold for constructing energy harvesting systems since their architectures roughly mimic that of natural photosynthetic centers in which light is harvested by chlorophyll chromophores encircling a reactive core. Energy transfer from the outer chlorophyll chromophores to the reaction center results in ATP production, which is essential for life. This chapter reviews artificial energy-harvesting systems that employ a dendritic architecture.

The role of dendrimers in light-harvesting systems varies substantially. First, the dendrimers can act as an inert spacer to separate two different chromophores (host/guest) from each other. In these cases, the dendrimers do not participate in the energy transfer between the chromophores, as their excited state energy is typically higher than both the host and guest. Also, increasing dendrimer generation increases host/guest distance and chromophore communication can begin to decrease (as above the Förster radius) as a result. Second, the dendrimer subunits can act as the host chromophores and are responsible for light harvesting. In this case, light absorbed by the dendrimer (acting as the host) is transferred to a core moiety (guest) from which emission ensues. Increasing generation allows for greater light harvesting capabilities since the number of dendrimers subunits increases.

In both cases (dendrimers as spacers or dendrimers as hosts), dendrimers offer a unique advantage over classical energy transfer between a chromophoric pair, in that the number of host chromophores can far exceed the number of guests chromophores, and is a strict function of generation. The effective molar absorptivity of the host chromophores can therefore be increased (predictably) and the dendrimer itself is a better light harvester then a single host chromophore. Better light harvesting and an increase in the amount of energy transferred to the guest results in a greater amount of emission from the dendrimers and is referred to as the 'antennae effect.'

In addition to the antennae effect, dendrimers are also effective insulators and exhibit the 'shell effect.' In providing a dense shell around the incorporated chromophores, dendrimers effectively prevent aggregation – which leads to non-emissive excimers and self-quenching that occurs when chromophores with small Stokes shifts are within short distances of one another. This 'shell effect' allows for increased photoluminescence efficiency of the enclosed chromophore, which is important for optoelectronic devices.

We have organized this review by type of chromophore and/or dendrimer structure.

2. METAL-CONTAINING DENDRIMERS

Light-harvesting metal containing dendrimers exist in two categories: (i) dendrimers where metals are at both the core and branching points[1-9] and (ii) dendrimers where metal cores are surrounded by aromatic dendrons.[10-18] In both cases, light-harvesting followed by energy transfer to the metal occurs. These two categories are addressed here.

2.1 Metal-containing Dendrons and Cores

In the early 1990's, Balzani and coworkers began to synthesize decanuclear homo- and heterometallic dendrimers which contain Ru(II) and Os(II) surrounded by nitrogen-containing ligands (bpy derivatives). The metals occupied sites in both the dendrons and at the core of the dendrimers.[1,6-9,19] Although the Balzani group had for some time been synthesizing multimetallic species in tri-,[20] tetra-,[20,21] hexa-,[22] and heptanuclear[23] complexes, this is the first work that displays these complexes in a dendritic fashion. The dendrimers are synthesized in a 'complexes as ligands/complexes as metals' manner in which metal complexes (instead of individual metals and ligands) are used as the reactive species. This approach has produced metallodendrimers in which the number of metals per dendrimer is as high as 22.[2,3,5] The dendrimers consist of bridging ligands 2,3- and 2,5-bis(2-pyridyl)-pyrazine (2,3-dpp and 2,5-dpp) that connect the metal centers, and terminal ligands, 2,2'-bipyridine (bpy), 2,2'-bisquinoline (biq) and 2-[2-(methylpyridiniumyl)]-3-(2-pyridyl) pyrazine (2,3-Medpp$^+$) that 'cap' the peripheral metals (Figure 1).

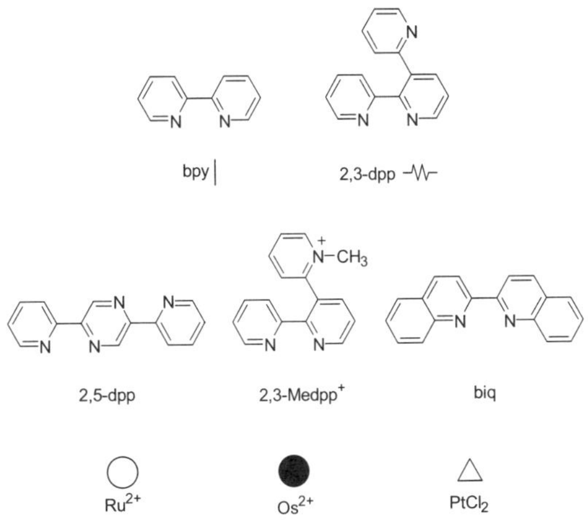

Fig. 1. Building blocks for multimetallic light harvesting dendrimers.[4,5]

Light harvesting in these systems is between ligand-centered (LC) transitions in the UV as well and several metal-to-ligand charge-transfer (MLCT) transitions in the visible region. Each metal center in the dendrimer is capable of absorbing light, so the molar absorptivity and amount of harvested light increases with increasing number of metal sites. Emission from these dendrimers is primarily radiative decay of the lowest lying metal-to-ligand charge-transfer triplet state, ^{3}MLCT, for both Ru(II) and Os(II) cases.[24] In the case of the homometallic Ru(II) dendrimers, emission occurs solely from the lower-energy peripheral Ru(II) complexes which are coordinated to the terminal ligands (either bpy or biq). In the Ru(II)/Os(II) mixed-metal systems, emission is expected solely from the lower energy Os cores (Figure 2). However, in these mixed metal dendrimers, replacing the Ru(II) at any of the metal sites with Os(II) results in only partial energy transfer from the Ru(II) sites (both interior and peripheral) to the Os(II) units and therefore emission from both.

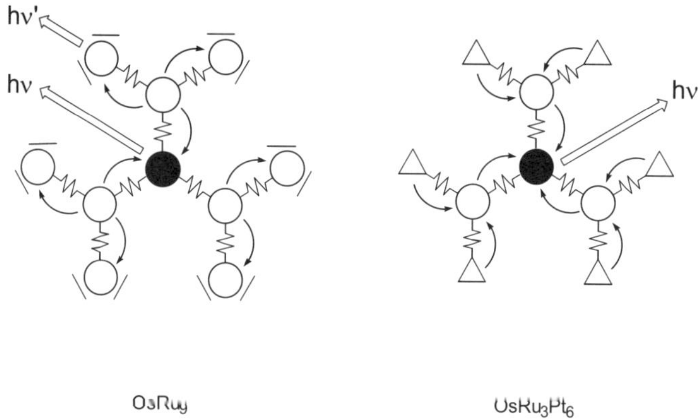

Fig. 2. Schematic of energy transfer in multimetallic dendrimers.[4,5]

Incomplete energy transfer from the Ru(II) sites to the Os(II) cores is thought to be due to the 'blocking' ability of the high-energy interior Ru(II) units. The energies of the lowest excited states follow in the order of Ru(II) (interior) > Ru(II) (exterior) > Os(II) (center), and therefore light harvested at the Ru(II) periphery (or transferred there from the interior) cannot progress completely to the center Os(II) by energy transfer due to the blocking interior Ru(II) units (Figure 2).[1] Hence, to create a true antennae species in which light harvested is transferred efficiently to the core, decanuclear dendrimers were prepared in which the peripheral Ru(II) units are replaced by Pt(II) whose lowest excited state resides above that of the interior Ru units.[4] Light harvested by an OsRu$_3$Pt$_6$ dendrimer gives rise to a sole emission band at 875 nm which is due to the emitting ^{3}MLCT of the Os core. These dendrimers represent the first metallodendrimers containing three different types of metals.

To introduce organic chromophores into multinuclear metallo-organic dendrimers, Campagna and coworkers have placed pyrene moieties on the periphery of a Ru(II)- and Os(II)-containing dendrimer (**1**).[25] An enhanced antenna effect occurs in the dendrimer because of intramolecular pyrene-to-ligand CT transitions which are seen in the visible region of the absorption spectra, in addition to the absorption bands due to the individual components of the dendrimer. The energies of the lowest excited-state are in the order: pyrene > Ru(II) interior > Os(II) core and, therefore, energy is funneled quantitatively to the Os(II) core from where emission is seen around 800 nm.

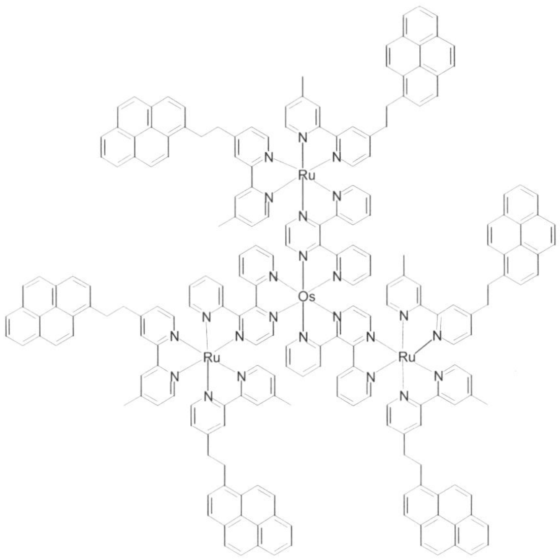

1

2.2 Metal-Cored Dendrimers with Organic Dendrons

The ability to alter the photo- and electrochemical properties of a reactive species by surrounding the unit with an organic dendrimer framework has been studied extensively.[26] In the case of metals, aromatic (bulky) dendrons have been used to insulate the metals from dioxygen and other excited-state quenchers,[10,27,28] prevent aggregation,[17,18] and in most cases, harvest light in the higher-energy dendrons which can be transferred to the lower-energy metal core.[10-12,17,29]

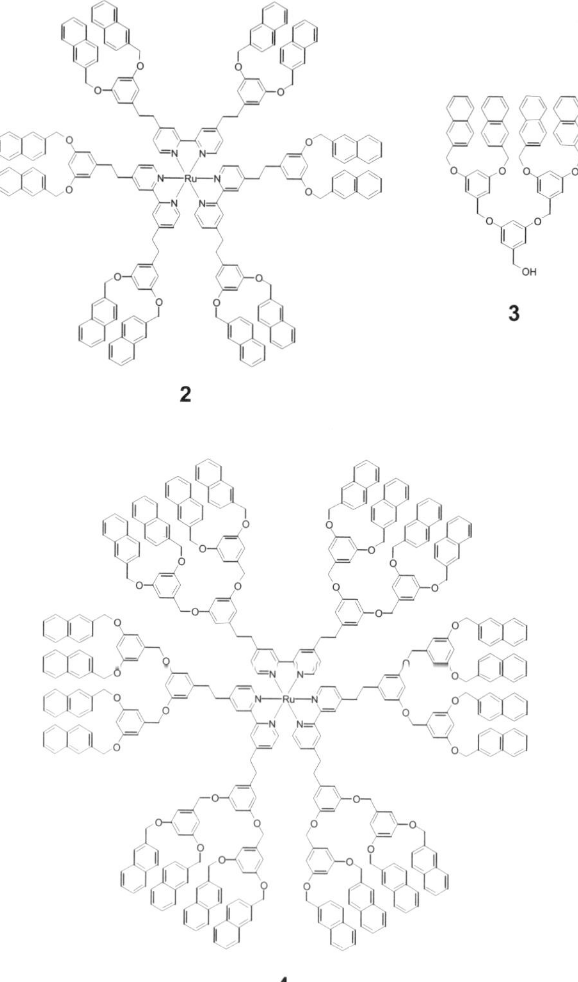

2

3

4

Balzani *et al.* synthesized first and second generation Ru(II) cored dendrimers with napthyl groups (12 or 24) at the periphery of benzyl(aryl ether) dendrons.[10] Dendrons were attached to bipyridine ligands (bpy) and complexed with Ru(II) to give the resulting dendrimers (**2** and **4**). Photoexcitation of the napthyl and dialkoxybenzyl groups of the dendron gives rise to an emission band near 600 nm which is attributed to the $[Ru(bpy)_3]^{2+}$ core complex (Figure 3).[24] The emission of the napthyl groups at 330 nm is nearly quenched, indicating an energy transfer from

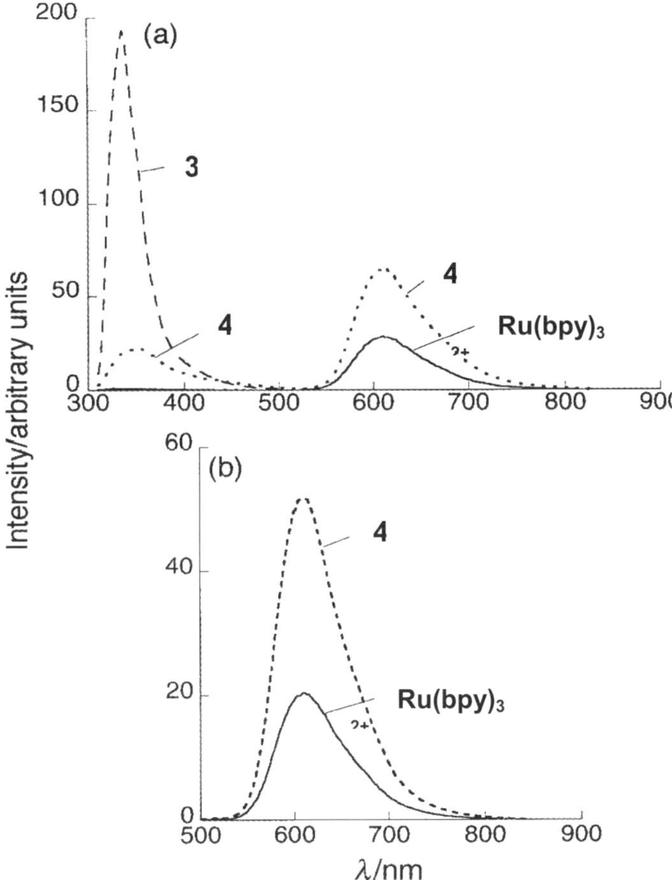

Fig. 3. (a) Emission spectra of dendrimer **4** (dotted), dendron **3** (dash) and $[Ru(bpy)_3]^{2+}$ (solid) when exciting into napthyl groups (270 nm). (b) Emission spectra of dendrimers (dotted) and $[Ru(bpy)_3]^{2+}$ when exciting into the $[Ru(bpy)_3]^{2+}$ core (450 nm).[10]

the peripheral napthyl groups to the $[Ru(bpy)_3]^{2+}$ core. Longer luminescence lifetimes of the dendrimers in aerated solutions (compared to the $[Ru(bpy)_3]^{2+}$ parent) indicate that the bulky dendrons prevent dioxygen from accessing the core and quenching the excited state, which has been shown for other Ru(II) containing dendrimers.[27,28]

A similar study by Castellano *et al.* placed coumarin 450 (C450) dye molecules on the periphery of 'reverse' benzyl aryl ether dendrons, first introduced by Hanson,[30] attached to a Ru(II) core moiety (**5**). The chromophores satisfy the requirements of a Förster energy transfer as the emission of the C450 significantly overlaps the absorption of the $[Ru(bpy)_3]^{2+}$ species and give a calculated Förster radius of 41.2 Å. Fluorescence spectra indicate that light absorbed by the C450 moieties on the periphery of the first generation dendrimer is accompanied by energy transfer and subsequent emission by the $[Ru(bpy)_3]^{2+}$ species at the core (Figure 4). Estimated energy transfer efficiency was calculated by comparing the absorbance and excitation spectra (observed at emission wavelength of $[Ru(bpy)_3]^{2+}$ complex, 610 nm) and determined to be around $95 \pm 10\%$. Preliminary excited state lifetime measurements

5

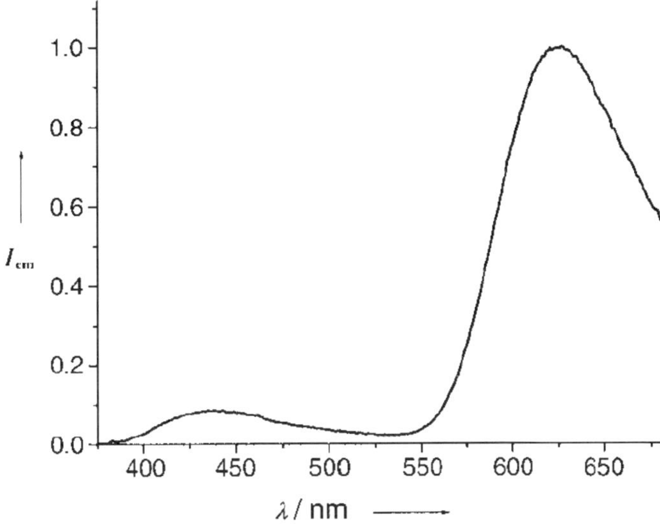

Fig. 4. Emission spectrum of dendrimer **5** in CH$_3$CN when excited into the coumarin 450 dyes (343 nm).[11]

indicated that dioxygen quenching was reduced by the presence of the 'shielding' dendrons.[11] However, further investigation of the bimolecular quenching rates for several additional quenchers indicated that dioxygen was an exception rather than the rule.[29] In all other cases the dendrons allowed access to the core by the quenchers (e.g. 9-methylanthracene and phenothiazine). Therefore, the reduced quenching in the dioxygen case was possibly due to the lower solubility of dioxygen in the microenvironment surrounding the metal core, not inability to diffuse to the core.

Dehaen and coworkers utilized carbazole dendrons attached to 1,10-phenanthroline ligands (phen) to construct a dendrimer around a Ru(II) core (**6**).[12] Absorbance transitions assigned to carbazole, phen, and carbazole-to-phen CT transitions all gave rise to emission from the Ru(II) core, providing yet another light-harvesting dendrimer which illustrates the antennae effect. Further examples of harvesting dendrimers that contain transition metals include poly(propylene imine) (PPI) dendrimers modified with 32 dansyl groups at the periphery (**7**)[13] whose fluorescence is quenched when a single Co^{2+} ion is coordinated to the interior amines, thereby creating a fluorescent chemosensor (Figure 5).[31,32]

The quenching arises from energy and electron transfer between the excited dansyl groups and the Co(II) amine complexes. Similar results have been obtained for lanthanide metals coordinated to the interior of polylysine dendrimers containing dansyl groups on the periphery.[14-16]

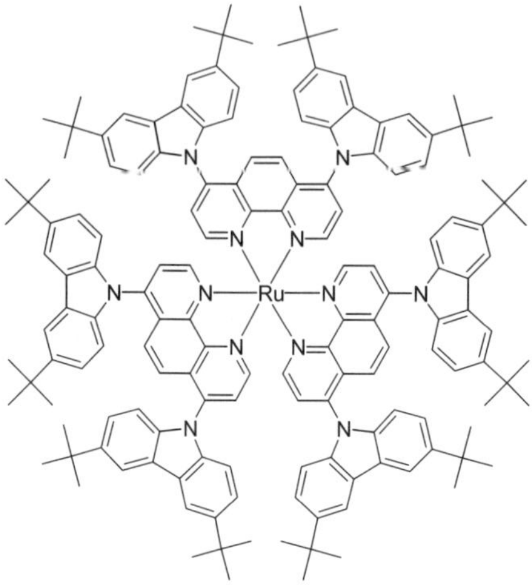

6

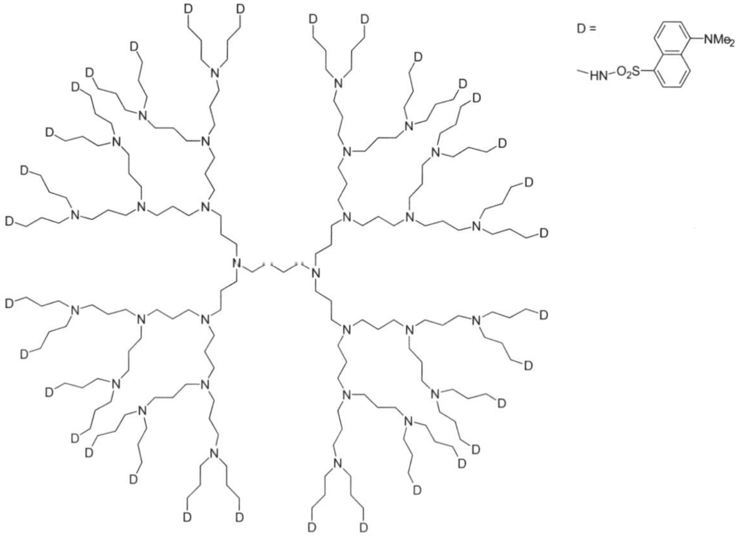

7

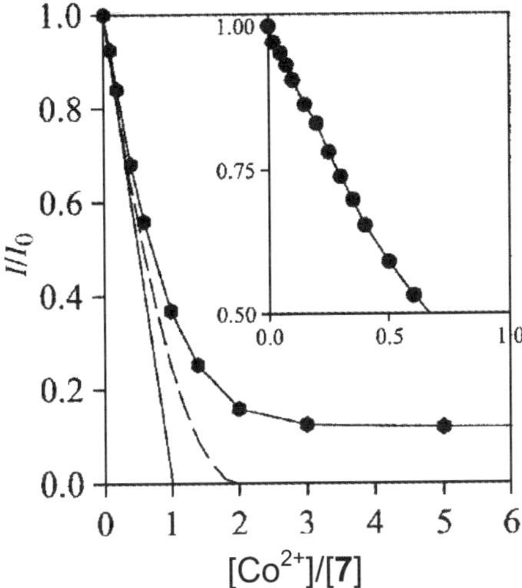

Fig. 5. Fluorescence intensity of dendrimer **7** with the addition of Co^{2+} (dark circles). Inset shows linearity at low concentration. Solid line is linear extension of the initial slope, dashed line is expected if two Co^{2+} are independently coordinated in the dendrimer.[32]

Because of their narrow emission bands, long luminescence lifetimes, and absorption profiles that span from the near-UV to the near-IR, lanthanides have been investigated for applications such as light-emitting diodes[33] and optical amplifiers.[34] However, the tendency of lanthanides to aggregate leads to self-quenching which limits their luminescence efficiency and lifetimes. Also, proximity to solvent molecules which can act as quenchers limits their potential use as opto-electronic materials.[35,36] In work in lanthanide containing dendrimers, Kawa and Fréchet[17,18] synthesized benzyl(aryl ether) dendrons with carb-oxylate focal points that formed 3:1 complexes with lanthanide ions Er^{3+}, Tb^{3+}, or Eu^{3+} (Figure 6). Emission spectra for **8b, 9b, 10b** at identical concentrations show an increase in emission from the Eu core when the dendrons are excited. This is due to both an increase in energy harvesters as generation increases (antennae effect) and an increase in site-isolation allowing for reduced self-quenching of the Eu cores (shell effect). Inter-estingly, when the substitution of the phenyl ring closest to the focal point is changed from 3,5- to 2,5-, emission enhancement is less pronounced.

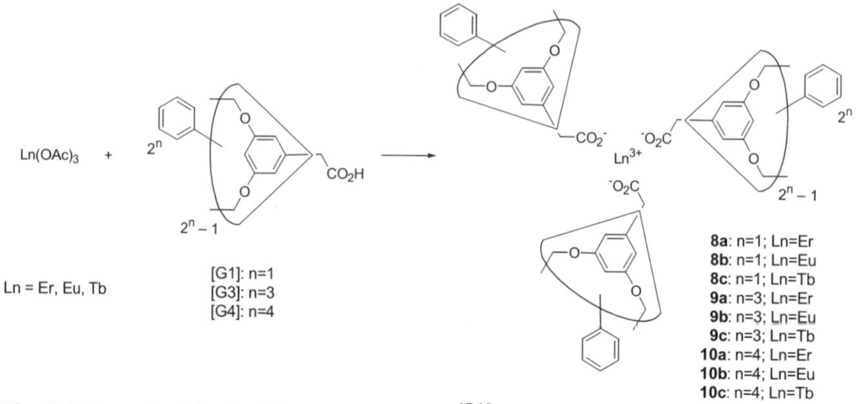

Fig. 6. Schematic of lanthanide-core dendrimers.[17,18]

3. PHENYLACETYLENE DENDRIMERS

Dendrimers consisting of phenylacetylene moieties exist as (i) 'compact' or 'extended' meta-conjugated phenylacetylene (PA) dendrimers,[37-51] and (ii) unsymmetrical branched phenylacetylene dendrimers.[52-55]

3.1 'Compact' and 'Extended' Phenylacetylene Dendrimers

Moore *et al.* have synthesized phenylacetylene dendrimers of two types, 'compact' (**11**) and 'extended' (**12**).[37-39] Both forms contain meta-substituted phenylacetylene subunits. However, the lengths of the phenylacetylene segments differ between the two types. In the 'compact' form, all the conjugated segments are effectively the same length and exhibit the same excitation energies; meta-substitution ensures that π-conjugation is blocked at each branching point, allowing for a localized excited state upon excitation.[40,43] In the 'extended' form, as the dendrimer proceeds from the periphery to the core, the conjugation length of each generation increases by one phenylacetylene moiety. Concomitant with the increase in π-conjugation is a stepwise decrease in the lowest excited state energy. This allows the dendrimer to act as an energy 'funnel'. The 'extended' dendrimer still contains localized excitations. However in each layer leading up to the core the excited state energy is more delocalized due to increasing conjugation. It is important to emphasize that neither type of dendrimer exhibits delocalized excitation over its entirety. Rather, energy is localized in each layer or generation.

Tris dendrimers (three dendrons attached to a 1,3,5-trisubstituted phenyl core) in 'compact'[37,39] and 'extended'[38] forms have been synthesized (figure 1a) as well as 'nanostars'[41,42] consisting of a 'compact' or 'extended' dendron attached to a perylene 'trap' (**13** and **14**). Their energy transfer rates and mechanisms have been studied extensively.[40,42-49]

11

12

13

14

Important features are observed when comparing the absorption spectra for a series of 'compact' and 'extended' dendrimers (Figure 7).[40,45] In the 'compact' case, the absorption maxima remain at a fixed wavelength throughout the series with only the intensity scaling with increasing number of absorbing chromophores. This indicates that not only are the excitations localized within the layers, but also that the chromophores,

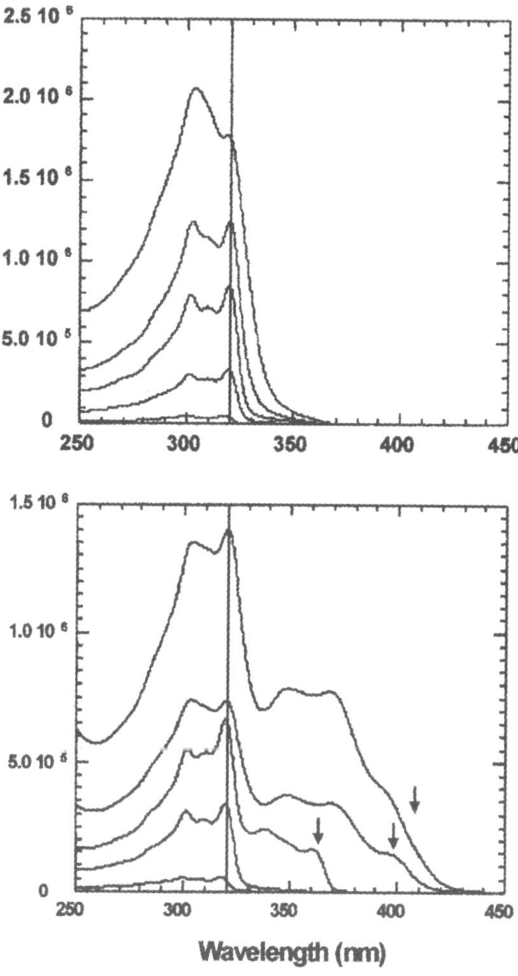

Fig. 7. Absorption spectrum for 'compact' dendrimers (top) and 'extended' dendrimers (bottom) in hexane. Arrows indicate the absorption bands for two, three and four phenyls in a phenylacetylene chain, respectively.[45]

regardless of location within the dendrimer, all absorb at the exact same wavelength. The spectra of the 'extended' series, however, contain additional peaks of lower energy with each generation, corresponding to the linear segments of extended conjugation. Broader absorption spectra for 'extended' dendrimers imply that these molecules will be more effective light harvesters. It should be noted that the absorption spectra of the 'compact' and 'extended' nanostars are similar to their respective dendrimers with an additional peak at 475 nm for perylene absorption.

The quantum yield for energy transfer Φ_{ENT} for 'compact' and 'extended' nanostars was measured by comparing absorption and excitation spectra normalized in the perylene absorption region (430-500 nm) (Figure 8).[42,44] For the 'extended' nanostar **13**, Φ_{ENT} was 98%. The

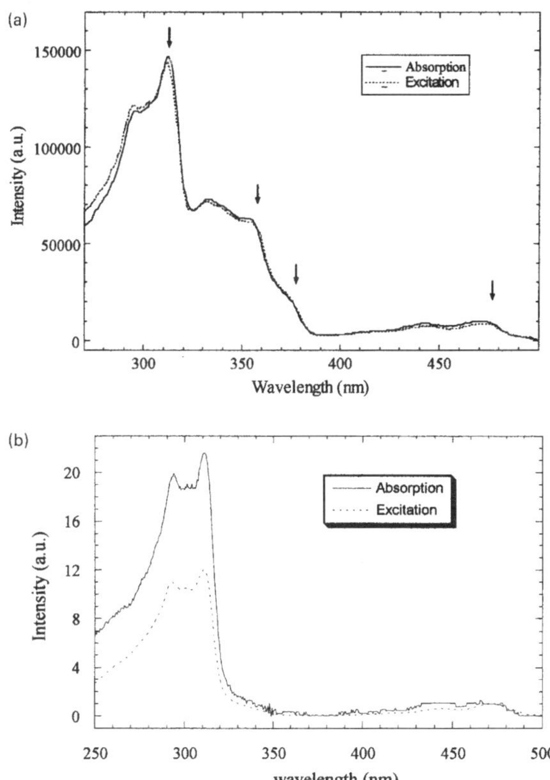

Fig. 8. Absorption and excitation spectra of an (a) 'extended' and (b) 'compact' nanostar.[45]

'compact' nanostars of varying generations give Φ_{ENT} of 95% for the lowest generation and 54% for the highest; residual emission from the phenylacetylene segments increases with generation. Fluorescence lifetime measurements also differ greatly for the 'extended' (270 ps) and 'compact' (2.3 ns) structures, yielding rate constants for energy transfer k_{ET} for the 'compact' molecules that are lower by two order of magnitude than those for the 'extended' structure. Although the Förster equation provides k_{ET} values that cannot be correlated to the observed ones, the reasons for lower efficiency in the 'compact' structures can be explained by the Förster model: as generation increases, chromophore separation becomes too great for efficient energy transfer. The energy 'funnel' inherent in 'extended' structures, on the other hand, allows for efficient stepwise transfer of energy to the perylene core, which is not hampered by the interchromophoric distance as in the 'compact' case.

Although at higher generations 'compact' structures have decreased Φ_{ENT} values, direct sensitization of the perylene cores by the diphenylacetylene segments in the dendrons can occur. In the 'extended' nanostar **14**, perylene excimer formation lends evidence that long-range energy transfer also takes place, in addition to a multistep process.[45,46] Although excimer concentration dependence was expected and observed, an unexpected dependence on the wavelength of excitation was also observed, as measured by steady-state fluorescence. Excimer fluorescence was highest when the innermost layer of phenylacetylene moieties was excited, indicating that a phenylacetylene excimer is initially formed which in turn excites a perylene excimer (Table 1).

Table 1. Excimer/monomer ratios for 'extended' nanostar **14**.[45,46]

Excitation Wavelength, nm	Dendrimer Generation	Excimer/Monomer Ratio
310	3	0.33
334	2	0.31
390	1	4.0
450	Perylene trap	0.14

Because the excimer emission was much lower when interior or periphery phenylacetylenes were excited, all excitons created in these outer regions do not always funnel to the interior ones, leaving the possibility that there is some amount of direct energy transfer that does take place between the periphery and the perylene trap.

Pu and coworkers are interested in using chiral dendrimers as enantioselective fluorosensors.[50,51] By utilizing the antennae effect available to light-harvesting dendrimers, fluorescence amplification can be achieved, implying that the dendrimers will be more sensitive fluorosensors to smaller amounts of a quencher. The dendrimer (**15**) consists of 'compact' phenylacetylene dendrons surrounding a chiral 1,1'-bi-2-naphthol (BINOL) core. Precise placement of the dendrons around the BINOL core creates a chromophore that has extended conjugation and therefore acts as an energy sink for energy harvested by the dendrons. Hydrogen bonds formed between the BINOL core and chiral amino alcohols in solution quench the fluorescence from the dendron/core chromophore. Stern-Völmer plots show a mirror-image relationship between the enantiomers of the dendrimers and the enantiomers of the quenchers (e.g. (*S*)-dendrimer quenches *(S)*-amino alcohol more efficiently and *(R)*-dendrimers quench *(R)*-amino alcohols more efficiently).

15

3.2 Unsymmetrical Phenylacetylene Dendrimers

If *meta* and *para* branching patterns are utilized, instead of the all-*meta* branching in Moore's PA dendrimers, dendrons now consist of a long linear segment that is attached to the core and from which PA segments of different length are attached (**16a-d**).[52-55] The overall effect is one of an energy 'short-cut' as now all segments varying in length are directly attached to the lowest energy PA moiety. In other words, the number of steps in this multistep energy transfer is diminished from Moore's 'extended' system. Energy transfer as measured by Φ_{ENT} is very efficient (85 – 95%) and does not seem to decrease with increasing generation as in the 'compact' symmetrical PA dendrimers. Also interesting is the minor effect that dendron geometry has on the Φ_{ENT}. Dendrons were attached to perylene 'traps' in two ways; either conjugated to the longest

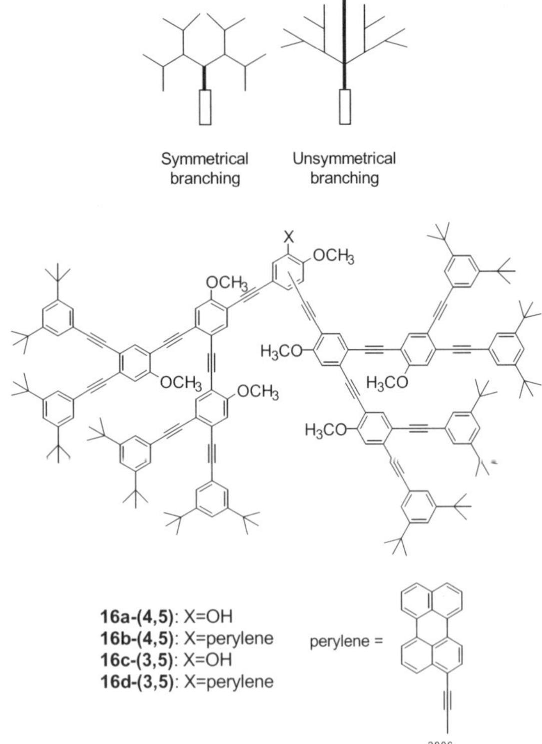

linear segment of the dendrons or attached by a *meta*-linkage, disrupting the conjugation. Only slightly lower Φ_{ENT} were obtained for the *meta*-linked [G1] dendrimers (difference of 3%) but the difference increased to 9% at the [G2] level.

4. DENDRIMERS CONTAINING DISTYRYLBENZENE OR STILBENE UNITS

Although the first examples of OLEDs included polymers, specifically poly (*p*-phenylene vinylene) (PPV), as the emissive material, conjugated oligomers can also be useful emitters as well as energy harvesters. Oligomers offer advantages over their polymeric parents such as ease of synthesis and functionalization allowing for easy introduction of various substituents. They can also be incorporated into dendrimers as both host and guest chromophores.

Distyrylbenzenes (**DSB**), also known as oligoPPVs, have been extensively functionalized to alter their band-gap properties as well as their two-photon cross-sections. By tailoring the end-capping styrene groups to be branching units in dendrons, several dendrimers have been synthesized containing **DSB** at the core. Kwok and Wong[56] have placed either one or two poly (aryl ether) dendrons around a core **DSB** to elucidate the influence of the dendritic wedge on aggregation of the cores. By placing a single electron-transporting oxadiazole on the exterior of the dendrons to offset the hole-affinity of the **DSB**, the dendrimers (**18** and **20**) then become more charge-balanced.

Fluorescence lifetimes and quantum yields of the symmetrical dendrons were higher then those of their asymmetric counterparts, indicating that there is more effective shielding of the **DSB** cores by two dendrons than one. In all other areas, such as energy transfer from the dendrons to the **DSB** core and OLED device performance, asymmetrical dendrimers seem to hold the advantage over the symmetrical ones. Performance is further enhanced in devices when a hole-transporting material, diphenylamine, is blended into the emissive layer of the oxadiazole asymmetrical dendrons, indicating that placing an oxadiazole on every dendron is excessive.

17: R$_1$ = OC$_3$H$_7$; R$_2$ = H

18: R$_1$ = H; R$_2$ =

19: R$_1$ = OC$_3$H$_7$; R$_2$ = H

20: R$_1$ = H; R$_2$ =

Samuel and Burn [57] have used **DSB** cores surrounded by conjugated dendrons as the emissive and charge transport layer in OLEDs. The conjugated dendrons consist of stilbene units in a meta-branching pattern, which prevents uninterrupted conjugation throughout the dendron. Each stilbene subunit therefore appears electronically isolated, although red-shifted absorbance and PL spectra indicate that there is some amount of delocalization. Excitation of the stilbene units does result in emission solely from the distyrylbenzene core, in both solution and thin films, giving evidence of efficient energy transfer. Light-emitting devices from all three generations of dendrimers were equal to (G1) or better than (G2 and G3) single-layer devices containing PPV. Dendrimers with **DSB** cores surrounded by electron transporting triazine groups (**22**) were found to produce comparable PL quantum yields but lower device external quantum efficiencies by a factor of 10.[58]

Burn and Samuel have used their stilbene dendrons to transfer energy to a number of different cores for color-tuning in OLEDs.[59] First

21a: n = 1
21b: n = 2
21c: n = 3

22

and second generation dendrons with an aldehyde at the focal point have been coupled with the bisphosphonate esters to form **DSB** and distyrylanthracene cored dendrimers. Dendron aldehydes and pyrroles were condensed to form porphyrin dendrimers. Proton NMR of the porphyrins-cored dendrimers showed a chemical shift of the *tert*-butyl groups on the periphery of the dendrons indicating that these groups were actually hovering over the porphyrins. This points to an orthogonal positioning of the dendrons to the porphyrin core, and the authors were interested as to whether this would hinder energy transfer from the dendrons to the porphyrin core. Interestingly, in the dendrimers with **DSB** and **DSA** cores, energy transfer was found to be around 65%. In the case of the porphyrin dendrimer, however, excitation of the dendrons resulted in quantitative energy transfer to the core as measured by comparing excitation and absorbance spectra.

X =

23a: n = 1
23b: n = 2
23c: n = 3

X =

24a: n = 1
24b: n = 2
24c: n = 3

R=

25a: n = 1
25b: n = 2

5. PORPHYRIN-CONTAINING DENDRIMERS

The three categories of porphyrin-containing dendrimers to be discussed include: (i) porphyrin cores surrounded by organic dendrons;[60-62] (ii) dendritic arrays of porphyrins at the both core and branching points;[63-65] (iii) organic dendrimers with porphyrins at the periphery.[66-68]

5.1 Dendrons Surrounding a Porphyrin Core

Work by Jiang and Aida on light-harvesting and energy transfer properties of porphyrin-containing dendrimers began as an extension of their investigation of azobenzene-containing dendrimers and their energy transfer phenomena.[69-71] Monoazobenzene-containing dendrimers **26a-d** exhibit remarkable generation-dependent energy harvesting capabilities. As expected, compounds **26a–26d** undergo ultraviolet photoinduced isomerization from the *E*-form to the *Z*-form of the core azobenzene and thermally isomerize in the reverse direction with typical half-lives. However, when individual solutions of the *Z*-forms of **26c** and **26d** were irradiated with infrared irradiation (75 W glowing nichrome source), isomerization to the *E*-form was accelerated over 250 times that of the thermal isomerization and, remarkably, over 20 times that of the rate found on irradiation with visible light (440 nm). Yet, the rates of *Z*→*E* isomerization of **26a** and **26b** were unaffected by the infrared irradiation. Apparently, spatial isolation of the azobenzene is crucial for this effect, as indicated by control experiments involving compound **27**.

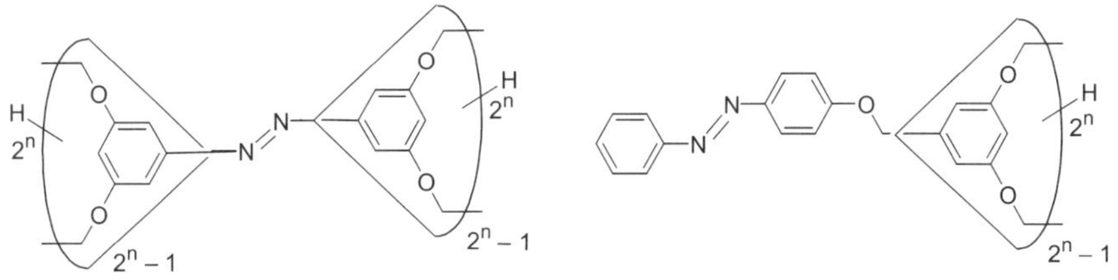

Fig. 9. Dendrimer system reported by Jiang and Aida to exhibit low-energy photon harvesting.[69-71]

Infrared irradiation of *Z*-**26d** was carried out using three specific wavelengths: (a) a stretching vibration for aromatic rings (1597 cm^{-1}), (b) a stretching vibration for CH_2–O (1155 cm^{-1}), and (c) a transparent region (2500 cm^{-1}). Only the 1597 cm^{-1} radiation accelerates the *Z*→*E* isomerization reaction. In addition, irradiation of *Z*-forms of the entire series of dendrimers (**26a**–**26d**) with 280 nm light (λ_{max} of the benzyl aryl ether dendrimer framework) accelerated the *Z*→*E* isomerization reaction, but only for **26c** and **26d** and not **26a** or **26b** (This result is curious in that the azobenzene itself has a finite absorbance at 280 nm and irradiation at this wavelength markedly accelerates *Z*→*E* isomerization.). These two results taken in concert strongly suggest that a matrix to core intramolecular energy transfer is partly responsible for the acceleration effect. Hence, the authors postulate that the dendrimer frameworks in **26c** and **26d** insulate the interior units from collisional energy scattering as well as serving as light harvesting antenna. Photon harvesting is necessary to account for how 1597 cm^{-1} light (0.2 eV) could accelerate a process that has an activation free energy ($\Delta G^{\ddagger}$) of 19.4 kcal mol^{-1} (0.84 eV) at 21 °C. Indeed, photon flux experiments indicated that 4.9 photons at 1597 cm^{-1} (0.98 eV total) are involved in this photochemical process.

Porphyrins were chosen as an alternative chromophore to elucidate whether this selective energy transfer was a feature of just the azobenzene chromophore or the morphology of the entire dendrimer.[60] A variety of symmetrical and unsymmetrical dendrimers were synthesized in which the four *meso*-positions on the porphyrin molecule, P, were substituted with methoxy-terminated poly(aryl ether) dendrons (3rd or 4th generation) and tolyl groups (Figure 10). When the **[G4]$_4$P** dendrimer was excited at 280 nm (dendron absorption), emission was observed primarily from the porphyrin core (656, 718 nm) with only weak emission seen from the dendrons (310 nm), giving an energy transfer quantum yield, Φ_{ENT}, of 80.3%. However, excitation at the same wavelength (280 nm) of the unsymmetrical dendrimers, (**[G4]$_2$P** (*syn*), **[G4]$_2$P** (*anti*), **[G4]$_3$P**) , resulted in emission primarily from the dendrons and only weakly from the porphyrin cores, yielding Φ_{ENT} values of 10.1, 19.7 (*syn*), 10.1 (*anti*), and 31.6%, respectively.

This increase in Φ_{ENT} with increasing number of dendrons suggests a cooperativity of dendrimer subunits involved in the energy transfer process, and further indicates the dependence of the process on morphology of the dendrimer. Fluorescence anisotropy experiments show that once excitation of the dendrons occurs in **[G4]$_4$P**, the energy migrates freely among the dendrimer subunits until it is transferred to the porphyrin core. However, in the unsymmetrical dendrimers, energy migration is less efficient and therefore energy transfer is lower. Temperature-dependent fluorescence measurements suggest that the lower energy transfer for unsymmetrical dendrimers is related to their conformational flexibility. Increasing the temperature results in lower values for Φ_{ENT}, yet the Φ_{ENT} of **[G4]$_4$P** remains essentially the same up to 80 °C.

	R_1	R_2	R_3	R_4
[G4]₁P	[G4]	tolyl	tolyl	tolyl
[G4]₂P (*syn*)	[G4]	[G4]	tolyl	tolyl
[G4]₂P (*anti*)	[G4]	tolyl	[G4]	tolyl
[G4]₃P	[G4]	[G4]	[G4]	tolyl
[G4]₄P	[G4]	[G4]	[G4]	[G4]

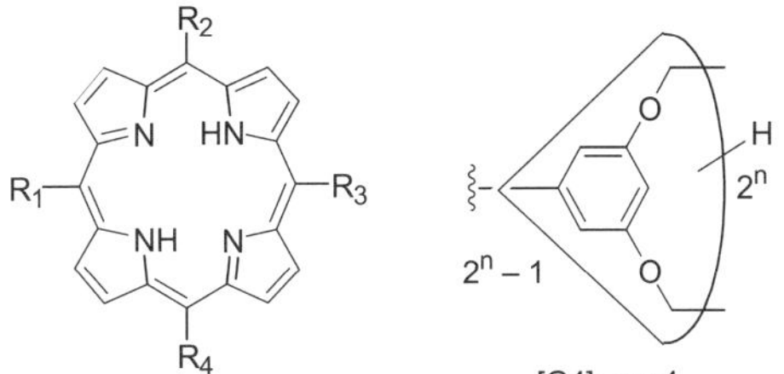

Fig. 10. Schematic of dendrimers containing a single porphyrin core.[60]

In a separate study by Shirai and coworkers, porphyrin dendrimers containing conjugated poly(phenylene) dendrons were synthesized (Figure 11) and exhibit Φ_{ENT} of the highest generation of 98%.[61] This

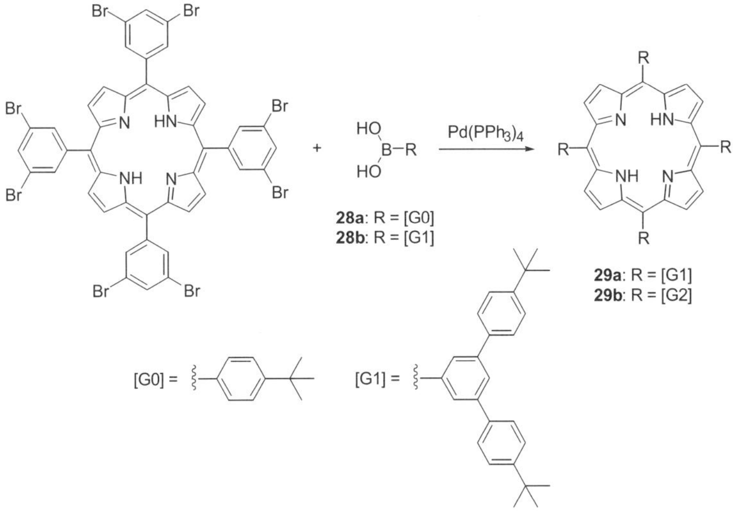

Fig. 11. Synthesis of porphyrin dendrimers containing poly(phenylene) dendrons.[61]

increase in efficiency over the poly(aryl ether) dendrons[60] was attributed to additional through bond (Dexter) energy transfer pathways from the conjugated dendrimer subunits to the porphyrin core. Fluorescence measurements on mixtures of [G2] boronic acid and the porphyrin core excited into the dendron absorption band (262 nm) exhibit emission only from the dendron. No energy transfer emission from the porphyrin appears. This again points to the important role that the dendrimer morphology plays in an energy-transfer process.

Fréchet and coworkers synthesized porphyrin cored dendrimers (Figure 12) and their architecturally isomeric linear analogues which were

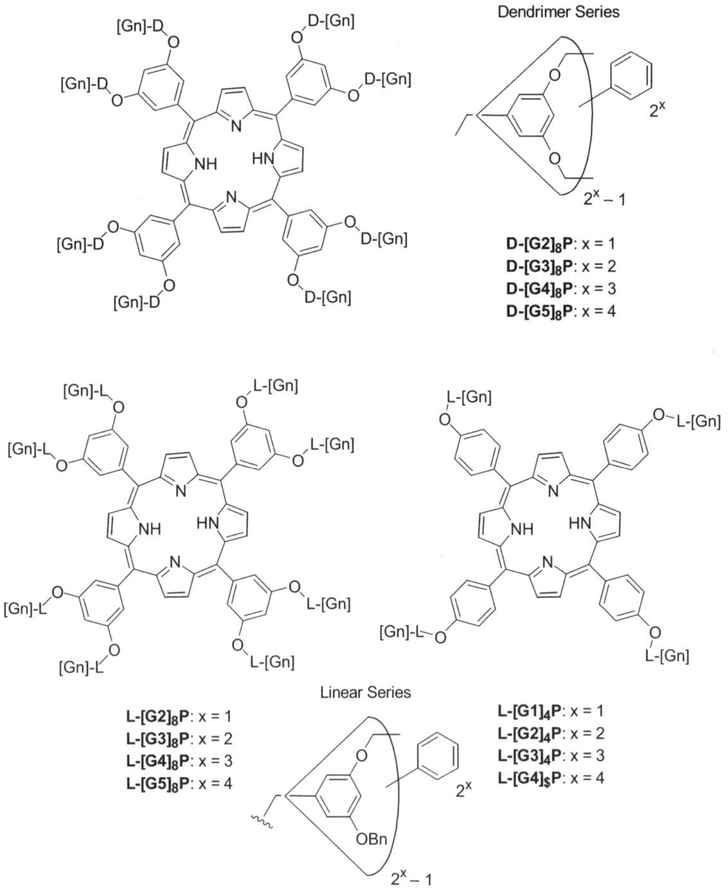

Fig. 12. Schematic of dendritic and four- and eight-arm porphyrins.[62]

then subjected to photophysical characterization.[62] Poly(aryl ether) linear oligomers were attached to a porphyrin core to give four-arm star and eight-arm star molecules. Corresponding eight-arm porphyrin cored dendrimers of generation 2 through 5 were exact architectural isomers, although their hydrodynamic volumes as measured by GPC vastly differed owing to the different conformations adopted by the linear and dendritic macromolecules. The dendrimer, with its more globular structure, gave a much lower hydrodynamic volume then its eight-arm isomer, whose molecular weight was closely predicted by linear polystyrene standards.

To investigate the effect that linear versus dendritic poly(aryl ether) architectures might have on their antennae effect, emission studies were performed to elucidate Φ_{ENT} values. While the Φ_{ENT} only slightly decreased with increasing generation in the dendrimer series (89.7% for [G3], 88.4% for [G4], and 83.9% for [G5]), Φ_{ENT} for the 8-arm linear analogs falls off more dramatically (87.8% for [G3], 74.1% for [G4], and 57.0% for [G5]) corroborating the morphological dependence on Φ_{ENT} that was observed in previously described systems.[60,61] While dendrimers provide a compact structure in which the distance between the dendron chromophores and the porphyrin core is within the Förster radius, the extended structure of the isomeric linear analogues can separate chromophores beyond a useful distance for energy transfer.

5.2 Dendritic Porphyrin Arrays

Star-shaped multi-porphyrin arrays have been constructed to mimic energy funneling in photosynthesis. These dendrimers contain free-base porphyrin cores (P_{FB}) connected to four dendrons consisting of 1, 3, or 7 zinc-metallated porphyrins (P_{Zn}) embedded in an organic matrix connected by ether linkages (Figure 13).[63,64] The periphery consists of methoxy-terminated poly(aryl ether) dendrons. For comparison, cone-shaped arrays consisting of P_{FB} cores monosubstituted with P_{Zn} dendrons were also synthesized.

Emission from the star-shaped arrays arises from excitation of the P_{Zn} followed by intramolecular energy transfer to the P_{FB} cores which act as energy traps. Measured Φ_{ENT} of the star-shaped arrays were 87%, 80%,

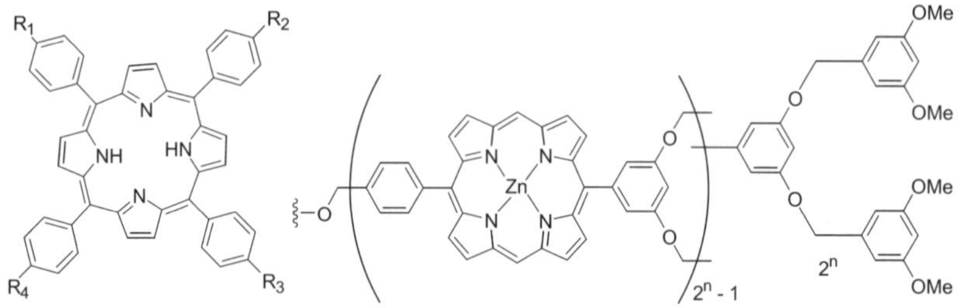

Fig. 13. Schematic of multiporphyrin array.[63,64]

	R_1	R_2	R_3	R_4
(7P$_{Zn}$)$_4$P$_{FB}$	n = 3	n = 3	n = 3	n = 3
(3P$_{Zn}$)$_4$P$_{FB}$	n = 2	n = 2	n = 2	n = 2
(1P$_{Zn}$)$_4$P$_{FB}$	n = 1	n = 1	n = 1	n = 1
(7P$_{Zn}$)$_1$P$_{FB}$	n = 3	OMe	OMe	OMe
(3P$_{Zn}$)$_1$P$_{FB}$	n = 2	OMe	OMe	OMe
(1P$_{Zn}$)$_1$P$_{FB}$	n = 1	OMe	OMe	OMe

and 71% for **(1P$_{Zn}$)$_4$P$_{FB}$**, **(3P$_{Zn}$)$_4$P$_{FB}$**, and **(7P$_{Zn}$)$_4$P$_{FB}$**, respectively. The drop-off of the cone-shaped arrays, however, was more dramatic with Φ_{ENT} of 86%, 66%, and 16% for **(1P$_{Zn}$)$_1$P$_{FB}$**, **(3P$_{Zn}$)$_1$P$_{FB}$**, and **(7P$_{Zn}$)$_1$P$_{FB}$**, respectively. The large gap in Φ_{ENT} between **(3P$_{Zn}$)$_1$P$_{FB}$** and **(7P$_{Zn}$)$_1$P$_{FB}$** indicates that there is an upper limit to the distance between communicating chromophores when those chromophores are not conjugated, consistent with the Förster energy transfer model. Fluorescence lifetimes, τ_D, (monitored at 585 nm, an emission maxima of P$_{Zn}$) of the star-shaped arrays were always shorter then those of their cone-shaped counterparts. However, the gap between the star- and cone-shaped lifetimes became larger with increasing dendrimer generation, demonstrating the inability of the higher generation cone-shaped arrays to efficiently transfer energy to the P$_{FB}$ core as corroborated by Φ_{ENT} values. Fluorescence anisotropies of the star-shaped arrays are lower then those of the cone-shaped molecules, which again indicates energy migration throughout the dendrimer framework followed by energy trapping by the P_{FB} core. Again, morphology of the dendrimer is critical for efficient light-harvesting porphyrins-containing dendrimers as the dendritic subunits show cooperativity in both light-harvesting and energy funneling to the core.

Dendritic porphyrin arrays have been synthesized in which P_{Zn} molecules are connected via conjugated diphenylethynyl linkages and subsequently attached to a P_{FB} core (Figure 14).[65] These arrays show similar energy transfer characteristics to those mentioned previously[63,64] (*i.e.* energy transfer occurs in a downhill fashion from the P_{Zn} to the P_{FB} core). However, the reported Φ_{ENT} values are much higher, presumably due to the increase in electronic communication by conjugation. Interestingly, bulk oxidation experiments show the removal of 20 electrons in $(\mathbf{P_{Zn}})_{20}\mathbf{P_{FB}}$, which yields a stable π-cation radical with hole-storage possibilities.

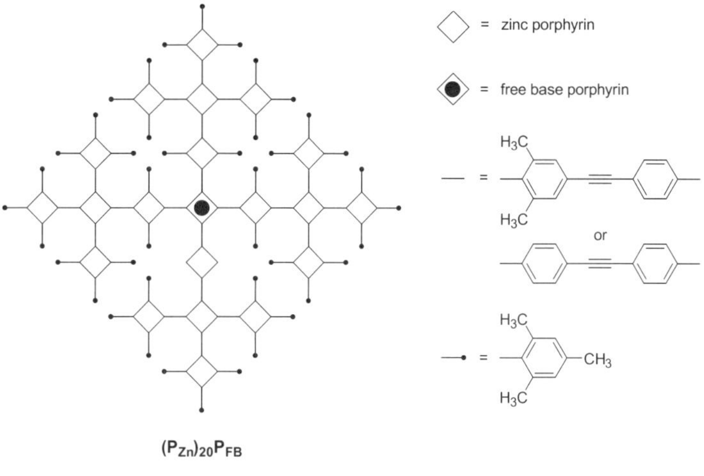

Fig. 14. Multiporphyrin array connected by diphenylethynyl linkages.[65]

5.3 Porphyrins as Peripheral Groups on Organic Dendrimers

Fifth generation poly(L-lysine) dendrimers have been modified at their periphery with 16 P_{Zn} in one hemisphere and 16 P_{FB} molecules in the other by using the orthogonal protection offered by Boc and Fmoc chemistry (Figure 15).[66] Energy transfer from the P_{Zn} to the P_{FB} was measured at 43% which corresponds to roughly 7 of the 16 P_{Zn} being in proximity to the P_{FB} groups. When the porphyrins are arranged in a random fashion on the periphery (not necessarily 16 of each), energy transfer is increased to 85%.[67]

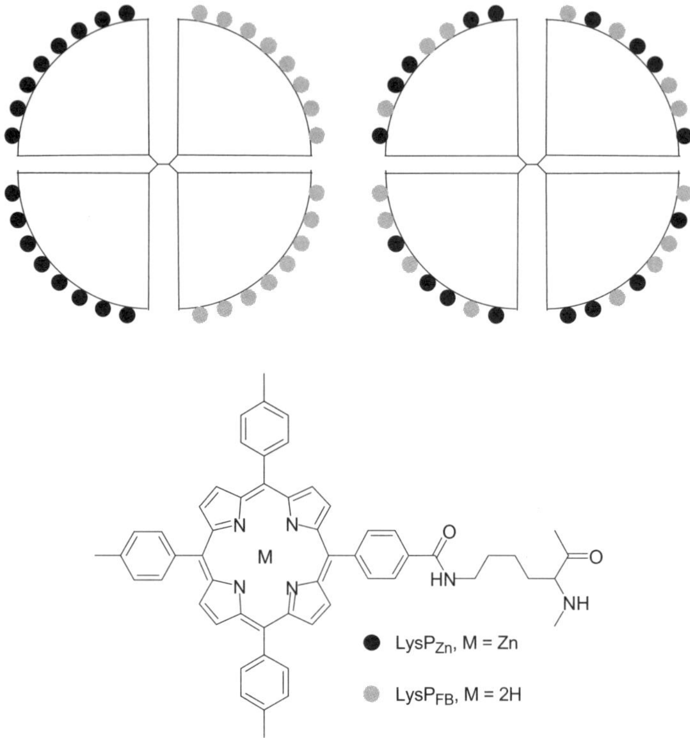

Fig. 15. Porphyrins on the periphery of poly(L-lysine) dendrimers.[66,67]

Meijer and coworkers have synthesized a family of poly(propylene imine) dendrimers of first, third, and fifth generations which contain 4, 16, and 64 P_{FB} moieties, at the periphery, respectively (Figure 16).[68] Fluorescence anisotropy values led to mechanistic models and equations to describe the energy transfer processes. The models take into account the different morphologies of the dendrimers: the disk-like nature of [G1] and the spherical natures of [G3] and [G5]. In the [G1] dendrimer, the Förster mechanism is sufficient to explain the energy transfer. Low anisotropy values in the [G3] dendrimer are explained by considering the dendrons as electronically separate from one another. Intra-dendron porphyrin energy transfer results in limited energy migration and therefore low anisotropy values. A segregated dendron explanation is not feasible for the crowded periphery of the [G5] dendrimer, as an intricate anisotropy decay indicates. The initial decay is fast followed by a slow return and leveling off. Rapid energy transfer between the dense porphyrins at the surface of the sphere, in addition to slower energy transfer between porphyrins further removed from the sphere, can explain the observed anisotropy.

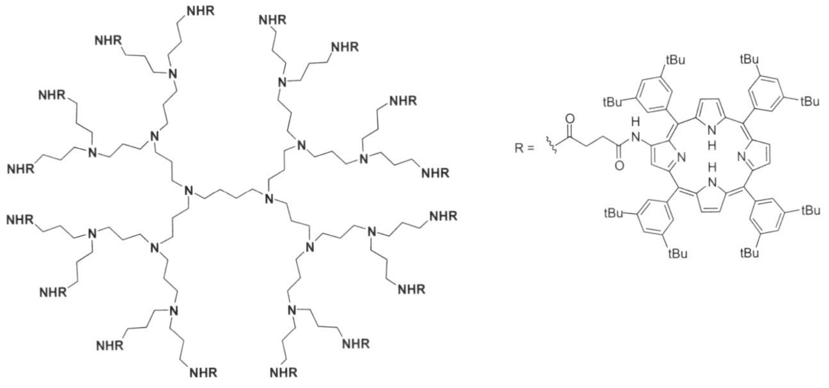

[G3]P$_{16}$

Fig. 16. Porphyrins at the periphery of PI dendrimers.[68]

6. COUMARIN DYE LABELED POLY(ARYL ETHER) DENDRIMERS

Fréchet and coworkers[72,73] have synthesized a series of dendrimers whose energy transfer mechanism is exclusively through-space. By designing dendrimers in which the donor periphery chromophores are effectively separated from the interior acceptors, the dendrimer architecture becomes simply a structural scaffold upon which chromophores can be placed. Chromophores are carefully chosen to satisfy the requirements of Förster energy transfer (i.e. emission of donor overlaps absorbance of acceptor), so that any photons absorbed by a molecule on the periphery undergo intramolecular singlet energy transfer to the core moiety and emission from that core ensues.

In these studies a pair of coumarin dyes was employed as the donor/acceptor pair. They were used for reasons that include: commercial availability at high purity, solubility in organic solvents, high fluorescence quantum yield (Φ_{fl}), sufficient spectral overlap (Figure 17), and a large Stokes shift of the peripheral coumarin-2 dye ensuring that energy transfer, rather then self-quenching, will be more probable following excitation.

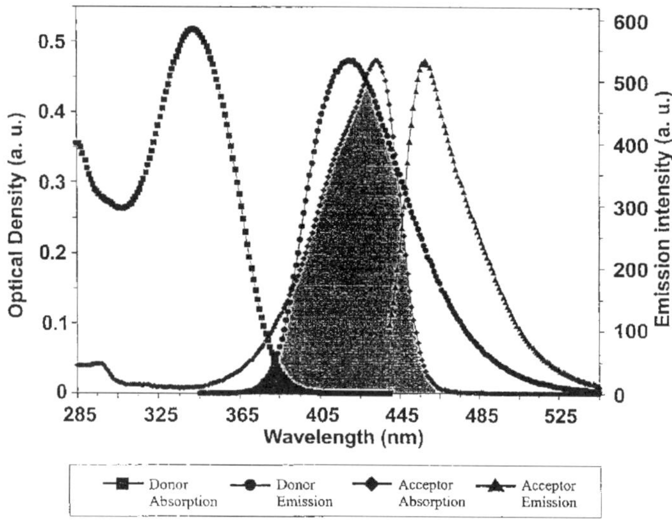

Fig. 17. Absorption and emission spectra of coumarin 2 and coumarin 343.[73]

To accommodate the nucleophilic nature of the coumarin-2 periphery dyes, 'reverse' dendrons[30] were used so that the dendrimer could be convergently synthesized (Figure 18). 'Reverse' refers to using 3,5-benzyl groups versus the typical 3,5-phenols, thereby switching the reactivity at the 3 and 5 positions from nucleophilic to electrophilic making them reactive towards the nucleophilic coumarin 2. The interior dye was coumarin 343 (C343), whose absorbance overlapped sufficiently with the emission of coumarin-2.

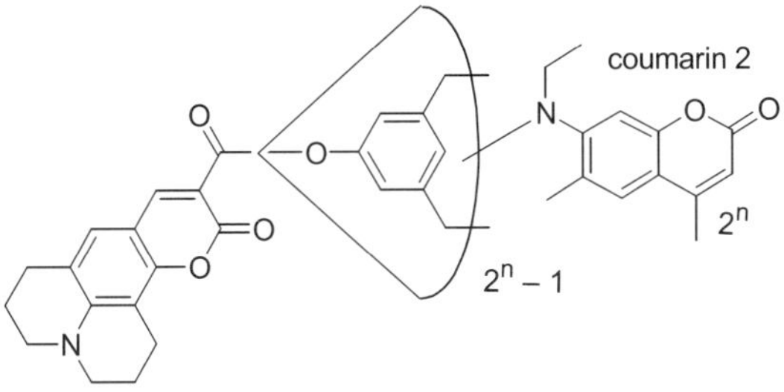

coumarin 2

coumarin 343

30a: n = 1
30b: n = 2
30c: n = 3
30d: n = 4

Fig. 18. 'Reverse' dendrons with coumarin 2 hosts and coumarin 343 guest.[72,73]

Energy transfer in the first through third generation dendrimers (**30a-c**) is nearly quantitative, as measured by comparing absorbance and excitation spectra and by studying fluorescence quenching of the donor by the acceptor. For the fourth generation (**30d**), the energy transfer efficiency decreases to 86% which is likely due to the increased interchromophoric distance. Energy transfer to the dendrimer scaffold is unlikely because its excited state lies at higher energy than both the coumarin donor and acceptor.

The significance of the dendrimers as effective light harvesters is most apparent in their 'amplified emission.' 'Amplified emission' occurs when the emission intensity from the core is greater when the donor is excited rather then the core itself, and is a direct result of both the light harvesting abilities of the donors and the energy transfer efficiency to the acceptor. Comparison of fluorescence spectra indicate that there is a pronounced 'amplified emission' for the higher generation dendrimers **30c** and **30d** which contain 8 and 16 donor chromophores, respectively (Figure 19). However, the increase in emission intensity does not scale with the absorbance increases for higher generations. The authors attribute this to an increase in non-radiative pathways for relaxation of excitons. Several generations of dendrimers containing coumarin-2 donor dyes on the periphery of 'reverse' dendrons with penta- and heptathiophene acceptors at the core were prepared and also exhibited similar energy harvesting capabilities to the coumarin-2/C343 systems.[74]

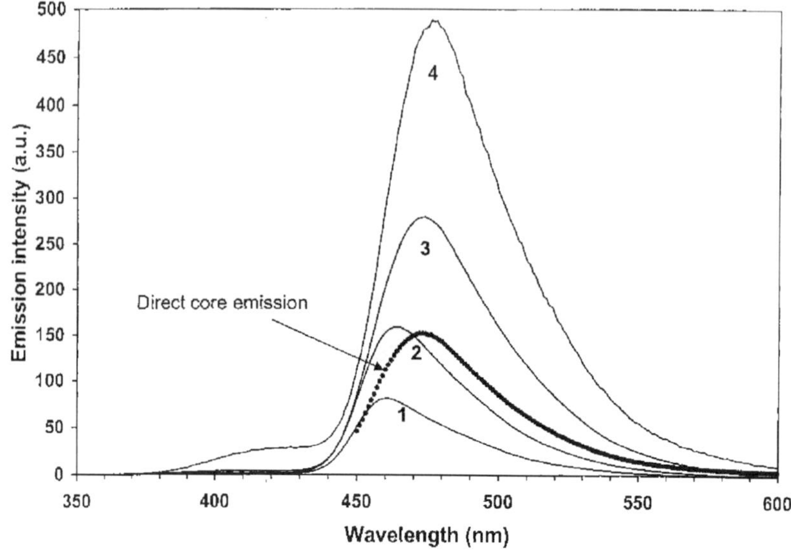

Fig. 19. Emission spectra of **30a-d** in toluene and direct core emission (dotted line).[73]

Dendrimers containing chromophore donor/acceptor pairs carefully placed on an inert scaffold have found a use as the emitting material in light emitting diodes.[75,76] If the donor/acceptor molecules also have hole and/or electron transporting capabilities, then a single layer device is possible. Fréchet and coworkers have synthesized dendrimers with hole-transporting triarylamines on the periphery and an energetically overlapping emissive acceptor on the interior (**31a** and **31b**).[75,76] These dendrimers, when mixed with electron-transporting oxadiazole and sandwiched between electrodes emit light with external quantum efficiencies of 0.012 % using C343 as the emitter, and 0.12 % when pentathiophene (T5) is used as the emitter. Mixed dendrimer films (both T5 and C343 dendrimers) showed light emission primarily from the T5 (lower energy) with only a small emission from the C343, indicating that energy transfer also takes place between the emitting groups in neighboring dendrimers since the Förster radius (35-38 Å) is smaller then the interchromophoric distance. Energy transfer is not quantitative, however, since the emission of light from the C343 dendrimers increases linearly as the percentage of C343 dendrimer increases in the blend.

31a, X=

31b, X=

To overcome this large Förster radius and create mixed dendrimer multicolor OLED's, larger dendrimers were needed so that energy transfer does not occur. However, poor solubility and increasing crystallinity of the dendrons prevented the synthesis of these target dendrimers.[75,76] To overcome these problems, Fréchet and coworkers[77,78] replaced every other triarylamine group on the periphery of the dendrons with a dialkyl substituted phenyl ring (**32** and **33**). Dendrimers up to the 4th generation could be synthesized with C343 cores, and up to the 5th generation for T5 cores when only two dendrons are attached to either end of the T5 core in a 'barbell' fashion. Site isolation of the individual cores, which implies that interchromophoric energy transfer is prevented, was probed by monitoring emission of C343 in films (thickness 1100–1300 Å) of mixtures of the dendrimers by photoluminescence (PL) and electroluminescence (EL). Results indicate that while site isolation for the C343 dendrimers is considerable at the fourth generation, it is only when the T5 dendrimers reach the fifth generation that their site isolation

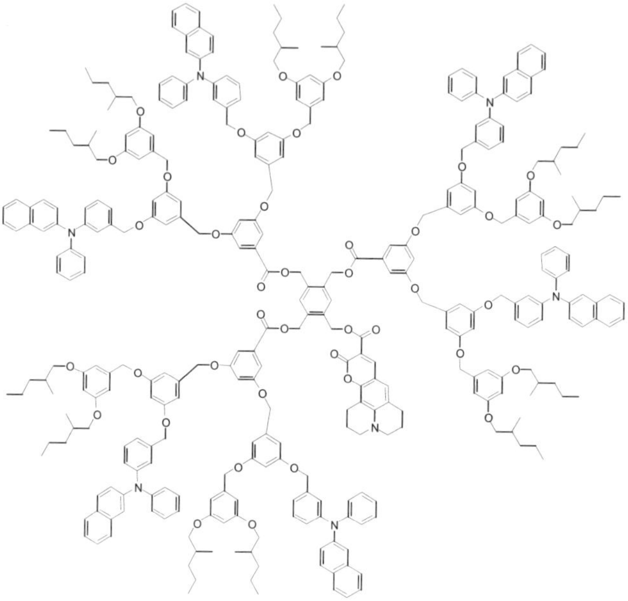

32

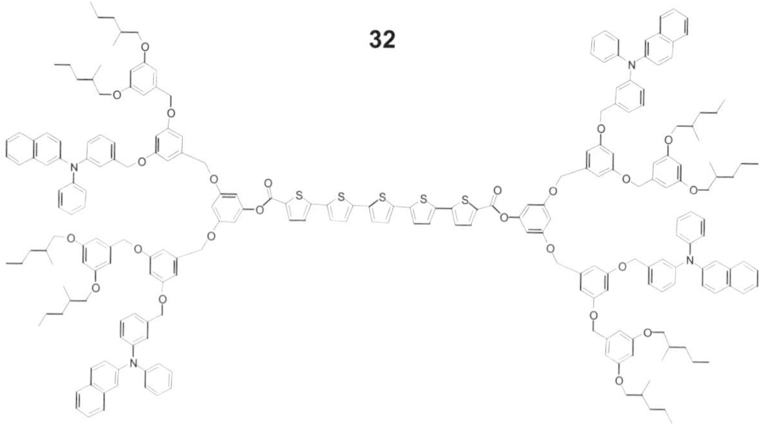

33

is significant. This is not surprising when considering the shape of the dendrimers. Presumably, the three-fold architecture of the C343 structures surrounds or encapsulates the core more effectively then the 'dumbbell' architecture in the T5 dendrimer.

Because of the site-isolation afforded by the higher generation dendrimers, there is a large increase in C343 emission (normalized to T5 emission) when the molar ratio of C343 to T5 is increased from 1:1 to 5:1, indicating a diminishing or lack of energy transfer. Upon dilution of mixtures of [G4] C343 and [G5] T5 by embedding in a polystyrene matrix, emission of C343 again increases, due partly to the reduced energy transfer to the T5 dendrimers but mostly to reduced self-quenching. The experiment also indicates that even at high generations of both the C343 and T5 dendrimers, complete site-isolation is still not achieved, although there is a significant improvement over lower generations. OLEDs were fabricated and showed external quantum efficiencies of 0.2% for mixed dendrimer devices and 0.76% for [G5] T5 devices alone. The higher efficiency for the [G5] T5 is attributed to its ability to trap electrons more effectively then C343 dendrimers which leads to exciton formation and subsequently light emission. Note the higher efficiencies then previously reported[75,76] (0.012% for C343 emitters and 0.12% for T5 emitters) with lower generation dendrimers.

Fréchet and coworkers[79] have also synthesized multi-chromophore cascade dendrimers whose excited state energies decrease towards the core. The authors have interested in elucidating whether energy transfer occurs in a step-wise fashion to the lowest energy acceptor (**Ac**), or whether there is a component of energy transfer that takes place from the highest energy donor (**D1**) to **Ac** *bypassing* the middle chromophore (**D2**), even though **D1** and **Ac** were spaced farthest apart in the dendrimer array (Figure 20). A series of model compounds were prepared combining only two chromophores so that energy transfer measurements could be made. FRET efficiencies for **D1** to **D2** (**35**; 99%), **D1** to **Ac** (**36**; 79%), and **D2** to **Ac** (**37**; 96%) were measured by comparing the emission of the donors with and without the presence of acceptors. Since the measured FRET in the dendrimer containing both donor chromophores, **D1** and **D2** (**34**), and the perylene acceptor, **Ac**, at the core was greater then 95% when exciting **D1** (342 nm), a cascade energy transfer from **D1** to **D2** to **Ac** seems likely since **D1** to **Ac** FRET efficiency was much lower (79%) then that of the multi-chromophore dendrimer (95%).

Mixed self-assembled monolayers (SAMs) which consist of coumarin-343 acceptors and dendrons substituted with coumarin-2 donors at the periphery have been constructed on silicon wafers and their energy transfer properties investigated by front-face fluorescence spectroscopy.[80] Energy transfer comparisons between mixed SAMs containing coumarin-2 donor dyes either in a dendritic array (**38**) or as a single molecule terminating a long chain mixed with acceptor C343 (**39**) shed light on the importance of the dendritic structure. In the case of the dendritic array **38**, energy transfer was efficient as noted from the lack of donor emission. Amplified emission was also apparent when the donors were excited rather then the acceptors. However, when the linear single molecules were mixed with the acceptors in a 4:1 ratio (to keep the molar ratio of donors to acceptors the same as in the dendritic case), there was significant donor emission. A likely explanation for the donor emission is that some chromophore pairs were too far apart (exceeding the Förster radius of 42 Å) due to phase separation of the donors into domains. A dependence upon donor/acceptor molar ratio was observed in mixed SAMs of dendrons containing only two coumarin-2 dyes and linear acceptors. It was found that if the molar ratio exceeded 4:1, emission resulted from both acceptor and donor, but at 4:1, emission occurred almost exclusively from the acceptors. Lower molar ratios (1:2) also showed complete quenching of donor emission. However, no amplified emission was observed.

7. TWO-PHOTON LIGHT HARVESTING AND ENERGY TRANSFER

Dendrons and dendrimers containing two-photon absorbing (**TPA**) chromophores as the donors and a Nile red dye (**NR**) as the core acceptor have been synthesized for potential uses in two-photon imaging and optical limiting (Figure 21).[81,82] Single-photon absorption and energy transfer from the **TPA** to the **NR** in dendrons consisting of one **TPA** and one **NR** resulted in a 3.4-fold increase in **NR** emission when exciting the

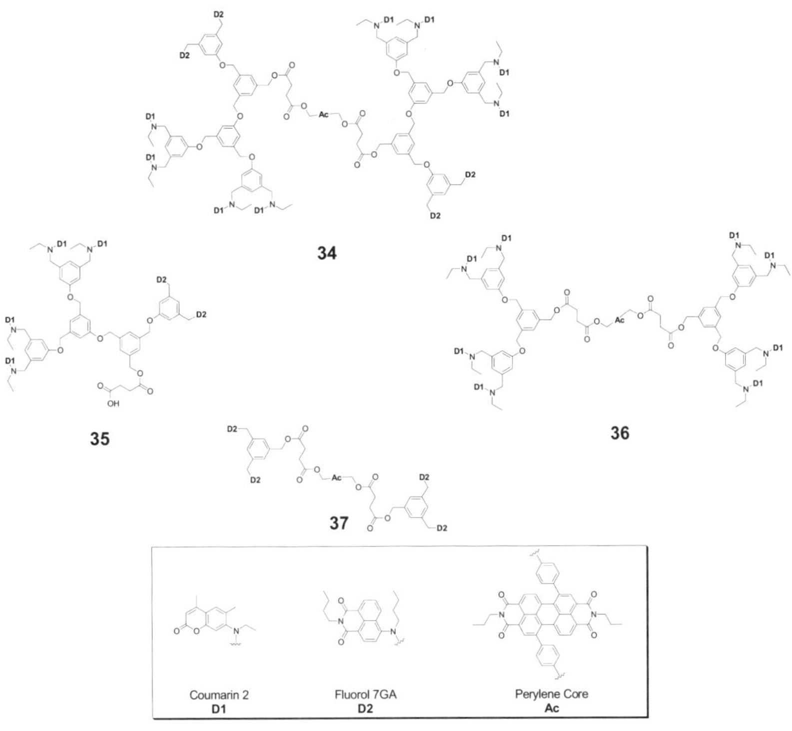

Fig. 20. Schematic of dendrimers with multiple donors and a single acceptor.[79]

TPA chromophores (405 nm) rather then the **NR** (540 nm).[81] This can be attributed to the antennae effect of the **TPA** moieties which effectively increases the extinction coefficient of the molecule at the **TPA** single-photon absorbing wavelength (405 nm). Energy transfer for single-photon absorption was found to be greater then 99%. Laser-induced two-photon absorption also resulted in energy transfer from the **TPA** groups to the **NR** core.[82] Three dendritic compounds consisting of 1, 2, and 4 **TPA** hosts surrounding a **NR** guest showed 8, 20, and 34-fold increases in **NR** emission, respectively, also attributed to the antennae effect. Compounds such as these which can absorb and transfer both single- and two-photon energy to a central core are important as energy harvesters because the window of wavelengths that can be used to excite the molecules is increased substantially over systems capable of only single-photon absorption.

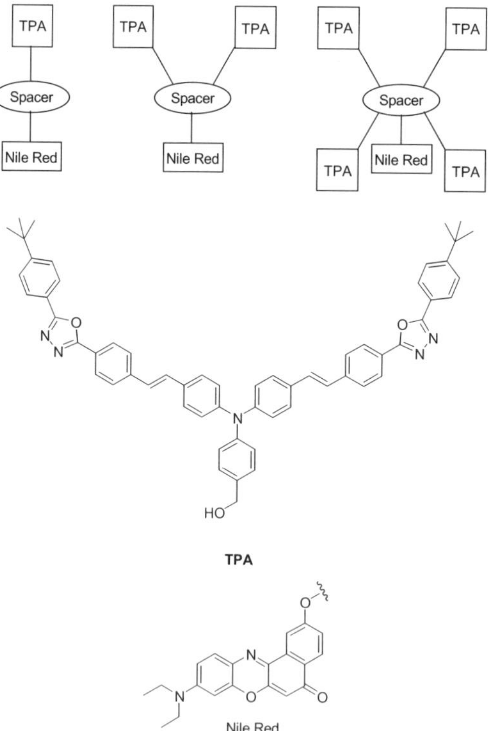

Fig. 21. Two-photon dendrimers with a nile red dye acceptor.[81,82]

8. POLYPHENYLENE DENDRIMERS

In the past several decades, many different dendrimer architectures have been introduced for a variety of purposes. One of the simplest designs are the polyphenylene dendrimers which consist only of branched phenyl groups. Since there are no bridging atoms between the rings, the phenyl groups of the dendrimer are twisted out of plane so that the dendrons themselves have limited conjugation. The bulky nature of the dendrons lends them as useful insulators for chromophores with aggregation, crystalinity, and solubility issues.[83] Also, as with other dendrimer architectures, polyphenylenes can be used as inert scaffolds onto which host/guest chromophores can be placed for Förster energy transfer.[84-86] These types of polyphenylene dendrimers have been studied extensively by Müllen and coworkers by a variety of techniques including fluorescence anisotropy and single-molecule spectroscopy.

The polyphenylene dendrimers were found to effectively isolate large chromophores such as perylene derivatives, when two polyphenylene dendrons (first through third generation) are attached to a perylenetetracarboxylic diimide core (PDI) (**40**).[83] Although the alkoxy substituents in the 'bay' area of the perylene twist the core out of planarity, the chromophore is only slightly blue-shifted when incorporated into the dendrimers. Therefore, energy transfer from the polyphenylene dendrons is still accomplished with high efficiency.

When a phenylene first generation dendrimer with a tetrahedral core is substituted on the periphery with three peryleneimides (PI) and one terryleneimide (TI) (**41**) to act as the lower energy acceptor, energy hopping between the PI molecules eventually results in energy transfer to the TI trap.[84,85] Since the dendrimer only contains the four equally spaced chromophores, no intramolecular excimers are formed, as evidenced by their localized fluorescence. Two different fluorescence anisotropy decay times measured for this system indicate two different energy pathways available for trapping by TI. The authors hypothesize that there are two different environments for the TI groups with respect to the PI donors. One environment places the chromophores close to one another, providing the initial close-proximity pathway. The two pathways, however, do not affect the overall energy transfer efficiency which is notably high at 96%.

40

41

By changing the position of the chromophores so that the terrylenediimide (TDI) is at the core and the PI are at the periphery of polyphenylene dendrons, Müllen and coworkers have created dendrimers which are incapable of intramolecular aggregation of chromophores due to their rigidity (**42** and **43**).[86] Additional high energy donors such as naphthaleneimides (NI) have been placed in a 2:1 ratio to PI on the periphery of the dendrons to create cascading dendrimers (**44**) which can undergo stepwise energy transfer from the NI to the PI and ultimately, to the TDI moiety at the core. Excitation of the PIs in the two-chromophore dendrimers (**42** and **43**) results in roughly 93% energy transfer to the TDI core in both **42** and **43**. Energy transfer in the three-chromophore system (**44**) also occurs, although a percentage is not reported for efficiency. Because of the lack of spectral overlap between the NI and TDI chromophores, any energy harvested by the NIs must be first transferred to the PI before ultimately arriving at the TDI core.

9. ENERGY TRANSFER TO ENCAPSULATED GUESTS

While several researchers have investigated energy transfer from a host (often the dendron itself) to a guest at the core of the dendrimer, there are few examples of guests that are not covalently bound but encapsulated in some fashion into the voids of the dendrimer. In an elegant example by Meijer and coworkers, oligo (*p*-phenylene vinylene)s (OPPVs) are covalently attached to the periphery of a PPI dendrimer and transfer energy to an encapsulated anionic dye, Sulforhodamine B (Figure 22).[87] The dye is first loaded into the dendrimer by extraction from an aqueous solution into a dendrimer-containing organic phase. An acid-base reaction between the tertiary amines in the dendrimer and the carboxylic acids of the dye holds the dyes creates electrostatically bound guests.

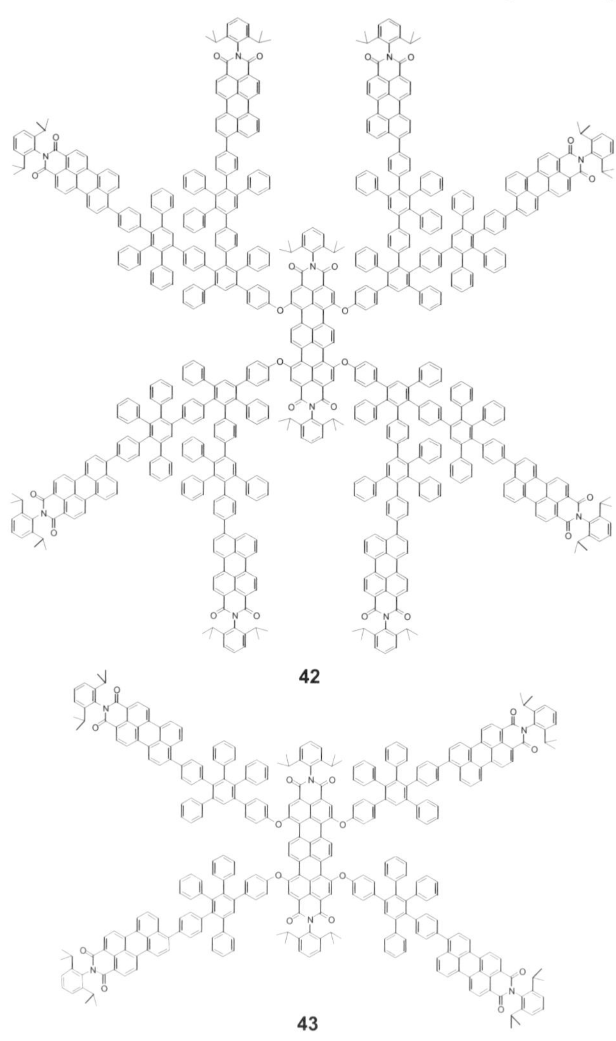

42

43

44

Seven dyes are loaded into the third generation dendrimer **45** while the fifth generation dendrimer is capable of extracting 26 dyes into its interior, occupying about half of the available tertiary amine sites located in each dendrimer. Solution photoluminescence measurements indicate that when the dye concentration in the dendrimer is at a maximum, energy transfer from the OPPV groups to the dye is approximately 40% for both third and fifth generation. Although the energy transfer percentage is lower than in most covalently incorporated guests, Meijer's system offers the unique opportunity of switching the wavelengths of emitted light if the dye can be removed from the dendrimer as easily as it is introduced.

Fluorescence titration curves show an initial steep increase in dye fluorescence when approximately one or two dyes are bound followed by only a gradual increase in fluorescence as each additional dye is bound to

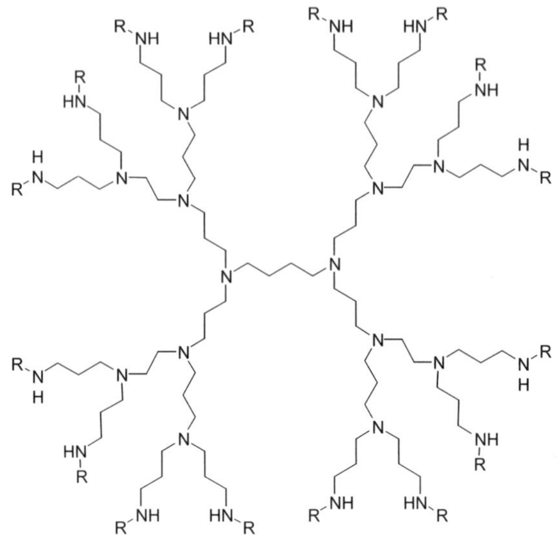

45

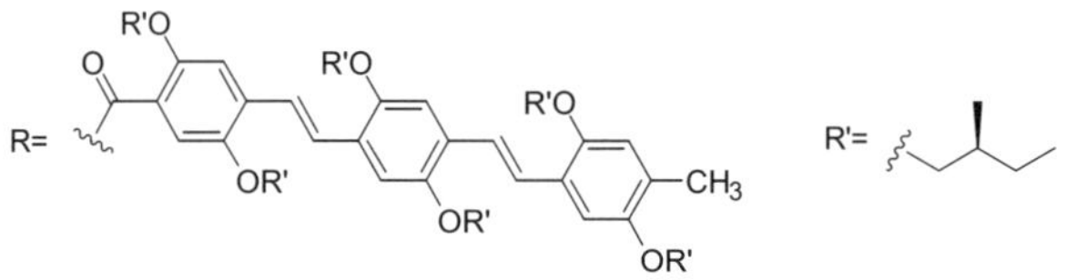

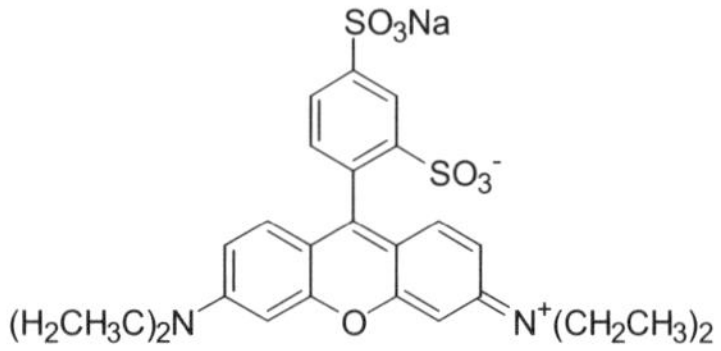

Sulforhodamine B

Fig. 22. Structures of third generation PPI-OPPV dendrimers and Sulforhodamine B.[87]

the interior of the dendrimers (Figure 23). This change in slope could be due to either self-quenching of the dyes which will compete with fluorescence once the concentration of dye is high enough, or just simply an antennae effect of the OPPVs. Thin films of the dye-loaded dendrimers showed a much higher energy transfer efficiency (greater than 90%) which could be due to either better spectral overlap (the film fluorescence is slightly red-shifted) or better orientation of the host and guest

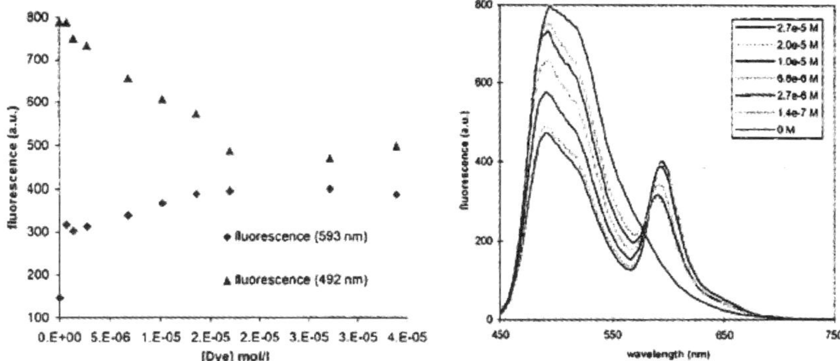

Fig. 23. (a) Fluorescence titration curve monitoring both the emission of the G3 dendrimer (492 nm) and the Sulforhodamine B dye (593 nm) in the CHCl$_3$ layer as the concentration of dye is increased in the water layer. (b) Fluorescence spectra of the CHCl$_3$ layer as concentration of the dye is increased in the water layer.[87]

molecules relative to each other within the film, a parameter in the Förster energy transfer equation. Additional dyes incorporated into the dendrimer all resulted in near-quantitative energy transfer in thin films.

Vögtle and Balzani have also explored light-harvesting and energy transfer to non-covalently linked guests.[88-90] They utilize PPI dendrimers modified with dansyl groups at the periphery (**46**) as the hosts for acidic dyes (i.e. eosin Y, fluorescein, rose bengal) as the encapsulated guest emitters.[88,89] Dendrimers of second through fifth generation were capable of extracting eosin Y (a diacid) into their cavities. The amount of eosin molecules incorporated into the dendrimers was both pH and concentration dependent. Fluorescence measurements performed mainly on the fourth generation dendrimer **46** show that when the 32 dansyl groups of the dendrimer are excited, the eosin guests quench their fluorescence. Emission from eosin, however, is also significantly quenched. This is thought to be due to non-radiative emission pathways made possible by the electrostatic bonds holding the eosin molecules in place. Encapsulation of one eosin molecule is enough to quench the fluorescence of all 32 dansyl groups.

A more recent example by Balzani and Vögtle provided an additional chromophore for cascade energy transfer.[90] Poly(aryl ether) dendrons with naphthyl groups at the periphery were attached to the

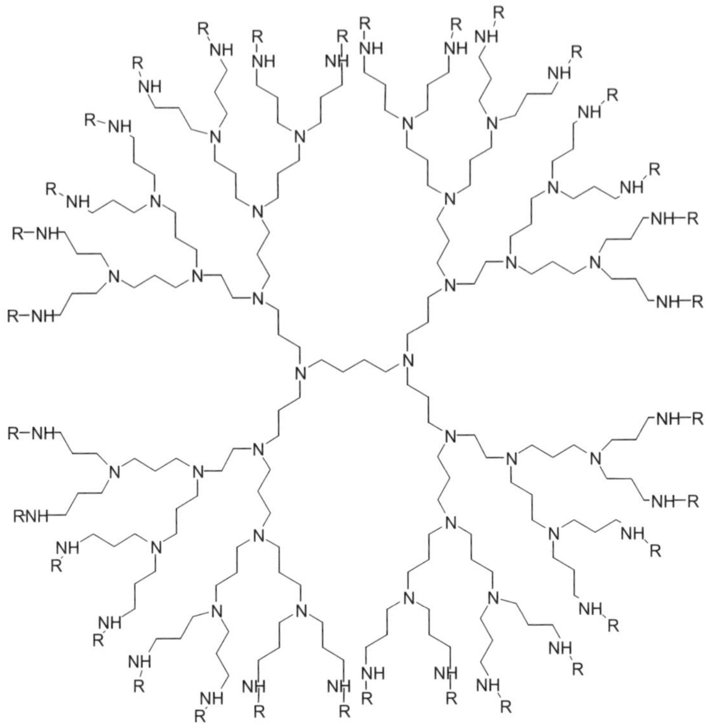

46

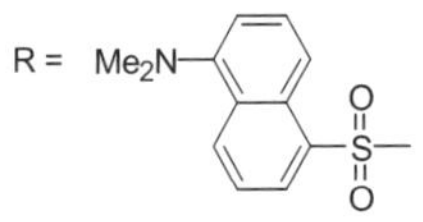

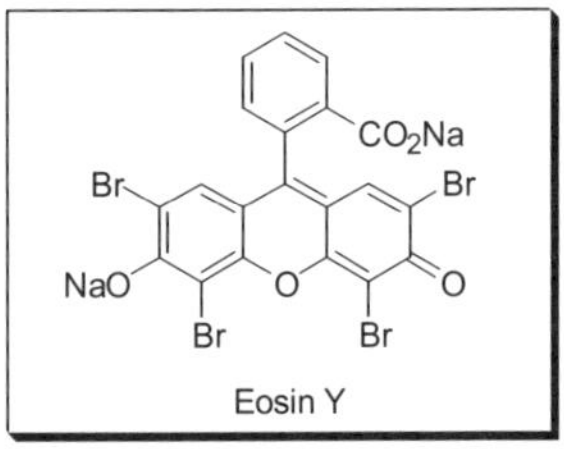

R =   (dansyl group)

Eosin Y

sulfonamides of the second generation dansyl terminated dendrimers to create a 'super' dendrimer **47** consisting of three different types of chromophores: 32 naphthyls, 24 alkoxybenzenes, and 8 dansyls. When eosin is extracted into the interior of the dendrimer and the dansyl groups are excited, emission from eosin has greater than 80% transfer efficiency. Any light harvested by the naphthyl or dimethoxybenzyl chromophores is thought to undergo energy transfer to the dansyl units and then to the eosin guest(s). Thus, a cascading energy transfer dendrimer was realized.

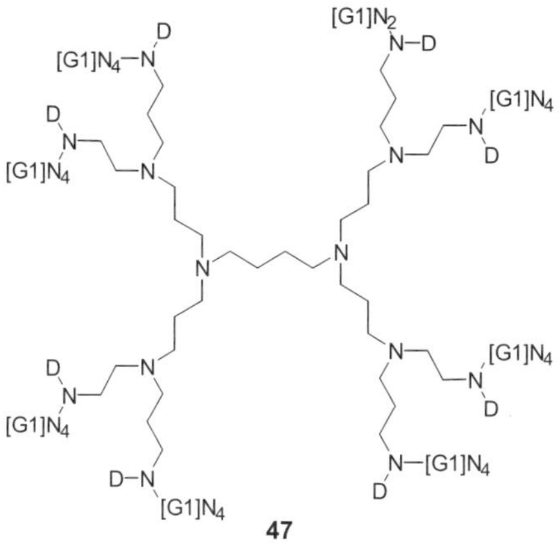

47

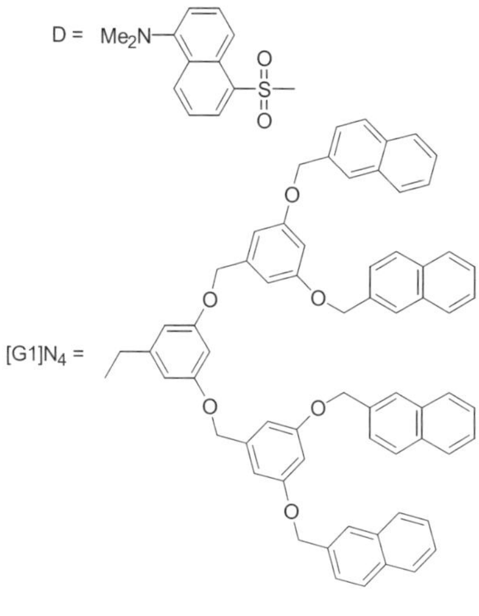

10.   CONCLUSION

Dendrimers have been shown to act as efficient light harvesters. Since they can be easily synthesized and modified to include chromophores, many routes to light-harvesting dendrimers have been explored and will continue to be developed. Energy transfer efficiency in many dendritic systems is near quantitative, establishing them as viable photosynthetic mimics. The high energy transfer efficiency in these systems has led to their use in optoelectronic devices such as OLEDs and fluorescent sensors. With new easier synthetic approaches to light harvesting dendrimers constantly emerging, it is envisioned that these dendritic systems will be competitive with polymers as organic materials for devices.

ACKNOWLEDGEMENTS

We gratefully acknowledge support from the NSF Science and Technology Center-Materials and Devices for Information Technology (DMR-0120967). DVM is a Camille Dreyfus Teacher-Scholar.

References

1. Denti G, Campagna S, Serroni S, Ciano M and Balzani V. *J. Am. Chem. Soc.* 1992; **114**: 2944-2950.
2. Serroni S, Denti G, Campagna S, Juris A, Ciano M and Balzani V. *Angew. Chem. Int. Ed.* 1992; **31**: 1493-1495.
3. Campagna S, Denti G, Serroni S, Juris A, Venturi M, Ricevuto V and Balzani V. *Chem. Eur. J.* 1995; **1**: 211-221.
4. Sommovigo M, Denti G, Serroni S, Campagna S, Mingazzini C, Mariotti C and Juris A. *Inorg. Chem.* 2001; **40**: 3318-3323.
5. Serroni S, Juris A, Venturi M, Campagna S, Resino IR, Denti G, Credi A and Balzani V. *J. Mater. Chem.* 1997; **7**, 1227-1236.
6. Balzani V, Campagna S, Denti G, Juris A, Serroni S and Venturi M. *Acc. Chem. Res.* 1998; **31**: 26-34.
7. Balzani V and Juris A. *Coord. Chem. Rev.* 2001; **211**: 97-115.
8. Campagna S, Di Pietro C, Loiseau F, Maubert B, McClenaghan N, Passalacqua R, Puntoriero F, Ricevuto V and Serroni S. *Coord. Chem. Rev.* 2002; **229**: 67-74.
9. Serroni S, Campagna S, Puntoriero F, Loiseau F, Ricevuto V, Passalacqua R and Galletta M. *Comptes Rendus Chimie* 2003; **6**: 883-893.
10. Plevoets M, Vögtle F, De Cola L and Balzani V. *New J. Chem.* 1999; **23**: 63-69.
11. Zhou X, Tyson DS and Castellano FN. *Angew. Chem. Int. Ed.* 2000; **39**: 4301-4305.
12. McClenaghan N, Passalacqua R, Loiseau F, Campagna S, Verheyde B, Hameurlaine A and Dehaen W. *J. Am. Chem. Soc.* 2003; **125**: 5356-5365.

13. Vögtle F, Gestermann S, Kauffmann C, Ceroni P, Vicinelli V, De Cola L and Balzani V. *J. Am. Chem. Soc.* 1999; **121**: 12161-12166.

14. Balzani V, Ceroni P, Gestermann S, Gorka M, Kauffmann C and Vögtle F. *J. Chem. Soc., Dalton Trans.* 2000; **21**: 3765-3771.

15. Vögtle F, Gorka M, Vicinelli V, Ceroni P, Maestri M and Balzani V. *ChemPhysChem* 2001; **12**: 769-773.

16. Vicinelli V, Ceroni P, Maestri M, Balzani V, Gorka M and Vögtle F. *J. Am. Chem. Soc.* 2002; **124**: 6461-6468.

17. Kawa M and Fréchet JMJ. *Chem. Mater.* 1998; **10**: 286-296.

18. Kawa M and Fréchet JMJ. *Thin Solid Films* 1998; **331**: 259-263.

19. Balzani V and Vögtle F. *Comptes Rendus Chimie* 2003; **6**: 867-872.

20. Campagna S, Denti G, Sabatino L, Serroni S, Ciano M and Balzani V. *Gazz. Chim. Ital.* 1989; **119**: 415-417.

21. Denti G, Serroni S, Campagna S, Ricevuto V and Balzani V. *Inorg. Chim. Acta* 1991; **182**: 127-129.

22. Campagna S, Denti G, Serroni S, Ciano M and Balzani V. *Inorg. Chem.* 1991; **30**: 3728-3732.

23. Denti G, Campagna S, Sabatino L, Serroni S, Ciano M and Balzani V. *Inorg. Chim. Acta* 1990; **176**: 175.

24. Juris A, Balzani V, Barigelletti F, Campagna S, Belser P and von Zelewsky A. *Coord. Chem. Rev.* 1988; **84**: 85-277.

25. McClenaghan ND, Loiseau F, Puntoriero F, Serroni S and Campagna S. *Chem. Commun.* 2001: 2634-2635.

26. Hecht S and Fréchet JMJ. *Angew. Chem. Int. Ed.* 2001; **40**: 74-91.

27. Vögtle F, Plevoets M, Nieger M, Azzellini GC, Credi A, De Cola L, De Marchis V, Venturi M and Balzani V. *J. Am. Chem. Soc.* 1999; **121**: 6290-6298.

28. Issberner J, Vögtle F, De Cola L and Balzani V. *Chem. Eur. J.* 1997; **3**: 706-712.

29. Tyson DS, Luman CR and Castellano FN. *Inorg. Chem.* 2002; **41**: 3578-3586.

30. Tyler TL and Hanson JE. *Polym. Mater. Sci. Eng.* 1995; **73**: 356-357.

31. Vögtle F, Gestermann S, Kauffmann C, Ceroni P, Vicinelli V and Balzani V. *J. Am. Chem. Soc.* 2000; **122**: 10398-10404.

32. Balzani V, Ceroni P, Gestermann S, Kauffmann C, Gorka M and Vögtle F. *Chem. Commun.* 2000: 853-854.

33. Kido J and Okamoto Y. *Chem. Rev.* 2002; **102**: 2357-2368.

34. Kuriki K, Koike Y and Okamoto Y. *Chem. Rev.* 2002: **102**: 2347-2356.

35. Haas Y and Stein G. *J. Phys. Chem.* 1971; **75**: 3677.

36. Stein G and Wurzberg E. *J. Chem. Phys.* 1975; **62**: 208-213.

37. Xu Z and Moore JS. *Ang. Chem. Int. Ed.* 1993; **32**: 246.

38. Xu Z; Moore, JS. *Ang. Chem. Int. Ed.* 1993; **32**: 1357.

39. Xu Z, Kahr M, Walker KL, Wilkins CL and Moore JS. *J. Am. Chem. Soc.* 1994; **116**: 4537-4550.

40. Kopelman R, Shortreed M, Shi Z-Y, Tan W, Xu Z, Moore JS, Bar-Haim A and Klafter J. *Phys. Rev. Lett.* 1997; **78**: 1239-1242.

41. Xu Z and Moore JS. *Acta Polymer* 1994; **45**: 83-87.

42. Devadoss C, Bharathi P and Moore JS. *J. Am. Chem. Soc.* 1996; **118**: 9635-9644.

43. Tretiak S, Chernyak V and Mukamel S. *J. Phys. Chem. B* 1998; **102**: 3310-3315.

44. Shortreed MR, Swallen SF, Shi Z-Y, Tan W, Xu Z, Devadoss C, Moore JS and Kopelman R. *J. Phys. Chem. B* 1997; **101**: 6318-6322.

45. Swallen SF, Kopelman R, Moore JS and Devadoss C. *J. Mol. Struct.* 1999; **486**: 585-597.

46. Swallen SF, Zhu Z, Moore JS and Kopelman R. *J. Phys. Chem. B* 2000; **104**: 3988-3995.

47. Zhu A, Bharathi P, White JO, Drickamer HG and Moore JS. *Macromolecules* 2001; **34**: 4606-4609.

48. Kleiman VD, Melinger JS and McMorrow D. *J. Phys. Chem. B* 2001; **105**: 5595-5598.

49. Gaab KM, Thompson AL, Xu J, Martinez TJ and Bardeen CJ. *J. Am. Chem. Soc.* 2003; **125**: 9288-9289.

50. Gong L-Z, Hu Q-S and Pu L. *J. Org. Chem.* 2001; **66**: 2358-2367.

51. Pugh VJ, Hu Q-S, Zuo X, Lewis FD and Pu L. *J. Org. Chem.* 2001; **66**: 6136-6140.

52. Peng Z, Pan Y, Yu B and Zhang J. *J. Am. Chem. Soc.* 2000; **122**: 6619.

53. Melinger JS, Pan Y, Kleiman VD, Peng Z, Davis BL, McMorrow D and Lu M. *J. Am. Chem. Soc.* 2002; **124**: 12002-12012.

54. Pan Y, Lu M, Peng Z and Melinger JS. *J. Org. Chem.* 2003; **68**: 6952-6958.

55. Pan Y, Peng Z and Melinger JS. *Tetrahedron* 2003; **59**: 5495-5506.

56. Kwok CC and Wong MS. *Chem. Mater.* 2002; **14**: 3158-3166.

57. Halim M, Pillow JNG, Samuel IDW and Burn PL. *Adv. Mater.* 1999; **11**: 371-374.

58. Lupton JM, Hemmingway LR, Samuel IDW and Burn PL. *J. Mater. Chem.* 2000; **10**: 867-871.

59. Pillow JNG, Halim M, Lupton JM, Burn PL, Samuel IDW. *Macromolecules* 1999; **32**: 5985-5993.

60. Jiang D-L and Aida T. *J. Am. Chem. Soc.* 1998; **120**: 10895-10901.

61. Kimura M, Shiba T, Muto T, Hanabusa K and Shirai H. *Macromolecules* 1999; **32**: 8237-8239.

62. Harth EM, Hecht S, Helms B, Malmstrom EE, Fréchet JMJ and Hawker CJ. *J. Am. Chem. Soc.* 2002; **124**: 3926-3938.

63. Choi M-S, Aida T, Yamazaki T and Yamazaki I. *Angew. Chem. Int. Ed.* 2001; **40**: 3194-3198.

64. Choi M-S, Aida T, Yamazaki T and Yamazaki I. *Chem. Eur. J.* 2002; **8**: 2667-2678.

65. del Rosario Benites M, Johnson TE, Weghorn S, Yu L, Rao PD, Diers JR, Yang SI, Kirmaier C, Bocian DF, Holten D and Lindsey JS. *J. Mater. Chem.* 2002; **12**: 65-80.

66. Maruo N, Uchiyama M, Kato T, Arai T, Akisada H and Nishino N. *Chem. Commun.* 1999: 2057-2058.

67. Kato T, Uchiyama M, Maruo N, Arai T and Nishino N. *Chem. Lett.* 2000; **29**: 144-145.

68. Yeow EKL, Ghiggino KP, Reek JNH, Crossley MJ, Bosman AW, Schenning APHJ and Meijer EW. *J. Phys. Chem. B* 2000; **104**: 2596-2606.

69. Jiang D-L and Aida T. *Nature* 1997; **388**: 454-456.

70. Aida T, Jiang D-L, Yashima E and Okamoto Y. *Thin Solid Films* 1998; **331**: 254-258.

71. Wakabayaski Y, Tokeshi M, Jiang D-L, Aida T and Kitamori T. *J. Lumin.* 1999; **83/84**: 313-315.

72. Gilat SL, Adronov A and Fréchet JMJ. *Angew. Chem. Int. Ed.* 1999; **38**: 1422-1427.

73. Adronov A, Gilat SL, Fréchet JMJ, Ohta K, Neuwahl FVR and Fleming GR. *J. Am. Chem. Soc.* 2000; **122**: 1175-1185.

74. Adronov A, Malenfant PRL and Fréchet JMJ. *Chem. Mater.* 2000; **12**: 1463-1472.

75. Freeman AW, Fréchet JMJ, Koene SC and Thompson ME. *Polymer Preprints* 1999; **40**: 1246-1247.

76. Freeman AW, Koene SC, Malenfant PRL, Thompson ME and Fréchet JMJ. *J. Am. Chem. Soc.* 2000; **122**: 12385-12386.

77. Furuta P, Brooks J, Thompson ME and Fréchet JMJ. *J. Am. Chem. Soc.* 2003; **125**: 13165-13172.

78. Furuta P and Fréchet JMJ. *J. Am. Chem. Soc.* 2003; **125**: 13173-13181.

79. Serin JM, Brousmiche DW and Fréchet JMJ. *Chem. Commun.* 2002: 2605-2606.
80. Chrisstoffels LAJ, Adronov A and Fréchet JMJ. *Angew. Chem. Int. Ed.* 2000; **39**: 2163-2167.
81. Brousmiche DW, Serin JM, Fréchet JMJ, He GS, Lin T-C, Chung SJ and Prasad PN. *J. Am. Chem. Soc.* 2003; **125**: 1448-1449.
82. He GS, Lin T-C, Cui Y, Prasad PN, Brousmiche DW, Serin JM and Fréchet JMJ. *Opt. Lett.* 2003; **28**: 768-770.
83. Herrman A, Weil T, Sinigersky V, Wiesler U-M, Vosch T, Hofkens J, De Schryver FC and Müllen K. *Chem. Eur. J.* 2001; **7**: 4844-4853.
84. Maus M, De R, Lor M, Weil T, Mitra S, Wiesler U-M, Herrman A, Hofkens J, Vosch T, Müllen K and De Schryver FC. *J. Am. Chem. Soc.* 2001; **123**: 7668-7676.
85. Weil T, Wiesler U-M, Herrman A, Bauer R, Hofkens J, De Schryver FC and Müllen K. *J. Am. Chem. Soc.* 2001; **123**: 8101-8108.
86. Weil T, Reuther E and Müllen K. *Ang. Chem. Int. Ed.* 2002; **41**: 1900-1904.
87. Schenning APHJ, Peeters E and Meijer EW. *J. Am. Chem. Soc.* 2000; **122**: 4489-4495.
88. Balzani V, Ceroni P, Gestermann S, Gorka M, Kauffmann C, Maestri M and Vögtle F. *ChemPhysChem* 2000; **4**: 224-227.
89. Balzani V, Ceroni P, Gestermann S, Gorka M, Kauffmann C and Vögtle F. *Tetrahedron* 2002; **58**: 629-637.
90. Hahn U, Gorka M, Vögtle F, Vicinelli V, Ceroni P, Maestri M and Balzani V. *Angew. Chem. Int. Ed.* 2002; **41**: 3595-3598.

Chapter 10

FULLERENES IN BIOMIMETIC DONOR-ACCEPTOR NETWORKS

Nazario Martín and Dirk M. Guldi

Biomimetic organization principles have inspired the design, synthesis and study of supramolecular fullerene architectures. The exceptional progress made to incorporate fullerenes into well-ordered arrays has been based on hydrogen-bonding, π-π stacking, metal-mediated complexation and electrostatic motifs, which are overviewed in this chapter. Owing to the presence of fullerenes as an integrative building block, the majority of the presented molecular architectures exhibit unique and remarkable features. Particular attention has been focused on those ensembles showing electron-donor/electron-acceptor interactions as a new class of energy harvesting materials.

Keywords: Fullerenes, supramolecular chemistry, energy and electron transfer processes.

1. INTRODUCTION

Since the coining of the term 'supramolecular chemistry' – defined as *the chemistry of molecular assemblies and of the intermolecular bond* by Lehn in 1978[1a] – supramolecular chemistry has undergone a spectacular expansion. A wide variety of chemical systems have been developed, which, although different in nature to the former systems, currently are also considered as real supramolecular ensembles. One of most remarkable examples is the field of supramolecular photochemistry, where, in a broader sense, a supramolecular compound is defined as a group of molecular components that contribute properties, which each component possesses individually, to the whole assembly.

The paradigm of natural processes involving the combining effect of photo- and redox-active chromophores is photosynthesis. Such photo-induced electron transfer processes have a great significance in nature, since they govern photosynthesis in plants and bacteria. This process, upon which our existence depends on this planet, occurs via a rapid charge separation at the reaction center, with 100% quantum yield, thus enabling the efficient transformation of sunlight into chemical energy. In this regard, during the last recent years the preparation of new artificial photosynthetic systems has been carried out as a fundamental task in chemistry. These artificial systems are constituted by electron-donor and electron-acceptor moieties, chemically connected through different spacers. Light irradiation promotes an electron from the donor to the acceptor unit, giving rise to the formation of a *charge-separated state.*

In nature, organization principles that regulate size, shape and function down to the nanometer scale include both covalently-bonded and self-assembled motifs.[1] Exceptional and esthetic illustrations for the sophistication of this course are protein shells – including those of the photosynthetic reaction center – with highly complex performances such as energy storage, protection and transport of inorganic or organic molecules.[2] Unquestionably, the weak molecular interactions, as found in natural donor-acceptor ensembles, bear a deep fascination as a superior means to control (*i*) the organization of photo- and redox-active components and (*ii*) their mutual, electronic coupling. Stimulated by this vision and considering the significant relevance to natural photosynthetic events, substantial efforts are devoted to the development of non-covalent linkages, which help associating donor and acceptor building blocks.[3]

A variety of biomimetic methodologies, ranging from hydrogen-bonding, to π-π stacking, metal-mediated complexation and electrostatic interactions, emerged as promising promoters to engineer stabilized arrays.[4] Common to all these methodologies is that they all guarantee control over integrating donor and acceptor composites and, simultaneously, over achieving predetermined architectures of controlled sizes and outer-shell structures. A central benefit of these biomimetic motifs is that they are reversible, and, in contrast to truly covalent bonds, their binding energies are highly dependent on the chemical environment and temperature.

Certainly, one of the major challenges that still lie ahead of us is to regulate the weak forces, on a molecular basis, which ultimately dictate size and shape in relation to the function of the resulting composite. Can molecular tailoring of, for example, fullerenes, which are rigid, conformationally restricted scaffolds, contribute to the induction of completely different kinds of assemblies? Owing to the spectacular advances in the chemistry of fullerenes[5] – especially in recent years – remarkable progress has been made in non-covalently bonded structures, in solutions and in crystals, which are to be highlighted in this chapter. Special attention will be devoted to those fullerene-based supramolecular systems, which have been studied in their excited states in the search of new materials of interest for the preparation of artificial photosynthetic systems and photovoltaic devices.

2. HYDROGEN BONDING MOTIFS

Non-covalent interactions play a leading role in controlling the secondary and tertiary structures of natural macromolecules such as peptides, polynucleotides and polysaccarides or, for example, to provide the double helix structure of DNA where the base pairing between guanine and cytosine takes place by means of a threefold H-bonding. However, it is only relatively recently that such interactions have been exploited in the molecular self-assembly of well-defined synthetic supramolecular structures and materials.

Regarding fullerene chemistry, whereas the covalent functional-isation of buckminsterfullerene C_{60} has seen dramatic development, the supramolecular aspects of its chemistry have not been explored to the same extent. In contrast to covalent bonds, the formation of H-bonds is reversible and their strength depends on the chemical environment, such as the solvent or temperature. This important difference has been skillfully used in the preparation of a variety of reversible donor-acceptor dyads.

Hydrogen bonding is recognized, among the many biomimetic methodologies, as the only motif that meets the criteria of high directionality and selectivity.[6] But its limitation should be considered: hydrogen bondings are insufficiently stable, particularly in polar or protic solvents. The presence of an array of highly directional multiple hydrogen bonds helps to overcome this deficiency. Multiple hydrogen bonding, for example, has been successfully applied for the preparation of very stable complexes in different areas.[7] A remarkably stable dimer ($K_a > 10^6$ M^{-1}) was prepared from 2-ureido-4-pyrimidone derivatives, *via* a self-complementary muster of four hydrogen bonds.

Although fullerenes have frequently been used as a building block in the preparation of supramolecular systems,[8] recently a few examples have been reported in which hydrogen-bonding motifs dictate the ensemble arrangement. In one of the first examples, the formation of a self-assembled monolayer (SAM), was carried out by self-assembly of fullerenes, endowed with a monobenzo 18-crown-6 ether moiety on a gold surface (*vide infra*).[9]

More recently, the synthesis of a C_{60}-dimer followed. Specifically, a C_{60}-dibenzylammonium adduct was threaded through the cavity of a crown ether macrocycle in C_{60}-dibenzo-24-crown-8 (DB24C8).[10] A combination of hydrogen bonding and ion-dipole interactions are responsible for the formation of the first supramolecular C_{60}-dimer with an association constant of 970 M^{-1}. The C_{60}-DB24C8 conjugate also forms a stable pseudorotaxane-like 1:1 complex, when dibenzyl-ammonium hexafluorophosphate is employed, for which an association constant of 820 M^{-1} was determined. Scheme 1 illustrates a similar scenario: the C_{60}-dibenzylammonium conjugate threads reversibly through the cavity of DB24C8 to form a pseudorotaxane-like complex ($K_a = 1.25 \times 10^4$ M^{-1}).

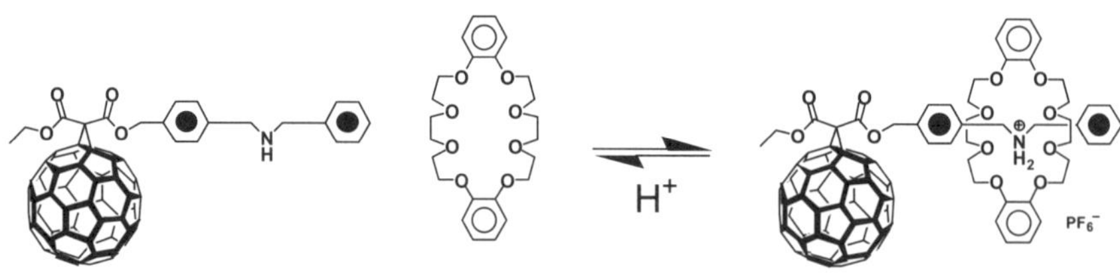

Scheme 1. Formation of a pseudorotaxane-like supramolecular complex.

Rotaxanes endowed with fullerenes are currently receiving a lot of attention as a consequence of the bulky nature of the C_{60} molecule, able to act as a stopper, as well as for its unique electrochemical and photochemical properties. Thus, a novel [2]rotaxane has been synthesized in which the C_{60} unit behaves as both a stopper and a photoactive unit,[11] as shown in Figure 1.

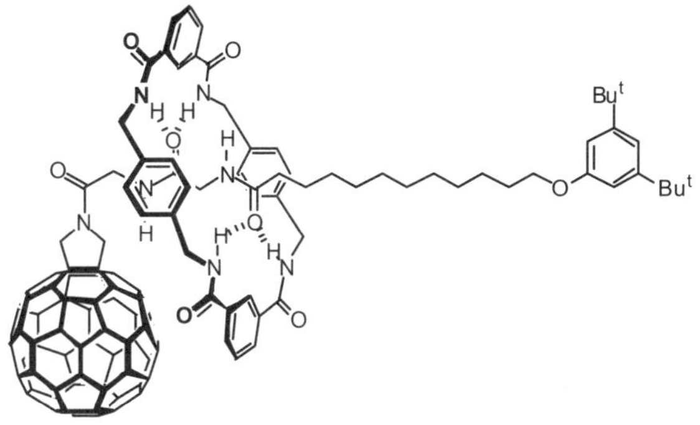

Fig. 1. C_{60}-based [2] rotaxane in which the location of the macrocycle depends on the solvent polarity.

Interestingly, the amphiphilic nature of the rotaxane thread can be used to shuttle the macrocyclic ring from close to the fullerene in non-polar solvents, to far away in polar solvents. Thus, in dichloromethane, the hydrogen bonding between the dipeptide moiety and the macrocycle brings the latter very close to the fullerene surface. On the contrary, in dimethylsulfoxide (DMSO) the macrocycle is located on the aliphatic portion of the thread and there are no significant interactions between the fullerene unit and the aryl groups of the macrocycle. Experiments carried out using fluorescence and time-resolved spectroscopy reveal that the proximity of the macrocycle to the C_{60} unit does not affect the fluorescence but it has a significant effect on the triplet-triplet spectra of the fullerene moiety.

The first example of a photoinduced intrarotaxane electron transfer between zinc porphyrin (ZnP) and C_{60} in benzonitrile has been reported very recently[12] (Figure 2). The aim of the authors was to use ZnP as a donor and C_{60} as acceptor with a rotaxane skeleton in order to mimic the photosynthetic reaction center. The rotaxane was synthesized by the hydrogen-bond assisted method for *sec*-amide-based rotaxanes. Photophysical studies reveal that charge separation takes place immediately after laser irradiation. The shortening of the fluorescence lifetimes indicates that charge separation occurs mainly from the singlet excited state of the ZnP moiety to produce a radical ion pair with lifetime values of 180 ns. The relatively low free energy of activation (57 meV) suggests that charge recombination takes place through space electron transfer with a superexchange mechanism.

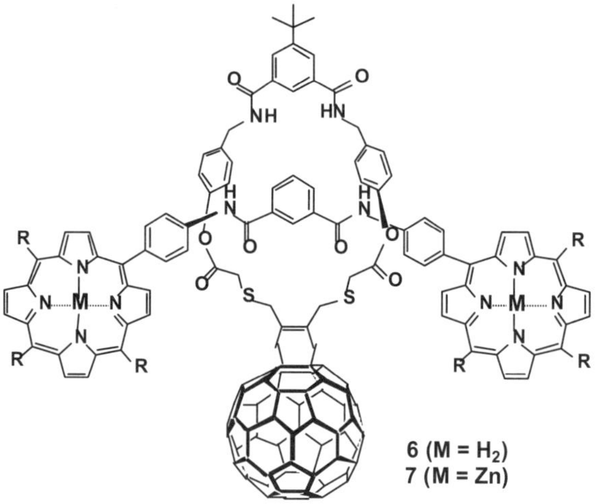

Fig. 2. Rotaxane formed by two porphyrin electron donors and a C$_{60}$ electron acceptor.

Slightly different is the approach that we reported on the synthesis of the first rigid non-covalent C$_{60}$-dimer, Figure 3: A self-complementary array of hydrogen bonding – donor-donor-acceptor-acceptor – is employed to link both molecular subunits (*i.e.*, fulleropyrrolidines).[13] In particular, the selected four-point hydrogen motif, due to the integration of a 2-ureido-4-pyrimidone moiety, guarantees the molecular recognition in solution. In a related C$_{60}$-dimer system, the 2-ureido-4-pyrimidone moiety was connected to the C$_{60}$ core through a cyclopropane ring.[14]

The electrochemical studies on these C$_{60}$-dimers indicate that there is – in the ground state – little, if any, electronic interaction between the

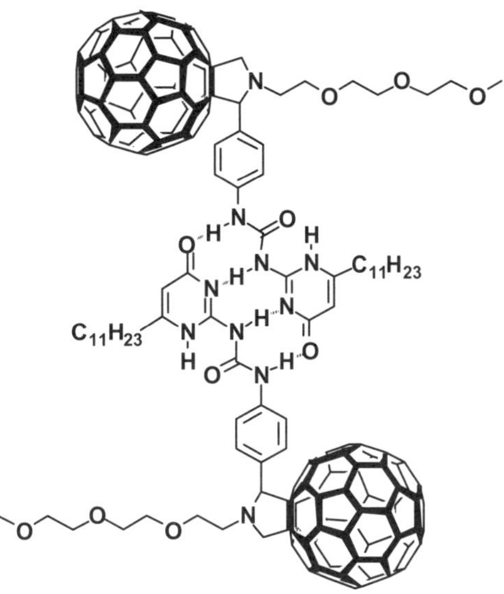

Fig 3. C_{60} dimer formed by a strong self-complementary array of four hydrogen bonding motif.

individual C_{60} subunits. In the excited state, the singlet excited state of the C_{60}-dimer is subject to an accelerated deactivation, resembling the trend seen for a covalently bridged C_{60}-C_{60}. Upon adding protic solvents to a CH_2Cl_2 solution, a progressive enhancement of the fullerene fluorescence was observed, reaching in the most protic solvent (hexafluoroisopropanol) a quantum yield similar to that observed for the reference 2',5'-dihydro-1'H-fulleropyrrolidine. Therefore, photochemical means bear a great potential, not only to distinct between monomer and dimer status, but also to monitor the gradual destruction of the hydrogen bonding motif.

Recently, the synthesis and spectroscopic characterization of the first hydrogen-bonded supramolecular array, formed by a methanofullerene bearing two coupling Meijer's self-complementary 2-ureido-4-pyrimidinones with a donor-donor-acceptor-acceptor (DDAA) hydrogen bonding motif, has been reported by Hummelen's group[15] (Figure 4). The dynamic behavior of the supramolecular polymer investigated by NMR spectroscopy reveals that the chemical integrity of the monomer as well as its redox and UV-Vis properties are fully preserved. This work

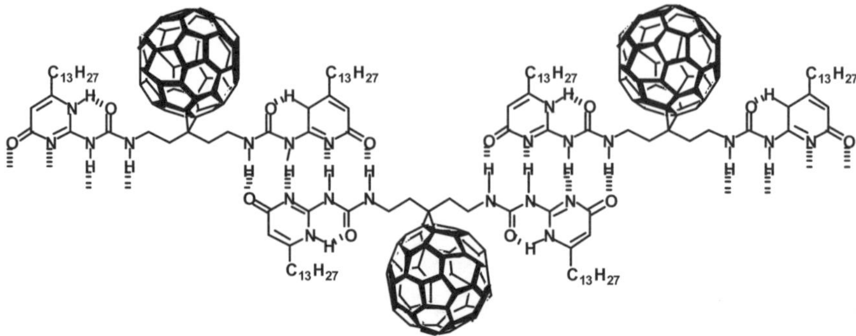

Fig. 4. Supramolecular fullerene polymer constituted by self-complementary 2-ureido-4-pyrimidones.

opens the supramolecular organization of fullerenes to a polymer level and, therefore, to the construction of supramolecular devices.

A different application of hydrogen bonding involving fullerenes has been the preparation of Langmuir films of C_{60}-uracil base paired complexes. Thus, a new C_{60}-uracil adduct capable of hydrogen bonding *via* complementary base pairing of adenine, adenosine and adenosine 5'-triphosphate (ATP) has recently been reported[16] (Figure 5). The determined isotherms of surface pressure *versus* area per molecule revealed the formation of 'expanded liquid' type films. The calculated area per molecule at infinite adduct dilution in the film was dependent on the composition of the sub-phase solution and increased in the order: water < adenine < adenosine < ATP, suggesting horizontal orientation of the complexes in the films. This study was completed with the formation of the Langmuir-Blodgett films by transferring the Langmuir films onto quartz slides, which were characterized by UV-Vis spectroscopy.

No doubt, hydrogen bonding motifs confer outstanding benefits for constructing photo- and electro-active donor-acceptor models. Studies performed with biomimetic model systems, wherein donor and acceptor molecules are tethered together non-covalently *via* hydrogen bonds, have shown that the electronic coupling through hydrogen bonds may, in fact, be larger than those mediated by either σ- or π-bonding networks. These findings pave the way for the preparation of novel C_{60}-based donor-

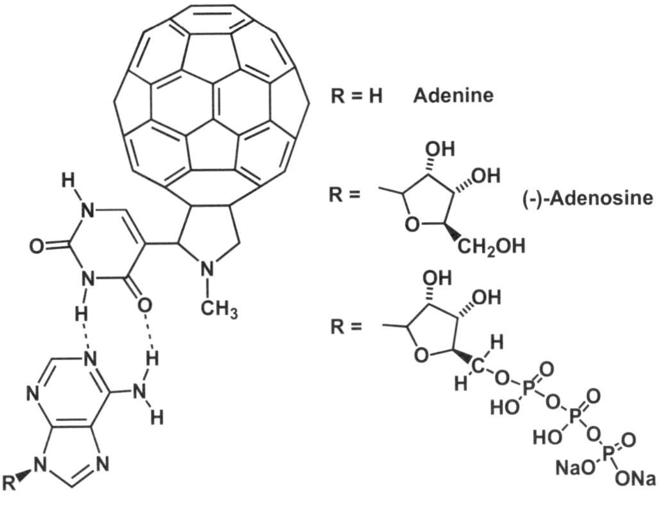

Adenosine 5'-triphospahte, disodium salt

Fig 5. C_{60}-uracil base paired complexes used in the preparation of Langmuir and Langmuir-Blodgett films.

acceptor systems, in which the electronic interaction of the C_{60} core with the electron donor unit will take place through a well-designed network of hydrogen bonding.

In this regard, important applications of this concept have recently been published. Using self-complementary 2-ureido-4[1*H*]-pyrimidinone units, two novel supramolecular dyads consisting of an oligo(*p*-phenylenevinylene) (OPV) donor and fullerene (C_{60}) acceptor were recently synthesized *via* quadrupole hydrogen bonding[17] (Figure 6). In these dyads, singlet energy transfer from the excited OPV unit to the fullerene causes a strong quenching of the OPV fluorescence. The high association constant of the 2-ureido-4[1*H*]-pyrimidinone quadruple hydrogen-bonding unit results in high quenching factors (≥ 90).

Although energetically possible, photoinduced electron transfer does not occur in these hydrogen-bonded dyads, even in polar solvents. The absence of charge separation has been accounted for by the low electronic coupling between the donor and acceptor in the excited state as result of the long distance between the chromophores.

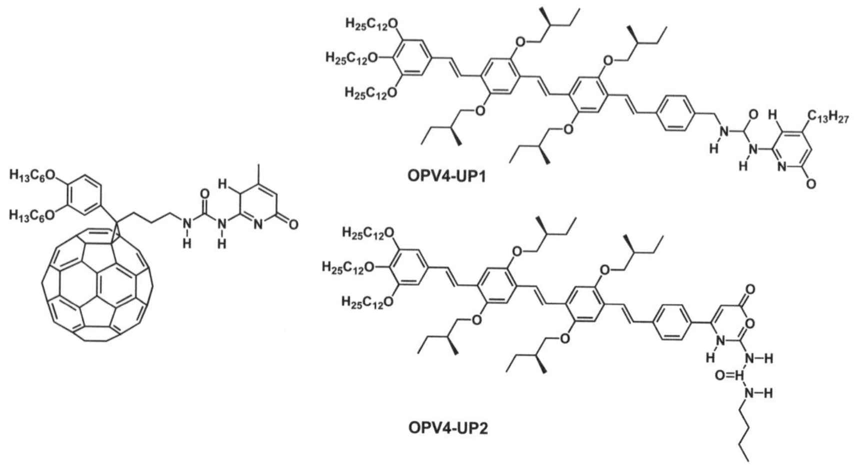

Fig. 6. Supramolecular dyads (OPV- C_{60}) connected by self-complementary 2-ureido-4 [1H]-pyrimidinones.

A new assembly between a conjugated *p*-phenylenevinylene (PPV) semiconducting polymer and an aziridino-functionalized fullerene has been carried out by means of complementary three-point hydrogen bonding motifs[18] (Figure 7). The formation of the hydrogen bonding between both electroactive units was confirmed by ^{1}H NMR spectroscopy and fluorescence quenching experiments clearly indicate a strong interaction between the PPV polymer and the fullerene derivative.

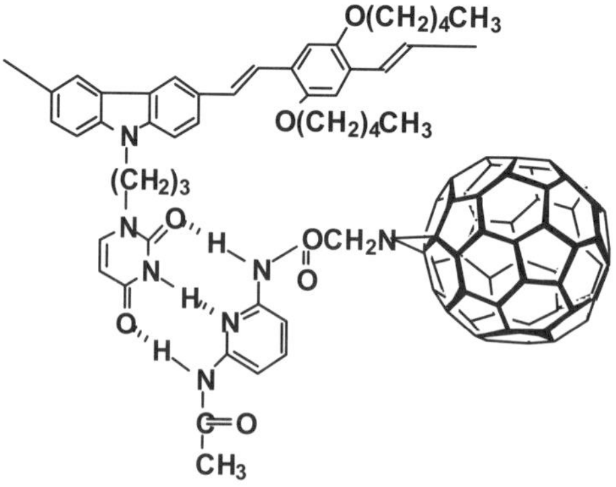

Fig. 7. Supramolecular assembly between a conjugated PPV and a functionalized fullerene constructed by groups of uracil and 2,6-diaminopyrimidine.

We have recently demonstrated that hydrogen bonding motifs confer outstanding benefits for constructing photo- and electro-active donor-acceptor models as novel artificial photosynthetic systems and possible photovoltaic applications. Thus, we have carried out the design, synthesis and physicochemical study of new supramolecular fullerene architectures built on highly directional and selective hydrogen-bonding as a biomimetic organization principle.[19] We started by synthesizing a series of novel TTF and C_{60} precursors that carry either a complementary guanidinium or carboxylate entity. A complementary array of hydrogen bonds – donor-donor-acceptor-acceptor (DD-AA) – opens the way to incorporate both moieties into well-ordered donor-acceptor arrays – **1a·5**, **1b·5**, **1a·6**, **1b·6**, **2a·4**, **2b·4**, **3a·4** and **3b·4** (Figure 8). Varying the nature and composition of the spacer connecting TTF and C_{60} – by using different functional groups (*i.e.*, ester *versus* amide) and spacer length (*i.e.*, phenyl *versus* biphenyl) – allowed the construction of topologically different ensembles. This provided meaningful incentives to differentiate between through-bond and through-space supported electron transfer interactions.

Steady-state and time-resolved emission studies showed a clear trend towards fluorescence quenching in the C_{60}•TTF dyads whose efficiency depends predominantly on the solvent polarity. Electron transfer, as the cause of fluorescence quenching, was confirmed independently by transient absorption spectroscopy. In particular, the rapid formation of the characteristic fullerene radical anion (1000 nm) and TTF radical cation (450 nm) transitions clearly attests $C_{60}{}^{•-}$•$TTF^{•+}$. Charge recombination rates in the C_{60}•TTF ensembles are typically around 10^6 s^{-1}, depending on the electronic coupling between donor and acceptor. Owing to the complexity of the network that connects C_{60} with TTF in our dyads, and the flexible nature of the spacer, through-space electron transfer interactions are operative to create the charge separated $C_{60}{}^{•-}$•$TTF^{•+}$ state. Addition of the hydrogen bonding-breaking solvent HFIP, on the other hand, resulted in a change of the photochemical reactivity, with no evidence observed for forming an intramolecular CS state. Alternatively, an intermolecular electron transfer process was observed.

This first example of electron transfer in C_{60}-based dyads, connected by strong hydrogen bonds, demonstrates that this approach can add outstanding benefits for the construction of artificial photosynthetic systems that bear closer resemblance to the natural one.

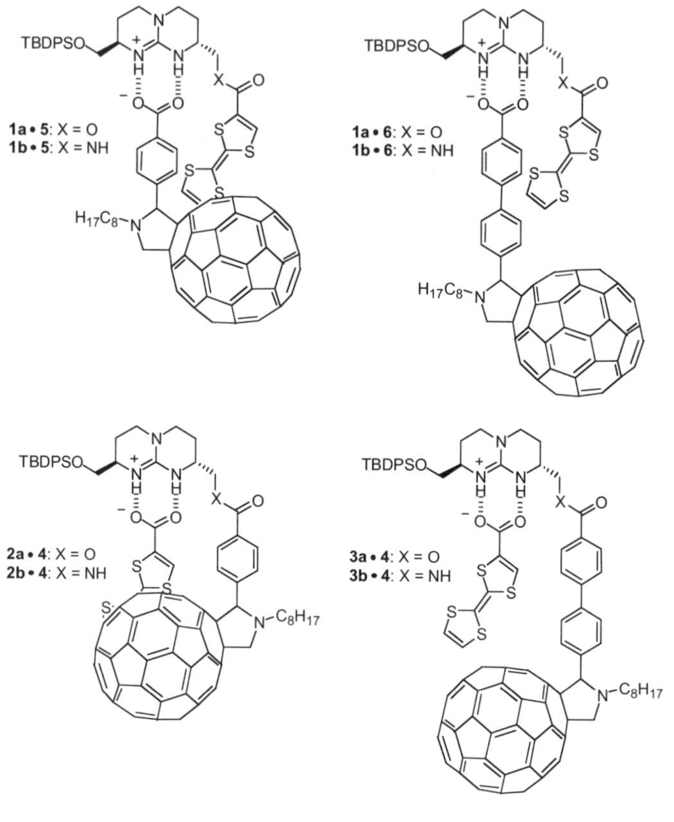

Fig. 8. Different photo-and electroactive donor-aceptor (TTF- C_{60}) supramolecular dyads based on complementary guanidinium and carboxylate units.

3. Π-STACK MOTIFS

The term 'π-π stacking' signifies a class of weak interactions between electron-rich and electron-poor aromatic rings. The attractive force is proportional to the contact surface between both π-systems. This interaction takes place between the negatively charged π-electron cloud of one molecule and the positively charged σ-framework of an adjacent molecule. The relative orientation of the two interacting molecules is determined by the electrostatic repulsions between the two negatively charged π-systems.

Only the introduction of five-membered (pentagonal) rings into fullerene structures accounts for their convex curvature.[20] Functioning like defects in a graphite structure, they govern the non-planarity of the fullerene's π-electronic arrangement. This structural topic also imposes a number of electronic consequences. Most importantly, they generate alternating areas of electron-richness (*i.e.*, hexagons) and electron-deficiency (*i.e.*, pentagons) in C_{60}. The anisotropic electron distribution is then the basis for predetermined interactions with substrates in general.[21] Depending on the electronic nature of the substrate – rich or deficient in electron density – the ambivalent structure of C_{60} can adapt and associate by maximizing the electronic / spatial overlap.

3.1. Concave-Convex

Matching the convex shape of C_{60} to a concave surface has been used to gain control over the inherently weak association between a host moiety and a molecular guest. A built-in shape- and size-specific receptor site ensures efficient binding, but it also exerts a major challenge to find the right match. The following examples form remarkably stable composites with fullerenes: nano-sized, bowl-shaped molecules or container molecules with curved, open-ended cavities, such as cyclodextrins,[22] benzotri(benzonorbornadienes),[23] calixarenes,[24] calixnaphthalenes[25] and cyclotriveratrylene.[26]

Alpha (α), beta (β) and gamma (γ)-cyclodextrins (CDs) are rigid, cyclic oligosaccharides with well-defined hydrophobic cavities.[22] A consideration regarding the right balance between cavity radius and fullerene size leads to the assumption that γ-CD should be the only candidate to host C_{60}. Nevertheless, molecular modeling suggests that a full incorporation – even in form of a 1:1 complex – is practically impossible to achieve. Despite this apparent size mismatch, incorporation of a single C_{60} molecule between the cavities of two γ-cyclodextrin molecules – in form of a 2:1 complex (Figure 9) – has been postulated and later confirmed by various spectroscopic techniques. In fact, it produces the most stable solutions, independent of the method of preparation, with C_{60} concentrations as high as 1.4×10^{-3} M.[22i] Parallel experiments attempting to incorporate (*i*) C_{60} into α- or β-cyclodextrin, or (*ii*) the larger, ellipsoidal C_{70} into γ-cyclodextrin, failed!

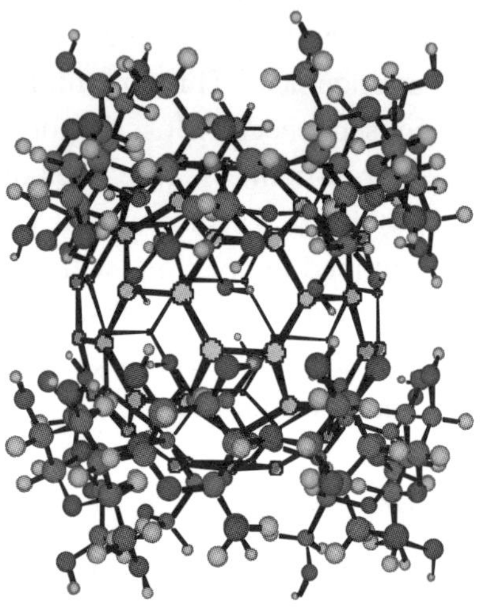

Fig. 9. A C_{60} molecule (convex shape) is complexed in the concave cavity of two γ–cyclodextrin molecules.

Another class of very rigid, cup-shaped molecular nanostructures are benzotri(benzonorbornadienes) which exhibit, once associated with C_{60}, a very distinct absorption spectrum.[23] Since the spectral features are uncharacteristic of both individual components in their unassociated status, it prompts to the existence of a strongly coupled inclusion complex. As of today, however, no information is available documenting the affinity and selectivity of benzotri(benzonorbornadienes).

In comparison with the aforementioned classes of containers, the structures of calixarenes[24] and calixnaphthalenes[26] allow for some more structural flexibility and impacts their affinity to bind C_{60}. Support for this view was lent from a comparison of the optical-active modes in the vibrational spectrum of the C_{60}/*p*-*t*Bu-calix[8]arene complex with those of the empty *p*-*t*Bu-calix[8]arene.[27] Typical association constants for

C_{60}/calixarenes ensembles in toluene are: 35 M^{-1} for *p-t*Bu-hexahomo-oxacalix[3]arene[28]; 2.1×10^3 M^{-1} for calix[5]arene[19e]; 8.3×10^3 M^{-1} for bridged calix[5]arene[25g]; 1.44×10^4 M^{-1} for hexamethoxy-*p-t*Bu-calix[6]-arene[25n]; 1.66×10^4 M^{-1} for octamethoxy-*p-t*Bu-calix[8]arene.[25n]

Probing calix[4]arene, homooxacalix[3]arene, calix[5]arene, calix[6]-arene and calix[8]arene (X = H) in toluene supports the fact that only calix[*n*]arenes can include C_{60}, if they hold a well-preorganized cone cavity.[29] Among those investigated, just homooxacalix[3]arene ($K = 35 \pm 5$ M^{-1}), calix[5]arene ($K = 330 \pm 10$ M^{-1}) and calix[6]arene ($K = 87 \pm 5$ M^{-1}) meet this requirements, that is, a cone configuration and a benzene ring inclination suitable for a multi-point interaction with C_{60}. By contrast, when the end-standing OH-groups are alkylated none of the calixarenes binds C_{60}. However *p-t*Bu-calix[n]aryl ester derivatives, which cannot interact with C_{60}, become excellent receptors in the presence of certain metal cations. Lithium and caesium, for example, induced favorable conformational changes of the *p-t*Bu-calix[n]aryl esters.

Besides the inclination of the benzene rings and the cavity of the varying temperature and / or solvent exerts similarly strong impact on the binding affinity. Increasing the temperature, for instance, from 294 to 315 K, led to smaller binding constants, with differences as large as two orders of magnitude.[25n] Benzene and toluene are generally found to yield the highest association constants, while variations of up to 10 were found in CS_2, CCl_4, *o*-dichlorobenzene and *o*-xylene.[25e,25g,25n] In retrospect, the association strength increases with decreasing the overall solubility of C_{60}, which reflects the ongoing competition between complex formation and solvation of the guest. In general the binding affinity for C_{60} is larger than for C_{70}.

Quite interestingly, the intensity of the narrow and broad linewidth EPR signals decreased, when $C_{60}^{\bullet-}$ was electrochemically generated in the presence of a complexing agent – *p*-benzylcalix[5]arene or cyclotriveratrylene.[29] This indicates that the π-π interactions were sufficiently strong to alter the EPR signals of $C_{60}^{\bullet-}$. Another profound impact was seen in the cyclic voltammetry experiments performed with C_{60} in the presence of these container molecules: the first four reduction processes of C_{60} split into two new processes upon complexation.

One of the water soluble complexes, namely C_{60}/homooxo-calix[3]arene, was subject to fluorescence and transient absorption measurements in aqueous media.[30] In comparison to C_{60} benzene and C_{60}/γ-CD aqueous solutions, very dramatic effects were noted: blue-shifted (from 700 nm to 531 nm) fluorescence and blue-shifted (from 750 nm to 545 nm) transient triplet maxima indicate rather strong electronic perturbations within the C_{60}/homooxocalix[3]arene complex.

In the solid state, the photophysical properties of C_{60}/*p-t*Bu-calix[8]arene reveal a triplet maximum at 780 nm, matching that of C_{60}/γ-CD under similar conditions.[31] As a consequence of the electronic interactions between C_{60} and *p-t*Bu-calix[8]arene, the lifetime at 1.7 μs is, however, substantially shortened relative to triplet lifetimes in solutions. Despite this notable impact, the short-lived triplet is efficiently deactivated in the presence of molecular oxygen, thus, photosensitizing the formation of singlet oxygen, $^{1}O_{2}$ ($^{1}\Delta_{g}$).

Inasmuch as C_{60} is particularly prone to co-crystallization with other molecules, the organization of the co-crystallizing components becomes useful for further refinement of the solid-state morphology.[32] C_{60} co-crystallizes with *p*-bromocalix[4]arene propyl ether in strictly linear, separated columns. The co-C_{60} / *p*-bromocalix[4]arene propyl ether crystal transformed, upon applying heat and pressure, into a linear [2+2] addition polymer – without, however, showing any evidence for notable crosslinking products. The photopolymer of a pristine C_{60} sample, on the other hand, is characterized by at least three polymer phases reported with characteristics ranging from those of rhombohedral and ortho-rhombic to tetragonal phases.[33] In retrospect, this elegant demonstration offers new stimuli in the field of crystal engineering.

One of the most important structural disparities between calixarenes and calixnaphthalenes is that the latter is a class of cavitands possessing deeper cavities.[26] Better contacts between host and guest and deeper penetration of the latter is made possible by the extra fused aromatic rings on each naphthalene unit. Of course, this effects their complexation properties, with values that range between 3.0×10^{2} M^{-1} and 7.08×10^{2} M^{-1} for *p-t*Bu-hexahomotrioxacalix[3]naphthalenes and *t*Bu-calix[4]naphthalenes, respectively. These values are recognizably larger than for the calixarene analogs. An unexpected observation was that the binding affinities in the different solvents were opposite to the trends reported for calixarenes – the binding for calixnaphthalenes in CS_2 is actually higher than in toluene or benzene. A detailed thermodynamic study reveals that both solvophobic effects (*i.e.*, a larger number of CS_2 molecules are displaced from the cavitand cavity upon C_{60} binding) and π-π interactions are cooperative driving forces for the complexation processes.[26b]

Similarly, cyclotriveratrylene forms an inclusion complex only with C_{60}, but not with C_{70}.[27] Again, the driving force is the creation of π-π charge transfer interactions. The X-ray structure is simplified in Figure 10 and shows that C_{60} adopts a nesting position above the concave surface of cyclotriveratrylene, with close contacts between the two entities at the van der Waals limits (3.34 – 3.51 Å). These short contacts compare well with the contacts in cubic close packed C_{60}.

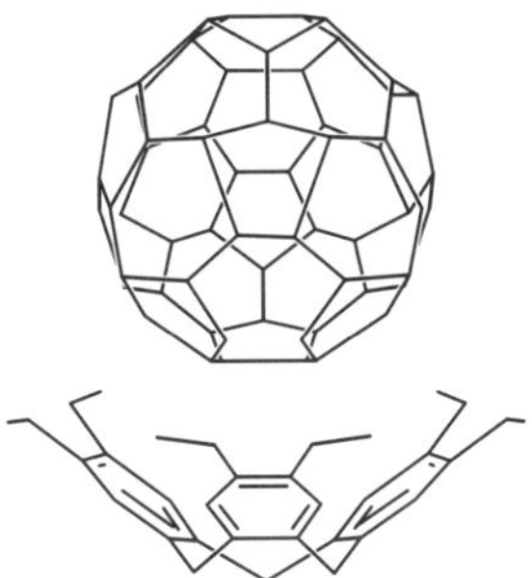

Fig. 10. Inclusion complex of a C_{60} molecule in the cavity of a cyclotriveratrylene.

Unequivocally, less constrains are encountered with adaptable nickel (II) macrocycles (*i.e.*, 5,7,12,14-tetramethyldibenzo[b,I]-[1,4,8,11]tetraazacyclotetradecinene – TMTAA and 5,14-dihydro-2,3,6,8,11,12,15,17-octamethyldibenzo[b,I]-[1,4,8,11]tetraazacyclotetradecinene – OMTAA).[34] Their shallow saddle-shaped voids give rise to two divergent concave surfaces, which are summarized in Figure 11. This structural issue is an important receptor asset for the formation of supramolecular arrays with globular hosts, since it confines molecular guests in either a bent sandwich or polymeric structure. In case of convex C_{60}, tightly bound 1:1 complex species are formed, since C_{60}'s favorable docking involves the phenyl groups of the macrocycle and, thus, blocks the association of the macrocycles. The latter motif is a key requirement for the realization of 2:1 complexes. Overall an entropically favored binding – relative to the binding of solvent molecules – prevails and no energy is required for the pre-organization prior to the association process. From the change in absorbance, prior to reaching stoichiometric equality, an association constant derived for the 1:1 complex exceeds 10^5 M^{-1}. Still, the curvature complementary, matching only the phenyl phase, imposes some restrictions to the applicability of the convex nickel (II) macrocycle.

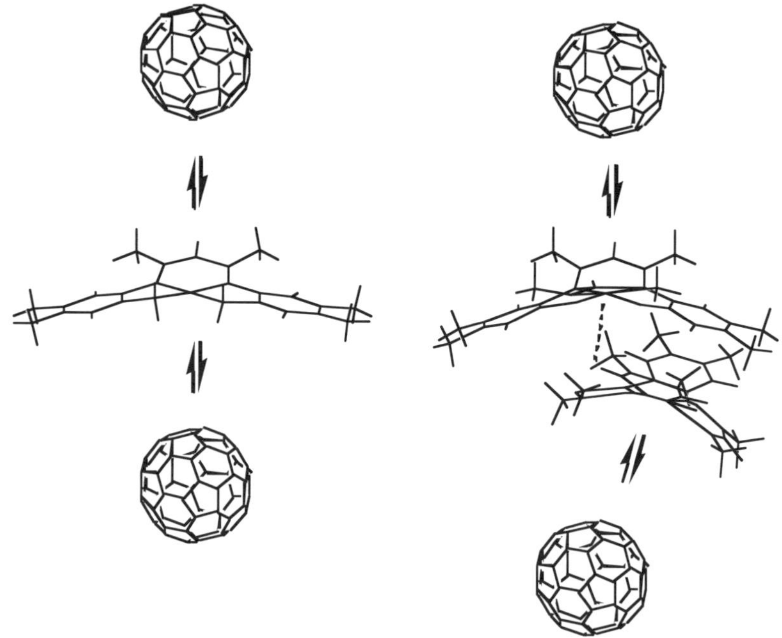

Fig. 11. Nickel (II) saddle-shaped macrocycles form supramolecular arrays with C_{60} in 1:1 and 2:1 stoichiometry

A wider synthetic flexibility is found in resorcarene-based materials to stabilize C_{60} within their framework.[35] Several metal ions, in conjunction with multiple units of a dithiocarbamate-resorcarene ligand, are the starting point for resorcarene-based nanostructures. In principle, the structures can be varied by the choice of metal ions and their oxidation states, and in case of cadmium and zinc, the resulting trimeric networks were shown to encapsulate C_{60} with binding constants ranging from 2.9×10^4 M^{-1} to 1.2×10^5 M^{-1} in toluene. The intramolecular distances, separating the opposing metal centers from each other are on the order of 14.7 Å. Importantly, in a copper-based tetrameric network, in which four ligands are placed at the apices of a distorted tetrahedron, distances across the tetrahedron can reach as high as 20.4 Å.

As the last example, a fundamentally different picture is summarized for phenylated triamino-*s*-triazines – see Figure 12.[36] In an attempt to more accurately mimic photosynthetic reaction centers and to better understand the physical implications of π-overlap between redox sites in supramolecular architectures, an electron donor which could strongly associate with the curved surface of C_{60} has been sought. Highly phenylated triamino-*s*-triazines are easily oxidizable highly fluorescent ($\Phi = 0.016$) electron donors, which are extremely suitable for intermolecular quenching studies. The convex surface of C_{60} is well-suited for an efficient and selective intermolecular association with the concave surface of triamino-*s*-triazines in toluene and *o*-dichlorobenzene *via* π-overlap. The value for this complexation is on the order of 10^5 M^{-1}, an interaction of unprecedented strength involving pristine C_{60}. Importantly, the association/dissociation process is reversible and acid/base controllable. Interestingly, in the C_{60}-triamino-*s*-triazine complex, a strongly emissive and long-lived charge-transfer state is produced upon photoexcitation. The advantage of this nanoarchitecture is its broad suitability associating with a large number of fullerenes, including C_{70} etc. Since for C_{70} a slightly lower association was registered ($\sim 5 \times 10^4$ M^{-1}), a selective interaction with the regions of highest reactivity, that is, the poles of the ellipsoidal C_{70}, might be postulated.

In retrospect, donor-acceptor interactions between the π-electron rich aromatic rings and the π-electron deficient domains of C_{60}, together with van der Waals forces, provide the major stabilization for the solution complexes. While the complementary of the curvature of the interacting species maximizes the number of intermolecular contact, formation of these concave – convex composites are entropically dis-favored due to a more ordered state. The above outlined examples manifest

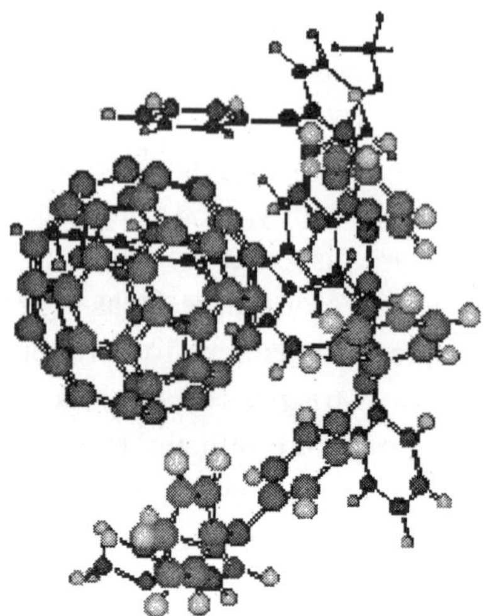

Fig 12. A self assembled C_{60} triamino-*s*-triazine complex.

that a broader applicability of a given nanostructure comes only at a price of an inferior selectivity. Notably, a mismatch in size generally results in the spontaneous and irreversible formation of aggregates, resembling the cluster phenomena of C_{60} in polar and aqueous media (*vide infra*) or even upon aging of surfactant-capped C_{60} solutions.[37] Scattered reports support this hypothesis. The *p-t*Bu-calix[8]arene complex of C_{60}, for example, exists in a stable configuration only in the solid state, while upon solubilization it dissociates into the free components or micelle-like composites with a trimeric cluster that is surrounded by three host molecules in a double cone formation.[38] Similarly, treatment of an intensively colored C_{70}/γ-CD reaction mixture by filtration or centrifugation becomes, in contrast to C_{60}/γ-CD, colorless. Again, aggregated C_{70}/γ-CD clusters, which are a direct product of the size-mismatch, are present in solution. Not unexpectedly, aggregation phenomena are also observed in the C_{60}/cyclotriveratrylene case, leading to a polymeric zigzag array of C_{60} in the solid state, each in the cavity of a cyclotriveratrylene molecule.[39] It is important to note that cluster or aggregates are ineffective probes for photoinduced electron transfer examinations, since the close packing expedites excited state deactivation processes.

3.2. Planar-Convex

The crystal structure involving a fulleropyrrolidine, H_2P-C_{60} (H_2P = free base tetraphenylporphyrin), which was grown as a chloroform solvate, gives way to a clear picture on the disposition of both moieties.[40] An unexpectedly close approach between C_{60} and the porphyrin is the basis for an appreciable intermolecular interaction. The distances of the closest C_{60} C-atoms to the mean plane of the inner core of the porphyrin are, with values of 2.78 Å and 2.79 Å, quite short. Upon considering the inter-layer separation in graphite (3.35 Å) and interfacial porphyrin / porphyrin separations (> 3.2 Å), a new donor-acceptor relationship can be formulated: favorable van der Waals attractions between the convex π-surface of C_{60} or C_{70} and the planar π-surface of MP (metallo-porphyrin), assist in the supramolecular recognition – overcoming, however, the necessity of matching a concave-shaped host with a convex-shaped guest structure (*vide supra*).

The remarkable results on H_2P-C_{60}[40b] led to further studies in a series of MP/C_{60} cocrystallates – Mn, Co, Ni, Cu, Zn to Fe were chosen (Figure 13).[41] In most C_{60}-based assemblies, electron-rich areas, namely, carbon atoms at hexagon-hexagon junctions, lie over the center of the porphyrin ring. C_{70}, on the other hand, adapts a configuration that brings the poles of the ellipsoidal framework – again carbon atoms located at the intersection of hexagon faces – into contact with the porphyrin. Again, complexes with unusually short contacts (2.7 – 3.0 Å), shorter than ordinary van der Waals contacts (3.0 –3.5 Å), are formed. This implies that MP/C_{60} associative forces constitute an important organization principle. A wide variety of crystal structures were found; honeycomb motifs, puckered graphite-like layers, zigzag-chains and columns.[42] However, experimental data from ESR, IR absorption and X-ray photoelectron spectroscopy suggest that no noticeable charge transfer prevails between the different MPs, which are all excellent electron donors, and the electron-accepting C_{60}.

Besides porphyrins, cocrystallates of C_{60} were also found with porphyrazines,[42] as another representative of the diverse family of π-extended macrocyclic complexes.[43] They reveal interesting supra-molecular structures: while, for copper (II) a C_I-symmetric sandwich

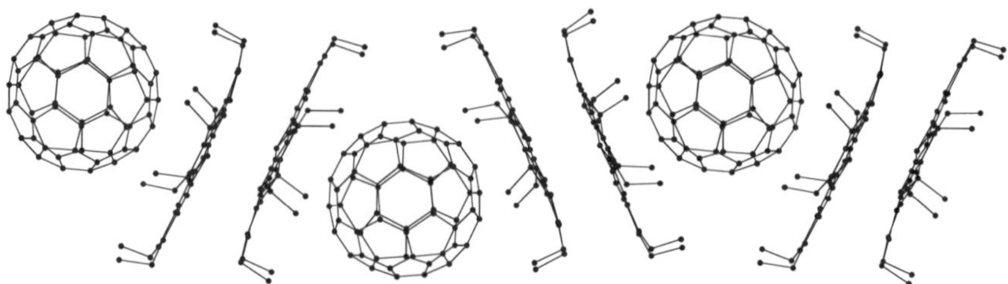

Fig. 13. Metalloporphyrin/ C_{60} cocrystallates formed by π–π weak interactions.

complex of two slightly dished porphyrazines units enclosing one C_{60} was found, nickel (II) features a non-centrosymmetric 1:1 complex with a strongly wrapped porphyrazine unit. Again, strong π-π associations are believed to promote these remarkable ordering principles.

At millimolar concentrations, when MP and C_{60} are either titrated or spontaneously mixed, NMR studies – ^{13}C and ^{1}H – reveal mutually upfield shifts. This clearly indicates the presence of complex formation. Absorption spectroscopy, on the other hand, proved to be insensitive to detect appreciable MP/C_{60} interactions, which may lead to the false presumption that the degree of MP/C_{60} association might be weak.

At micromolar concentrations, MP / C_{60} interactions are also inferred on the following grounds.[44] The rate constants for electron transfer from various MP π-radical anions (M = Zn, In, Ge, Al, Ga, Sn, Sb) to C_{60} are found to be in the range of $(1\text{-}3) \times 10^{9}$ $M^{-1}s^{-1}$, which corresponds to nearly diffusion-controlled processes. The lack of dependence, despite the large variation in one-electron reduction potentials for the examined MPs between $E_{1/2}$ ZnP/ZnP$^{\bullet-}$ = -1.35 V and $E_{1/2}$ SnP/SnP$^{\bullet-}$ = -0.8 V *versus* SCE, might reflect the fact that the investigated MPs and C_{60} already experience electronic interactions in the ground state. Again, no ground state charge transfer interactions were detectable, at least, in the form of a perturbation of the ground state transitions.

Attractive van der Waals forces between porphyrins and C_{60} are definitely appreciable, when both moieties are linked to each other, rather than just mixed. Importantly, whenever affirmed possible by the molecular topology of the system, these moieties spontaneously tend to achieve close spatial proximity relative to each other. As illustrative examples, a series of flexibly-spaced dyads[45] should be considered in comparison to a π-stacked dyad,[46] in which two tethers ensure a locked configuration with the two π-systems sitting tightly on top of each other. Notable imprints – reflecting perturbation of the *Soret* and *Q*-band transitions, charge-transfer absorption and charge-transfer emission – are clearly registered, whose extent, although not reaching the magnitude seen for a true π-stack, is appreciably stronger than in donor-acceptor structures – where the rigid, constrained architecture opposes those interactions. In MP-C_{60} ensembles, ultraviolet-visible absorption spectroscopy emanates as an easy and sensitive probe for assessing the binding and electronic coupling.

Let us turn back to truly intermolecularly organized ensembles. Complementarity of size and maximizing the number of points of interaction are key factors in devising stable fullerene architectures, at least in the absence of alternative motifs such as hydrogen bonding, electrostatic and metal coordination. The tedious control over the competition between host-host, guest-host and host-host interactions, which is particularly evident in fullerene chemistry where, for example, C_{60}-C_{60} interactions play a major role, is important in determining the structure of supramolecular ensembles. Thus, following similar incentives, that is, the utilization of topological controlled π-π associations, a porphyrin 'cylic-dimer'[47] and a porphyrin 'jaw',[48] depicted in Figure 14, were developed. The electron-rich walls of the porphyrins and their considerable contact with incumbent C_{60} encouraged experiments, and strong interactions were indeed detected. In both constructs, discrete van der Waals complexes are realized with a core of two porphyrins, *i.e.* PdP (palladium 3-pyridiyltriphenyl-porphyrin)[46] or ZnP (zinc biphenyltetrahexylporphyrin)[44a], controlling the selective C_{60} incorporation.

Strong π-electronic donor-acceptor interactions stem from the close proximity of the MP and C_{60} π-systems in the 'cyclic-dimer' and 'jaw', and the effects are detectable in shifts of the absorption bands. Typically, red shifts of the *Soret* and *Q*-band transitions, accompanied by lower extinction coefficients – relative to the model porphyrin systems – are observed. Concomitantly, the chromophore's emission gives rise to a progressive quenching after addition of variable C_{60} concentrations.

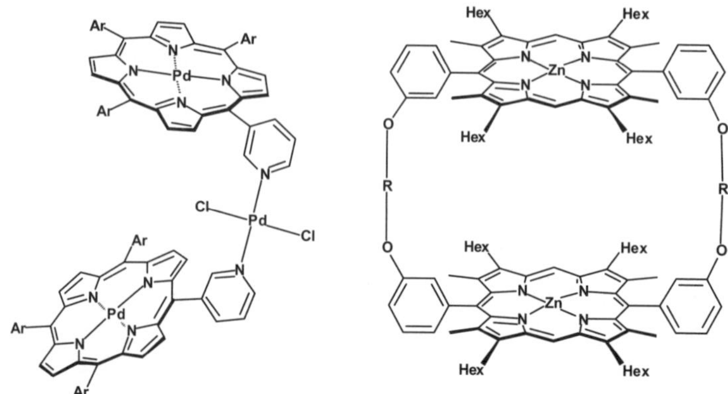

Fig. 14. A metalloporphyrin 'jaw' (left) and a metalloporphyrin 'cyclic-dimer' (right) as host to bind the C_{60} molecule.

To shed light onto the evident C_{60} encapsulation within the porphyrin 'cyclic-dimer' a variety of metal centers were probed (*i.e.*, Co(II), Rh(III), Ni(II), Cu(II), Ag(II) and Zn(II)).[48b] Based on metal-to-fullerene-charge-transfer interactions, the Rh(III) led to an unprecedentedly high association constant – for binding C_{60} – of 2.4×10^7 M^{-1}. Inferior rates were derived for the remaining metal centers, with values typically on the order of $\sim 10^6$ M^{-1}.

Weaker interactions and smaller association constants (7×10^5 M^{-1}) are determined for the PdP-based porphyrin 'jaw'. The difference in association rates can be rationalized in terms of the flexible framework of the cyclic-dimer, which ensures perfect encapsulation. For example X-ray crystallographic measurements reveal, in the C_{60}-cyclic-dimer complex, shortest zinc – carbon distances of 2.765 and 2.918 Å, notably shorter than the sum of the van der Waals radii (3.09 Å). In addition, the hexamethylene spacers are folded and the ZnP's planarity is slightly distorted to maximize the π-overlap with the convex C_{60} surface.

Based on these cyclic dimers, a supramolecular oscillator composed of carbon nanocluster C_{120} and a rhodium (III)porphyrin cyclic dimer has recently been reported[49] (Figure 15).

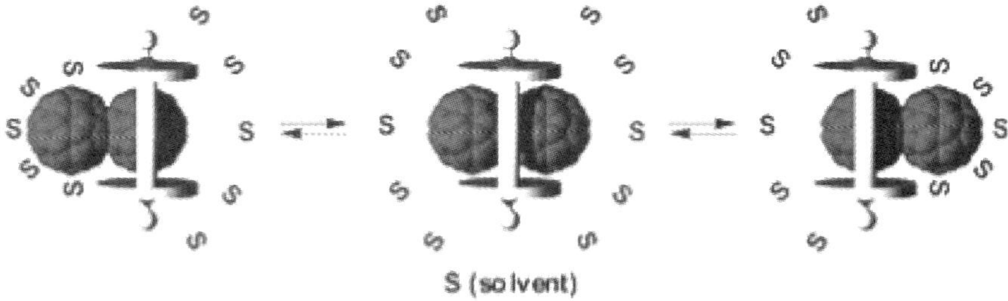

Fig. 15. Supramolecular oscillator constructed from C_{120} and a rhodium (III) porphyrin cyclic dimer.

An inclusion complex of the fullerene dimer (C_{120}) in the cyclic porphyrin dimer showed pairs of two singlet signals in the ^{1}H NMR spectra both for the meso and for the pyrrole-β-methyl protons. These signals are due to the protrusion of one of the C_{60} moieties of C_{120} from the cavity of the porphyrin cyclic dimer. However, upon increasing the temperature, a coalescence of these signals was observed – which has been accounted for, not by the expected dissociation/association dynamics of the inclusion complex, but by the oscillation of the included C_{120}. This oscillation has been visualized by the solvation/desolvation dynamics of carbon nanoclusters. This result has been recognized as an interesting potential of the supramolecular chemistry of fullerenes for molecular sensing.

Strong association constants in combination with well-defined geometries were achieved when a multi-point interaction approach was probed in form of a self-assembled porphyrin 'box' (*i.e.*, four side wall ruthenium porphyrins, RuP, and one central zinc porphyrin, ZnP) – Figure 16.[50] Large area of contacts render these systems useful for expediting the supramolecular interactions of the porphyrins with a suitable C_{60} electron acceptor at the molecular level. The systematic variation of the ZnP(4)-(RuP)$_4$ / ZnP(3)-(RuP)$_4$ geometry, namely, straight *versus* tilted, facilitated or hindered the incorporation of a 3-D C_{60} moiety into an ensemble, constituted by four side-wall RuP and one central ZnP. Evidence for the efficient engagement is given by transient absorption spectroscopy, revealing the nearly quantitative formation of the C_{60} triplet excited state.

Fig. 16. Self assembly of a ZnP(4)- (RuP)$_4$ box

In an alternative method, supramolecular nanoarchitectures have been devised using polybenzyl ether dendrimer hosts to complex C_{60}.[51] A phloroglucinol or meso-tetraphenylporphyrin core provides the correctly sized space for the C_{60} inclusion, a hypothesis confirmed by ^{13}C-NMR. More important is the observation that the dendrimer's affinity to host C_{60} increased significantly as the generation number of the surrounding dendritic substituents increased (*i.e.*, 1st: 5 ± 2 M^{-1}; 2nd: 12 ± 2 M^{-1}; 3rd: 68 ± 4 M^{-1}) – mimicking the function of natural globular proteins. In the porphyrin analog, the expected decrease of the *Soret*-band was seen in the presence of C_{60}, but without the obligatory red-shift of the transition. Due to the poor solubility, determination of the association constant could not be performed. However, comparing the different dendrimer generations gave rise to the following conclusions: firstly, the dendritic branches are crucial to provide the important complexation. Secondly, C_{60} is in close vicinity to the meso-tetraphenylporphyrin core.[52]

3.3. Convex-Convex

Pristine fullerenes are virtually insoluble in aqueous environments. Although the attachment of hydrophilic, solubilizing functionalities to the hydrophobic core emerged as a probate means to overcome the insolubility,[6,9g,38,53] a new and interesting feature, namely aggregation/clustering, transforms the convex spheres into hierarchical mesoscopic structures. Hereby, the driving force, *i.e.* a dominant intrinsic geometry constraint, evolves from the uniquely rigid hydrophobic C_{60} core.

A fundamental and imperative challenge is to gain control over the cluster size, leading to two different strategies. The first one implements the rapid injection of a polar – non-fullerene-like – solvent (*i.e.*, acetonitrile or water) into a non-polar solution of the respective C_{60}, C_{70}, etc. (*i.e.*, toluene or THF).[54-63] In an alternative approach a dispersion of a hydrophobic fullerene sample, which carries, for example, a hydrophilic ammonium group in distilled and filtered water or acetone, is immersed in an ultrasonic bath for several time intervals.[64-68]

Following the first approach, C_{60}- and C_{70}-clusters were generated in a room temperature toluene – acetonitrile solvent mixture.[55] Reversible solvatochromics turned out to be good indicators for the cluster formation: for C_{70} the color changed from orange to reddish purple, while for C_{60} the color changed from magenta to brownish yellow. Upon reverting the solvent composition – namely, decreasing the overall acetonitrile content – the solution color and absorption spectrum changed back to those of the corresponding monomers. Photon-correlation spectroscopy of quasi-elastic light scattering showed that the average size of the clusters is 186 nm for C_{70}, while for C_{60} the clusters fall in the relatively wide range of 140 to 270 nm. The large variation of the C_{60}-cluster sizes were correlated with the initial C_{60} concentration (5.0×10^{-6} – 1.0×10^{-4} M) and the toluene – acetonitrile composition (40 – 90 %).

Water as an additive to solutions of C_{60} and C_{70} – instead of THF – yielded fairly monodisperse clusters. The diameters of the monodisperse clusters agree well with the average hydrodynamic diameter obtained by dynamic light scattering.[56] Values of 62.8 nm for C_{60} and 63.0 nm for C_{70}, are significantly smaller than the values reported previously for fullerene dispersions in water (*i.e.*, mean diameter of 300 nm) and those in toluene – acetonitrile (mean diameter 300 nm)[57]. The electrostatic repulsion between the similarly charged clusters is believed to be important for the stability of the dispersions.[58]

Larger clusters were generated from 1,2,5-triphenylfulleropyrrolidine in toluene – acetonitrile (1:3 *v/v*).[61] Dynamic light scattering revealed a mean diameter of 180 nm. Surprisingly, electron transfer quenching rates of the triplet excited state ($\tau = 2$ µs) with ferrocene, *N*-methylphenothiazine and *N,N*-dimethyl-*p*-anisidine were found to be orders of magnitude faster than that of the monomeric analogue, in the range of diffusion-controlled limits. This is attributed to the entrapment of donor molecules within the porous cluster network.

A similar mean diameter (*i.e.*, ~ 170 nm) was found for C_{60}-aniline clusters in toluene – acetonitrile (1:3 *v/v*).[62] This cluster formation exerted an improving impact on the performance of C_{60}-based donor-acceptor dyads. For example, in aggregates of a C_{60}-aniline dyad the radical pairs have a lifetime of 60 µs, while no detectable charge-transfer intermediates were noted for the isolated dyad. The close and condensed network in the cluster composites facilitates the hopping of electrons from the parent fullerene to an adjacent one etc., progressively increasing the spatial separation between radical ions of the charge-separated state.

Globular particles of 17 nm diameter, by far the smallest spheres detected so far, were seen by AFM and dynamic light scattering measurements upon addition of water to a THF solution of a potassium salt of pentaphenylated C_{60} ($Ph_5C_{60}K$).[66] The resulting anion, $Ph_5C_{60}^-$, associates into spherical bilayers.

In line with the second approach, a first report on ultrasonication described the treatment of an *N,N*-dimethylfulleropyrrolidinium salt in an aqueous solution.[67] After subsequent filtration through a 0.45 µm filter, and centrifugation to remove possible suspended solid, the resulting solution was transferred to a TEM grid. Representative images clearly reveal perfectly round shapes. The spheres have very similar sizes, with diameters ranging from 10 to 70 nm and wall thicknesses of 3 – 6 nm. Much larger spherical objects of mean diameters reaching 1.2 µm (Figure 17) were formed upon sonicating a dispersion of a fulleropyrrolidine, in which an ethylammonium group is linked to the pyrrolidine's nitrogen.[68]

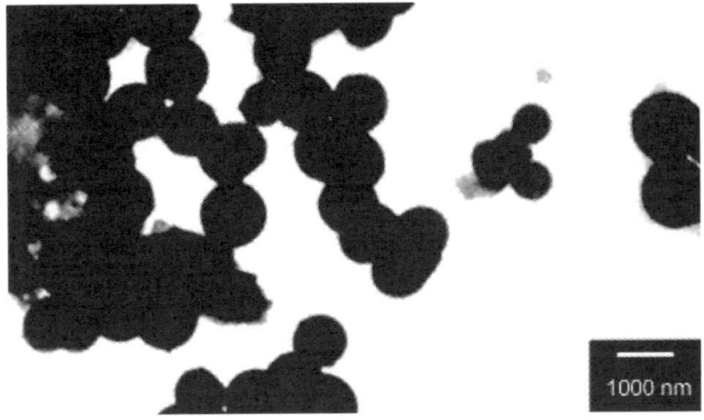

Fig. 17. TEM images of the spheres formed by a fulleropyrrolidine carrying an ethylamonnium group.

Finally, intermediately sized spheres were found, when a fulleropyrrolidine ammonium chloride solution of chlorobenzene methanol (10: 1 *v/v*) was left at room temperature for a few days and then dried *in vacuo*.[69] But no sonication was applied to these samples. TEM images reveal diameters that are in the range of 80 – 130 nm, while complementary AFM led to smaller sizes, 75 – 95 nm. The size discrepancy was ascribed to the different nature of the substrates used for the imaging techniques.

The size of all these spheres falls within a wide range, reaching from 17 nm to 10 μm. The size of the spheres is determined by:

1. the different methods of preparation (*i.e.*, time and type of sonication, presence of co-solvents, deposition, *etc.*);
2. the hydrophobic area left on the fullerene core (*i.e.*, the number and nature of the functional groups);
3. the side chain appendage of the fullerene spheroid (*i.e.,* the balance between attractive and repulsive forces).

4. CROWN ETHER COMPLEXATION MOTIFS

Crown ethers are very simple macrocyclic ligands constituted by a cyclic array of ether oxygen atoms connected through carbon atoms, which have been successfully used as appealing hosts in supramolecular chemistry for cations as well as neutral molecules. Recently, crown ethers in association with fullerenes have led to novel supramolecular ensembles exhibiting interesting properties.

Crown ether complexation emerged as a versatile molecular recognition principle to realize molecularly-organized thin film assemblies and nanoarchitectures.[70] Self-assembled monolayers (SAM), for example, were successfully employed to gain control over the organization of a C_{60} derivative – bearing a crown ether functionality – and an ammonium-terminated alkanethiolate that was attached irreversibly to a gold electrode (Figure 18).[10] As a consequence, the SAM approach is a viable alternative to deposition techniques such as Langmuir-Blodgett. On the basis of Osteryoung Square Wave Voltammetry experiments, determination of the surface coverage yielded a value (1.4×10^{-10} mol/cm^2), which is well in accord with a fcc close-packed packing of C_{60} ($\sim 1.9 \times 10^{-10}$ mol/cm^2). Complementary desorption experiments confirmed quantitatively the reversibility of this organization principle: there is no covalent and irreversible linkage of the fullerene moiety to the modified surface.

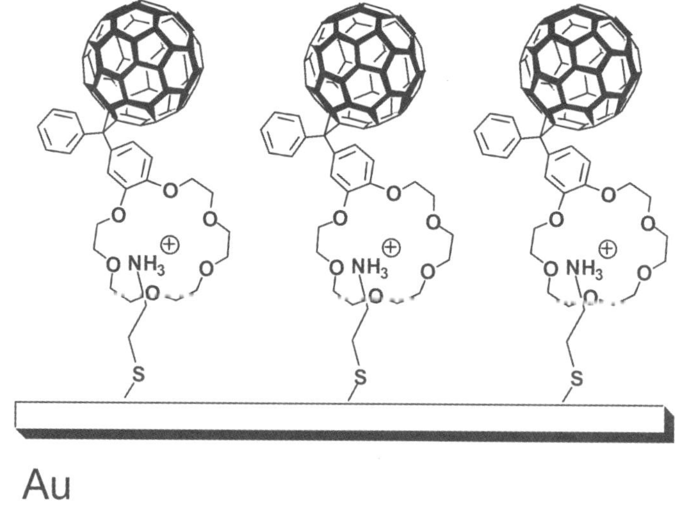

Fig. 18. A self-assembled fullerene-crown ether conjugate monolayer.

Not only playing a leading role in SAM constructs, fullerene-crown ether conjugates are also employed as molecular probes in complexation assays with potassium, sodium, cesium and lithium cations.[71a] Significant perturbations of the fullerene's electronic structure were, however, only registered when the bound cation was in a position close and tight relative to the C_{60} surface. These requirements were clearly guaranteed in the *trans-1*-bisadduct and, to a somewhat lesser extent, in a *trans-2*- and *trans-3*-bisadduct. Placing the crown ether conjugate, on the other hand, at greater distance from the C_{60} surface eliminated the cation-mediated effects.

When linking the crown ether conjugate to photoactive C_{60}-TTF donor-acceptor dyads – but adjacent to the TTF moiety – upon complexation of potassium, sodium and lithium, electronic effects were only exerted onto the TTF features.[72b] C_{60} with its highly delocalized π-system is not very perceptive of electronically induced changes. The electroactive probe must be placed in close proximity to ensure strong coupling. Illustrations are found in the following donor-acceptor dyads: Fc-C_{59}N, π-stack ZnP-C_{60}, π-stack H_2P-C_{60}, ZnPC-C_{60} and H_2PC-C_{60} (ZnPc and H_2PC signify zinc phthalocyanine and metal-free phthalocyanine, respectively), where the strong coupling between electron donor and electron acceptor is the only probate promoter for perturbing the fullerene's π-system.[47,72]

A fullerene-crown ether conjugate, similar in structure to that used for building SAM associates, forms in solution a 1:1 complex with 3-aminomethyl-(2,2,5,5-tetramethylpyrrolindin-1-oxyl).[73] Using visible light, which engages mainly with the fullerene core, a radical-triplet pair in the quartet excited state has been recorded. Strong electronic interactions between the triplet excitations and an ammonium aminoxyl free radical are responsible for the radical-triplet character.

Threading a dibenzylammonium unit attached to a fullerene, through the crown ether of an unsymmetrically substituted phthalocyanine which contains a dibenzo-24-crown-8, led to the assembly of a supramolecular ZnPc-C_{60} dyad.[74] Both components, which contain different electroactive subunits, can be assembled in $CHCl_3$ or CH_2Cl_2 leading to the typical stable pseudorotaxane-like complex structure. The ^{1}H NMR spectrum of a 1:1 mixture in $CDCl_3$ displays the characteristic high field shift and splitting of the resonances assigned to the 1,2-dioxybenzene unit. Diagnostic signals for both free and complexed subunits on the fullerene component were identified and allowed determination of the association constant ($K_a = 1.53 \times 10^4$ M^{-1}). When titrating ZnPc with variable C_{60} concentrations and on excitation at the 680 nm ground-state maximum, a decrease in intensity of the ZnPc emission was observed, eventually reaching a plateau value. From the fluorescence / fullerene concentration relationship an association constant of 1.8×10^4 M^{-1} was derived, matching that derived from the ^{1}H NMR data. Complementary transient absorption spectroscopy confirmed the presence of a charge-separated radical pair, ZnPc$^{\bullet+}$-$C_{60}^{\bullet-}$, as the product of the instantaneous fluorescence quenching.

5. METAL MEDIATED MOTIFS

The use of metal ions as synthetic templates has been widely used in supramolecular chemistry as an excellent method to bring about the organisation of a number of reacting components, in order to control the geometry of the product. Since some metal ions, such as the transition metals, usually present preferred coordination geometries, changes in the metal ion may have a strong effect on the nature of the templated product. In the following we show some of the most remarkable metal mediated supramolecular structures involving fullerenes.

5.1. Polypyridyl Precursors

Coordinating copper (I) to two separate bidentate 6,6′-disubstituted 2,2′-bipyridine ligands, attached to an o-quionodimethane derivative of C_{60}, afforded a novel dimeric form of C_{60}, separated by an intervening [Cu(bpy)$_2$]$^+$ unit.[75a] On account of the efficient and selective formation of copper (II) terpyridines, the corresponding C_{60}-2,2′:6′,2″-terpyridine ligands were also synthesized, which can be viewed as promising building block for supramolecular chemistry and nanoscience.[76b]

Similar, but more sophisticated, is the copper (I) templated approach bearing two C_{60}'s as end-terminating stoppers – Figure 19.[76] A three component precursor ensemble – a coordinating ring, a redox-active copper (I) center and a bisfunctionalized fragment threaded inside the ring – is reacted with a C_{60}-derivative. The MLCT state of $[Cu(phen)_2]^+$ and the singlet excited state of C_{60} are both substantially quenched, but the outcome of these rapid intramolecular deactivation routes is very different in character and product. Deactivation of the fullerene excited state, for example, follows a mechanism of energy transfer to the adjacent $[Cu(phen)_2]^+$. In contrast, the MLCT state of $[Cu(phen)_2]^+$ is mainly quenched by electron transfer to form the charge-separated radical pair comprising an oxidized metal center, namely, $[Cu(phen)_2]^{2+}$ and the one-electron reduced fullerene π-radical anion, $C_{60}^{\bullet-}$.

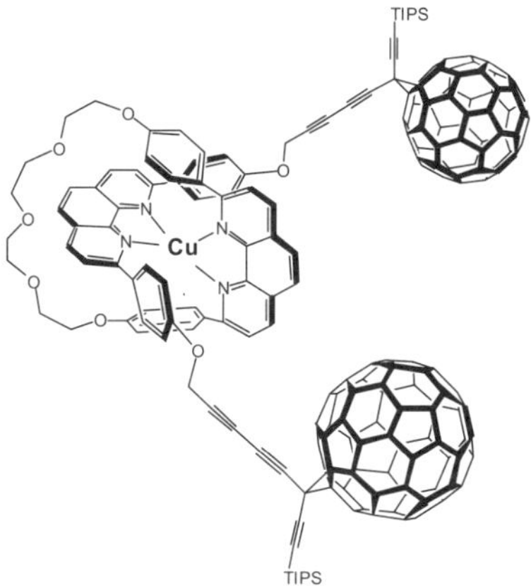

Fig. 19. A rotaxane bearing C_{60} units as electroactive stoppers.

The design of bis(phenanthroline) complexes with C_{60}-based dendrimers of different size was pursued in parallel work – Figure 20.[77] In the 1st, 2nd and 3rd generations, 4, 8 and 16 C_{60} moieties, respectively, surround a bis(phenanthroline) copper (I) core, $[Cu(phen)_2]^+$. The 16 C_{60} moieties create in the 3rd generation-based ensemble cluster as a black box around the copper (I) complex, shielding it electronically from the environment: ultraviolet light fails to penetrate through the densely packed C_{60}-periphery, and does not reach the copper (I) core. The indirect route, that is, transduction of singlet or triplet excited state energy from the C_{60}-periphery to the core, is also unsuccessful. This funnel effect is unlikely to happen, since the excited states of fullerenes are in general very low in energy, typically around 1.75 eV and 1.5 eV for the singlet and triplet manifold, respectively. Thus, an exothermic energy transfer to the copper (I) MLCT excited state (1.85 eV) is energetically unrealizable.

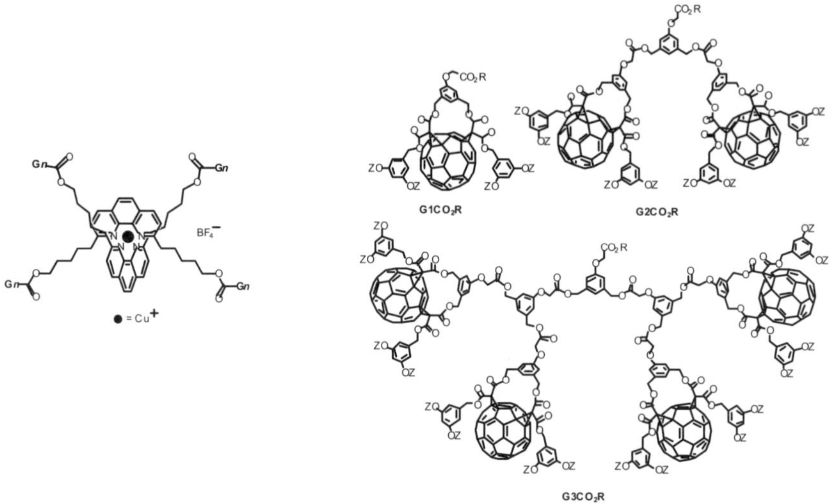

Fig. 20. Bis (phenanthroline) copper (I) complexes with C_{60}-based dendrimers of different size.

The product of the ruthenium (II) mediated organization of C_{60}-bipyridyl (*i.e.*, electron acceptor) and phenothiazine-bipyridyl (*i.e.*, sacrificial electron donor) building blocks is the photo- and electro-active C_{60}-[Ru(bpy)$_3$]$^{2+}$-PTZ triad ensemble.[78] The correspondingly formed ruthenium (II) complex, [Ru(bpy)$_3$]$^{2+}$, constitutes an important and widely used redox-active chromophore. Photoexcitation of the ruthenium (II) chromophore leads to ^{3}*MLCT[Ru(bpy)$_3$]$^{2+}$, from which a sequence of short-range intramolecular electron and charge transfer reactions evolve, at whose end a long-lived charge-separated state, $C_{60}^{\bullet-}$-[Ru(bpy)$_3$]$^{2+}$-PTZ$^{\bullet+}$, is formed. This radical pair exhibits in deoxygenated dichloromethane a lifetime of 1290 ns, and it deactivates completely back to the initial ground state.

Much simpler is the organization of C_{60}-$[Ru(bpy)_3]^{2+}$ dyads by a reaction of suitable C_{60}-bipyridyl precursors, bipyridyl ligands (*i.e.*, in a 1:2 stochiometry) and ruthenium (II) chloride.[79] Spacers such as androstane, polyglycol, crown ester and hexapeptide were employed as molecular rulers to separate a C_{60} acceptor unit from the bipyridyl ligand, yielding innovative donor-acceptor ensembles with diverse topographies – Figure 21.[80a,c,e-g]

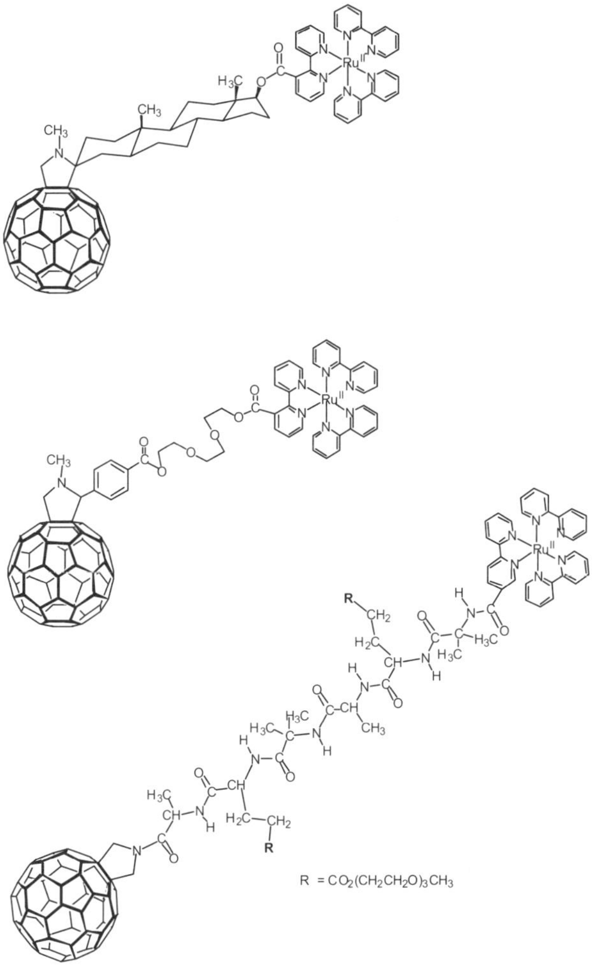

Fig. 21. C_{60}- $[Ru(bpy)_3]^{2+}$ dyads linked by different (rigid and flexible) spacers.

A common feature of all these C_{60}-$[Ru(bpy)_3]^{2+}$ systems is that upon photoexcitation a long-lived charge-separated state, $C_{60}{}^{\bullet-}$-$[Ru(bpy)_3]^{3+}$, evolves from an intramolecular electron transfer quenching of the $^3*(MLCT)$ state. Owing to the diverse topologies of these dyads, the lifetimes of the $C_{60}{}^{\bullet-}$-$[Ru(bpy)_3]^{3+}$ turn out to be quite different. For example, in dichloromethane solutions the rigidly spaced C_{60}-*androstane*-$[Ru(bpy)_3]^{2+}$ and C_{60}-*hexapeptide*-$[Ru(bpy)_3]^{2+}$ dyads yield lifetimes of 304 ns and 608 ns, respectively, while no appreciable lifetime was noted for the flexibly spaced C_{60}-*polyglycol*-$[Ru(bpy)_3]^{2+}$ analogue.[80]

Important to this work were the solvent-dependent conformational changes occurring within peptide bridges which were shown, for example, to influence the charge-separation process in the C_{60}-*hexapeptide*-$[Ru(bpy)_3]^{2+}$ dyad, upon photoexcitation of the ruthenium chromophore.[81e] A strong protic solvent disrupts the helical secondary structure of the peptide spacer that locates donor ($[Ru(bpy)_3]^{2+}$) and acceptor (C_{60}), at the N- and C-termini of the peptide chain, respectively. Our results strongly support the view that, upon disruption of the 3_{10}-helical structure, the separation between the two components of the dyad, C_{60} and $[Ru(bpy)_3]^{2+}$, tends to increase to a point that eventually disfavors their mutual electronic interactions. Thus, an unfolding of the 3_{10}-helix leads to a statistically unordered conformation, and consequently to a greater average distance between the two termini. Despite the general flexibility of the peptide backbone, the experimental data fail to support any short- or long-lived products, stemming from intramolecular charge-separation.

An intriguing feature of the peptide backbone is that the more randomized configuration can be reversibly transferred into the starting 3_{10}-helical conformation. After careful removal of the protic component from a binary solvent mixture the luminescence intensity of the $[Ru(bpy)_3]^{2+}$ chromophore becomes comparable again to that for the original non-protic solution, prior to the addition of the protic solvent. The successful repetition of the activation / deactivation cycle shows that the luminescence serves as a sensitive probe for the secondary structure of peptides.

The metal-mediated organization of another intriguing fullerene receptor is illustrated in Figure 22.[81] Silver (I) complexation holds together two calix[5]arene precursor units ($K_a = 5.7 \times 10^3$ M^{-1}) – each carrying a bipyridine ligand – creating a cavity sufficiently large to bind C_{60} and C_{70}. The assembly of π-conjugated oligo(*p*-phenylene vinylene) (OPV) donors and [60]fullerene acceptor moieties into a supramolecular donor-acceptor system has recently been achieved by means of ruthenium complexation[82] (Figure 23).

Oligomers have received a lot of attention since their precise chemical structure and conjugation length allow defined optoelectronic properties and facilitate enhanced control over their molecular architectures. OPVs have been shown to exhibit highly luminescent and electron donor properties. Therefore, these π-conjugated systems are very appealing to interact with fullerenes and, in fact, these two electro-

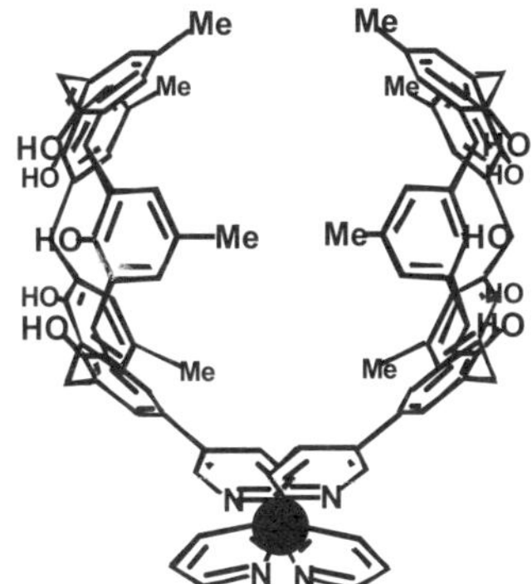

Fig. 22. Metal [silver(I)] mediated organization of a C_{60} receptor.

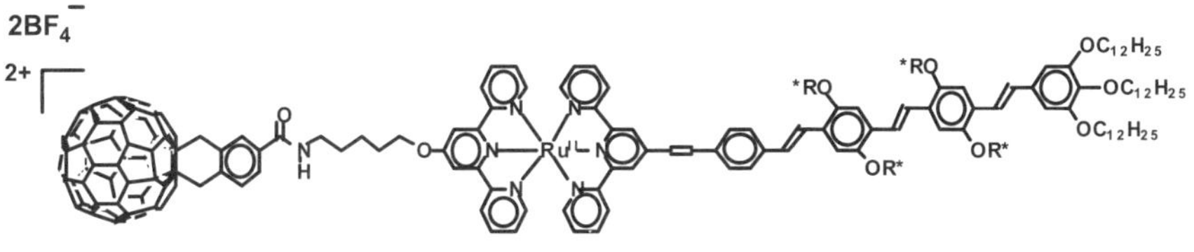

Fig. 23. Electroactive triads formed by metallo-supramolecular chemistry.

electroactive molecules (OPV and C_{60}) have recently been linked by metallo-supramolecular chemistry. In particular, the exploitation of non-covalent metal-ligand interactions by using 2,2′:6′,2″-terpyridine (tpy) as effective chelating agent with ruthenium metal, has allowed the preparation of new triads in which OPV and C_{60} are connected through a photoactive $Ru(tpy)_2$ moiety. Preliminary photophysical studies indicate that a charge-separated state can be formed in these complexes, resulting in an OPV radical cation and fullerene radical anion.

5.2. Porphyrin Precursors

A first demonstration was presented in form of a ZnP-*pyridine*-C_{60} complex. In the latter the reversible coordination of a pyridine functionalized fullerene ligand (*pyridine*-C_{60}) to the square-planar zinc center constitutes a labile, but nevertheless measurable, $(K \sim 5,000 \; M^{-1})$ binding motif, explored by three different research teams simultaneously.[83] The ground state features of ZnP in the visible (*Q*-bands) were employed as sensitive aids to monitor the progression of the ZnP-*pyridine*-C_{60} complexation: red-shifted transitions and the observance of clear isosbestic points. In a chain of events – triggered by light – the excited donor activates a rapidly occurring electron transfer to the electron accepting C_{60} within the ZnP-*pyridine*-C_{60} complex.[33] The weak equilibrium between dissociation and association of the 'metal-pyridine' bond then facilitates, in the final step of the sequence, the crucial break-up of the radical pair, before the competing charge-recombination starts to become a restriction. In ZnP-*pyridine*-C_{60} the free radical ions $(ZnP^{\bullet+}/C_{60}^{\bullet-})$ live for tens of microseconds in THF and benzonitrile, and any processes that may take place are exclusively governed by *inter*-molecular diffusion.

In this regard it appeared attractive to us that complexation of *pyridine*-C_{60} to a ruthenium tetraphenylporphyrin (RuP) produces the quite stable RuP-*pyridine*-C_{60} complex.[84d] Here the overriding principle is that utilization of the strong π-back-bonding strengthens the 'metal-pyridine' bond relative to that found in the ZnP-*pyridine*-C_{60} analog, in which the bonding is limited to a weak σ-character. As a consequence, the *intra*molecular charge-separated $RuP^{\bullet+}$-*pyridine*-$C_{60}^{\bullet-}$, as observed in polar solvents, recombines rapidly on the picosecond timescale (<4000 ps), since the diffusional splitting of the radical pair is largely suppressed.

The X-ray structure of ZnP-*pyridine*-C_{60} supports several key features.[84e] First, the tilting of the C_{60} unit towards the porphyrin is clearly discernible. Secondly, the edge-to-edge distance, that is, the closest distance between the porphyrin π-ring carbon and the C_{60} carbon of the axially linked fulleropyrrolidine, is 3.51 Å. Thirdly, the zinc to axially coordinated pyridyl nitrogen distance is 2.158 Å. Finally, the center-to-center distance between the porphyrin zinc ion and C_{60} is ~9.53 Å.

A more linearly aligned supramolecular architecture was assembled, based on linking a heterofullerene acceptor (*pyridine*-$C_{59}N$) to the central zinc atom of a zinc tetra-(*p*-tert.-butylphenyl)porphyrin (ZnP) donor – Figure 24.[84] The linear structure was achieved by attaching the donor pyridine ring to the $C_{59}N$ moiety, yielding *pyridine*-$C_{59}N$, by Mannich functionalization method of the dimer, $(C_{59}N)_2$. The ZnP-*pyridine*-$C_{59}N$ adduct is also ideally suited for devising integrated model systems to transmit and process solar energy. Depending on the solvent, either photoinduced singlet-singlet energy transfer or electron-transfer was observed. The latter process takes place in *o*-dichlorobenzene as solvent and leads to the corresponding *pyridine*-$C_{59}N$ π-radical anion and ZnP π-radical cation.

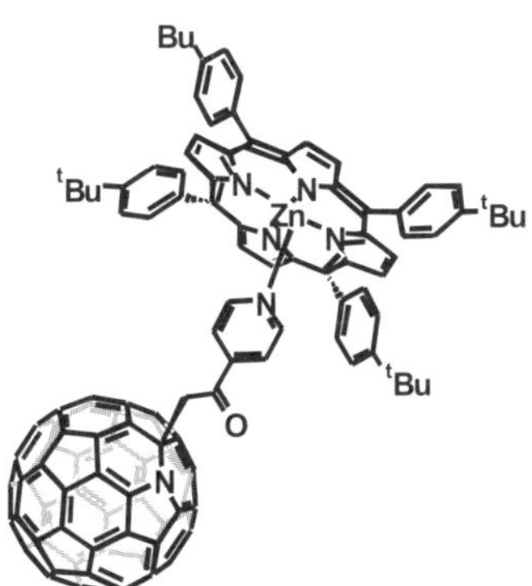

Fig. 24. Linearly aligned ZnP- C_{60} supramolecular architecture.

Similarly, novel dyads formed by axial coordination of zinc tetraphenylphorphyrin (TTP)Zn, and fulleropyrrolidine bearing either pyridine or imidazole coordinating ligands, have been synthesized and electrochemically and photochemically investigated.[85] As expected, the spectroscopic studies revealed a 1:1 stoichiometry between the donor, (TTP)Zn and the fulleropyrrolidine acceptor. The determined association constants follow the order: *o*-pyridyl $\ll$ *m*-pyridyl $\cong$ *p*-pyridyl $\ll$ N-phenylimidazole entities of the fulleropyrrolidine. Therefore, the constants are controlled by the nature of the axial ligand and the associated steric factors. The results of steady-state and time-resolved emission, and transient absorption studies revealed the occurrence of electron transfer mainly from the singlet excited state zinc porphyrin to the fullerene core in the non-coordinating solvent *o*-dichlorobenzene. However, in the presence of a coordinating solvent such as benzonitrile, the main quenching occurs via an intermolecular electron transfer from the triplet excited (TTP)Zn to the C_{60} unit.

Notably, this metal-mediated concept, namely, coordination of a fullerene-pyridine ligand by a macrocyclic π-system, is very general and has been successfully extended to zinc complexes of phthalocyanines, porphycenes and corrphycene macrocycles.[84] Interestingly, the binding strength reveals a close resemblance to the oxidation potential of the macrocyclic π-system: porphycene > porphyrin.

A fascinating modification of the ZnP-*pyridine* coordination focuses on the modulation of the donor-acceptor proximity by controlling a 'tail-on / tail-off' binding mechanism.[86] In this work, the *pyridine*-C_{60} ligand is covalently attached to the phenyl group of ZnP and the nitrogen of the pyrrolidine through a flexible chain (Scheme ?). The defined spatial organization is controlled *via* temperature variation or replacement of the axial ligand with 3-picoline. As far as charge-separation rates and efficiencies are concerned in the 'tail-off' status (*i.e.*, ZnP/*pyridine*-C_{60}) both parameters are slightly changed in comparison with the results of the 'tail-on' status (*i.e.*, ZnP-*pyridine*-C_{60}), suggesting through-space interactions in the former. In line with this assumption is the observation that an acceleration of the charge-recombination also comes to light in the 'tail-off' form.

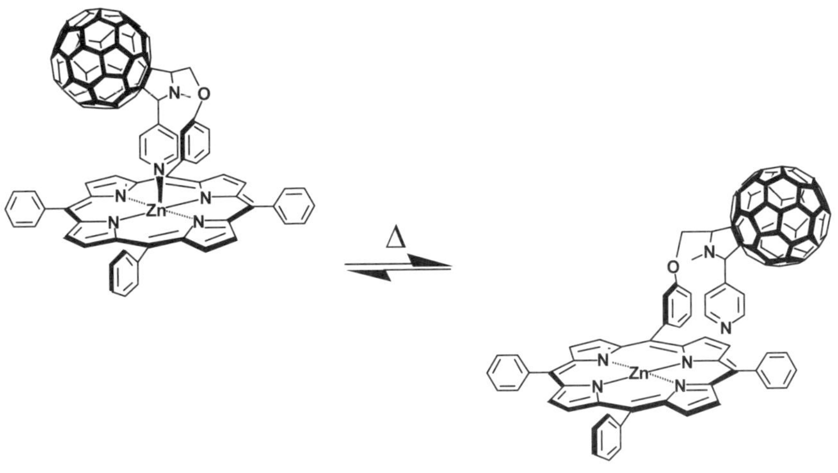

Scheme 2. Temperature-dependent 'tail-on / tail-off' equilibrium.

With the objective to devise linear architectures of higher complexity, *i.e.* triads, tetrads and pentads, several ZnP/C_{60} containing ensembles were subjected to complexation assays with diazabicyclo-octane (DABCO).[87] The bidentate DABCO ligand exhibits a number of appealing features: it not only forms square pyramidal 1:1- or 2:1-complexes with, for example, ZnP, but it is also a good electron donor. A prerequisite for successful construction restricts use to non-coordinating solvents. Suitable polar media, on the other hand, set up a competition between DABCO and solvent complexation and subsequently lead to dissociation into the free components. Importantly, charge-recombination kinetics in the primary building block (*i.e.*, *trans-2*-$ZnP-C_{60}$, etc.) of these complexes reveal that the large $-\Delta G_{CR}°$ values in toluene are extremely helpful to stabilize the $ZnP^{•+}-C_{60}^{•-}$ radical pair. Considering these facts in concert, it is clear that among the many unique fullerene features the small reorganization energy guarantees appreciable affects in these photoactive architectures, especially in the light of retarding charge-recombination. Varying the relative concentrations of DABCO and *trans-2*-$ZnP-C_{60}$, the precursor $ZnP-C_{60}$ dyad was transformed step by step into DABCO-$ZnP-C_{60}$ (micromolar concentrations) and C_{60}-ZnP-DABCO-$ZnP-C_{60}$ (millimolar concentrations) (Figure 25). The lifetimes

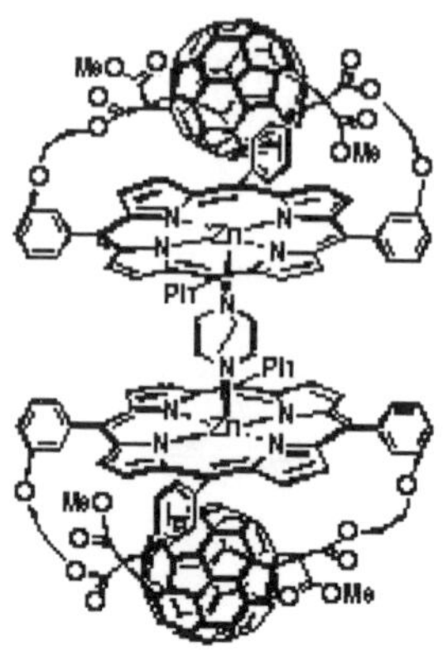

Fig. 25. A linear C_{60}- ZnP-DABCO-ZnP- C_{60} stack.

of the charge-separated states in these newly formed ensembles change markedly with the ensemble constitution. A significant improvement is seen upon going from *trans*-2-ZnP$^{\bullet+}$-$C_{60}^{\bullet-}$ (toluene: $\tau = 619$ ps) and DABCO$^{\bullet+}$-ZnP-$C_{60}^{\bullet-}$ (toluene: $\tau = 1980$ ps) to C_{60}-ZnP-DABCO$^{\bullet+}$-ZnP-$C_{60}^{\bullet-}$ (toluene: $\tau = 2280$ ps).

A different example involves a simplistic but powerful means to regulate donor-acceptor separations and orientations. More specifically, rigid, confined model ensembles are self-assembled, starting from a flexible ZnP-C_{60}-ZnP system and DABCO.[88] Similar to the simpler dyad ensembles (*m*-ZnP-C_{60} and *p*-ZnP-C_{60}) a photoinduced electron transfer evolving from the ZnP singlet excited state to the electron accepting fullerene governs the photophysics of the *m*-ZnP-C_{60}-ZnP and the more electron-rich *p*-ZnP-C_{60}-ZnP. The resulting ZnP$^{\bullet+}$-$C_{60}^{\bullet-}$-ZnP states decayed on a timescale of a few hundred nanoseconds to regenerate the ground state. For example, in *o*-dichlorobenzene the actual values are 150 ns and 290 ns in the *m*-ZnP-C_{60}-ZnP and *p*-ZnP-C_{60}-ZnP isomers, respectively. Addition of DABCO to toluene or *o*-dichlorobenzene solutions of *m*-ZnP-C_{60}-ZnP or *p*-ZnP-C_{60}-ZnP evoked a strong reactivation of the ^{1}*ZnP fluorescence (factor of ~ 3). A similar impact was concluded from the transient absorption measurements, monitoring the decay and grow-in kinetics of $^1{*}$ZnP and ZnP$^{\bullet+}$/C$_{60}$$^{\bullet-}$ features, respectively. The sum of these effects evokes a model that infers the successful complexation of DABCO to the vacant sites of the two ZnP (dz^2-orbitals) to expand the donor-acceptor separation considerably (*i.e.*, triad *versus* tetrad). In a polar medium (*o*-dichlorobenzene) these readily available and stable ensembles are subject to a rapid *intra*molecular electron transfer to yield a long-lived radical pair with lifetimes close to a microsecond (~ 700 ns). Again, the simple addition of a suitable component leads to a nearly five-fold improvement of the radical pair stability, besides an overall higher quantum yield of formation. In retrospect, the reversible coordination of DABCO to ZnP-C$_{60}$-ZnP creates a simple photoswitch, in which either an electron or energy transfer pathway deactivates the photoexcited ZnP.

Linear arrays (Figure 26) are built when millimolar concentrations of DABCO are added to ZnP-C$_{60}$-ZnP. In the millimolar concentration regime, precipitation of a poorly soluble oligomeric material is observed, which, however, can be re-suspended in *o*-dichlorobenzene. It is interesting to note that the lifetime of the radical pair in these linear structures gives rise to a further improvement (7.5 µs): a sufficiently fast separation of charges along the longitudinal axis is probably responsible for this consequence.

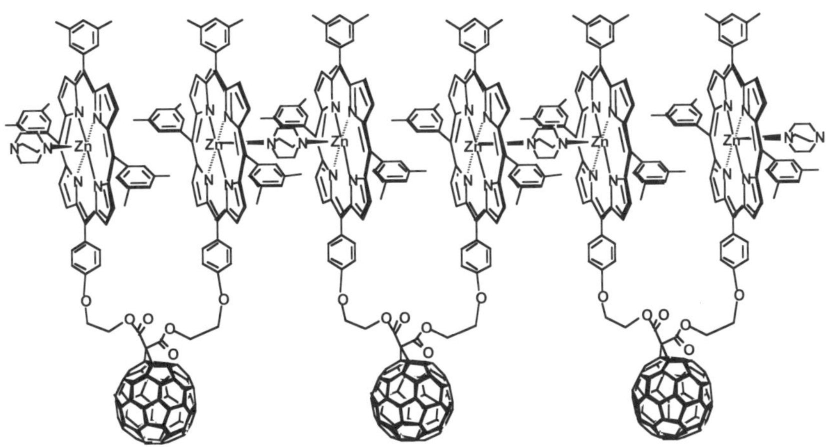

Fig. 26. Linear supramolecular arrays from ZnP- C$_{60}$- ZnP and DABCO.

6. ELECTROSTATIC MOTIFS

Exploration of electrostatic interaction as a probate means to bind C_{60} – with cationic headgroups – to a suitable template was first reported in binding assays with duplex DNA.[89] An *N,N*-dimethylfulleropyrrolidinium salt and three isomeric fulleropyrrolidines carrying pyridinium moieties were found to bind to double stranded DNA. Two major contributions control the binding modes; electrostatic and also hydrophobic interactions with the phosphate groups along the DNA backbone and the DNA grooves, respectively. All derivatives cleaved double stranded DNA under photoirradiation, which promotes the action of singlet oxygen. The latter species evolves as a product of the reaction of triplet excited C_{60} with molecular oxygen. Importantly, preferential binding to the DNA grooves brings the C_{60} photosensitizer closer to the nucleic bases, which enhances the cleavage efficiency.

More recently, a 'two-handed' C_{60} derivative – carrying two protonated diamine side chains – was shown to bind through electrostatic interaction to duplex DNA.[90] It also condenses and allows the complexed DNA to be delivered into, and transiently expressed in, the target cell. It is an interesting fact that, despite the intrinsic photoactivity of C_{60}, no differences were noted when the transfection assays were carried out under ambient light or black light.

7. SUMMARY AND OUTLOOK

This article has demonstrated the effectiveness of employing fullerenes in biomimetic strategies for devising thermodynamically stable but kinetically labile 1-D, 2-D, or 3-D networks. These strategies (highly directional hydrogen bondings; π-stack motifs including concave-convex, planar-convex and convex-convex interactions; crown ether complexation; metal-mediated and electrostatic interactions) provide the means for an evident trend towards the facile preparation of precise structures never before accomplished by conventional synthetic chemistry. The most important aspect in this area is the regulation of the inherently weak forces seen in biomimetic organization principles on a molecular basis. In this context, the current concepts illustrate that relating size and shape to the function of the resulting composites leads to composites with new and original properties. Unquestionably, the peculiar shape and unique electronic properties of fullerenes – as stiff molecular scaffolds – has an important bearing on the design of novel, well-ordered supramolecular arrays.

Despite some remarkable recent successes, it is clear that the examples discussed in this chapter represent only the tip of the iceberg. More research in this relatively new area is needed to fully explore the possibilities offered by these materials, for example, in the production of active as well as passive devices. This underlines more than ever the great need for creative synthetic chemistry, which will ensure that fullerenes may eventually become an important building block of future technologies, such as optoelectronics, batteries, photovoltaics and, in a more general sense, as a new and realistic class of energy harvesting materials.

ACKNOWLEDGMENTS

This work was supported by the Office of Basic Energy Sciences of the Department of Energy and the MCyT of Spain (Project BQU2002-00855). This is document NDRL-4484 from the Notre Dame Radiation Laboratory.

References

1. a) Lehn, J.-M. *Supramolecular Chemistry – Concepts and Perspectives*, Weinheim: VCH; 1995. b) McDermott G, Priece SM, Freer AA, Hawthornthwaite-Lawless AM, Papiz MZ, Cogdell RJ and Isaacs NW. *Nature*, 1995; **374**: 517-521. c) Barber J. *Nature*, 1988; **333**: 114-114. d). Balzani V and de Cola L (Ed): *Supramolecular Chemistry*, NATO ASI Series. Dordrecht: Kluwer Academic Publishers; 1992.
2. Deisenhofer J and Norris JR (Ed): *The Photosynthetic Reaction Center*. New York: Academic Press; 1993.
3. a) Atwood JL, Davies JED, MacNicol DD, Vögtle F and Lehn J-M (Ed): *Comprehensive Supramolecular Chemistry* Vol. 1-10. Oxford:

Pergamon/Elsevier; 1996. b) Lindoy LF and Atkinson IM, *Self-Assembly in Supramolecular Systems*: Cambridge: Royal Society of Chemistry; 2000.

4. Fendler J.H. (Ed): *Nanoparticles and Nanostructured Films* Weinheim: Wiley; 1998.

5. a) Diederich F and Kessinger R. *Acc. Chem. Res.* 1999; **32**: 537-545. b) Hirsch A (Ed): *Fullerenes and Related Structures*, Topics in Current Chemistry, Vol. 199. Berlin: Springer; 1999. c) Prato M and Maggini M. *Acc. Chem. Res.* 1998; **31**: 519-526. d) Taylor R. *Lecture Notes on Fullerene Chemistry* London: Imperial College Press; 1999. e) Guldi D and Martín N. *Fullerenes: From Synthesis to Optoelectronic Properties*. Dordrecht: Kluwer Academic Publishers; 2002.

6. a) Philp D and Stoddart JF. *Angew. Chem. Int. Ed. Engl.* 1996; **35**: 1155-1196. b) Lawrence DS, Jiang T and Levitt M. *Chem. Rev.* 1995; **95**: 2229-2260. c) Sherrington DC and Taskinen KA. *Chem. Soc. Rev.* 2001; **30**: 83-93.

7. For a recent review, see: Schmuck C and Wienand W. *Angew. Chem. Int. Ed.* 2001; **40**: 4363-4369

8. For recent reviews on fullerene containing donor-acceptor ensembles and fullerene anions see: a) Imahori H and Sakata Y. *Adv. Mater.* 1997; **9**: 537-546. b) Prato M. J. *Mater. Chem.* 1997; **7**: 1097-1109. c) Martín N, Sanchez L, Illescas B and Perez I. *Chem. Rev.* 1998; **98**: 2527-2547. d) Imahori H and Sakata Y. *Eur. J. Org. Chem.* 1999; 2445-2457. e) Diederich F and Gomez-Lopez M. *Chem. Soc. Rev.* 1999; **28**: 263-277. f) Guldi DM. *Chem. Commun.* 2000; 321-327. g) Guldi DM and Prato M. *Acc. Chem. Res.*, 2000; **33**: 695-703. h) Reed CA and Bolskar RD. *Chem. Rev.* 2000; **100**: 1075-1119. i) Gust D, Moore TA and Moore AL. J. *Photochem. & Photobiol. B*, 2000; **58**: 63-71. k) Gust D, Moore TA and Moore AL. *Acc. Chem. Res.* 2001; **34**: 40-48.

9. Arias F, Godinez LA, Wilson SR, Kaifer AE and Echogoyen L. *J. Am. Chem. Soc.* 1996; **118**: 6086-6087.

10. Diederich F, Echegoyen L, Gomez-Lopez M, Kessinger R and Fraser Stoddart J. *J. Chem. Soc., Perkin Trans. 2*, 1999; 1577-1586.

11. Da Ros T, Guldi DM, Morales AF, Leigh DA, Prato M and Turco R. *Org. Lett.* 2003; **5**: 689

12. Watanabe N, Kihara N, Furusho Y, Takata T, Araki Y and Ito O. *Angew. Chem. Int. Ed.* 2003; **42**: 681.

13. González JJ, González S, Priego EM, Luo C, Guldi DM, De Mendoza J and Martín N. *Chem. Commun.* 2001; 163-164.

14. Rispens MT, Sánchez L, Knol J and Hummelen JC. *Chem. Commun.* 2001; 161-162.

15. Sánchez L, Rispens MT and Hummelen JC. *Angew. Chem. Int. Ed.* 2002; **41**: 838-840

16. Marczak R, Hoang VT, Noworyta K, Zandler ME, Kutner W and D'Souza F. *J. Mater. Chem.* 2002; **12**: 2123-2129

17. Beckers EHA, van Hal PA, Schenning A-PHJ, El-ghayoury A, Peeters E, Rispens MT, Hummelen JC, Meijer EW and Janssen RAJ. *J. Mater. Chem.* 2002; **12**: 2054-2060

18. Fang H, Wang S, Xiao S, Yang J, Li Y, Shi Z, Li H, Liu H, Xiao S and Zhu D. *Chem. Mater.* 2003; **15**: 1593-1597

19. Segura M, Sánchez L, de Mendoza J, Martín N and Guldi DM. *J. Am. Chem. Soc.* 2003; *in press*

20. a) Haddon RC, Brus LE and Raghavachari K. *Chem. Phys. Lett.* 1986; **131**: 165-169. b) Prassides K, Kroto HW, Taylor R, Walton DRM, David WIF, Tomkinson J, Haddon RC, Rosseinsky MJ and Murphy DW. *Carbon* 1992; **8**: 1277-1286. c) Kelly MK, Etchegoin P, Fuchs D, Krätschmer W and Fostiropulos K. *Phys. Rev. B* 1992; **46**: 4963-4968. d) Haddon RC. *Science* 1993; **261**: 1545-1550.

21. Mirkin CA and Caldwell WB. *Tetrehedron* 1996; **52**: 5113-5130 and references cited therein.

22. a) Andersson T, Nilsson K, Sundahl M, Westman G and Wennerstroem O. *Chem. Commun.* 1992; 604-606. b) Sundahl M, Andersson T, Nilsson K, Wennerstroem O and Westman G. *Synth. Met.* 1993; **55**: 3252-3257. c) Andersson T, Westman G, Wennerstroem O and Sundahl M. *J. Chem. Soc. Perkin Trans. 2* 1994; 1097-1101. d) Constable E. *Angew. Chem. Int. Ed.* 1994; **33**: 2269-2271. e) Priyadarsini KI, Mohan H, Tyagi AK and Mittal JP. *J. Phys. Chem.* 1994; **98**: 4756-4759. f) Yoshida Z, Takekuma H, Takekuma S and Matsubara Y. *Angew. Chem. Int. Ed.* 1994; **33**: 1597-1599. g) Kuroda Y, Nozawa H and Ogoshi H. *Chem. Lett.* 1995; 47-48. h) Marconi G, Mayer B, Klein CT and Koehler G. *Chem. Phys. Lett.* 1996; **260**: 589-594. i) Komatsu K, Fujiwara K, Murata Y and Braun T. *J. Chem. Soc. Perkin Trans. 1* 1999; 2963-2966. k) Buvari-Barcza A, Rohonczy J, Rozlosnik N, Gilanyi T, Szabo B, Lovas G, Braun T, Samal S and Geckeler

KE. *Chem. Commun.* 2000; 1101-1102. l) Samu J and Barcza L. *J. Chem. Soc. Perkin Trans. 2* 2001; 191-196.

23. Zonta C, Cossu S and DeLucchi O. *Eur. J. Org. Chem.* 2000; 1965-1971.

24. a) Williams RM and Verhoeven JW. *Recl. Trav. Chim. Pays-Bas.* 1992; **11**: 531-532. b) Atwood JL, Koutsantonis GA and Raston CL. *Nature* 1994; **369**: 229-231. c) Suzuki T, Nakashima K and Shinkai S. *Chem. Lett.* 1994; 699-702. d) Isaacs NS, Nichols PJ, Raston CL, Sandova CA and Young DJ. *Chem. Commun.* 1997; 1839-1840. e) Haino T, Yanase M and Fukazawa Y. *Angew. Chem. Int. Ed.* 1997; **36**: 259-960. f) Atwood JL, Barbour LJ, Raston CL and Sudria IBN. *Angew. Chem. Int. Ed.* 1998; **37**: 981-983. g) Haino T, Yanase M and Fukazawa Y. *Angew. Chem. Int. Ed.* 1998; **37**: 997-998. h) Tsubaki K, Tanaka K, Kinoshita T and Fuji K. *Chem. Commun.* 1998; 895-896. i) Yanase M, Haino T and Fukazawa Y. *Tetrahedron Lett.* 1999; **40**: 2781-2784. k) Atwood JL, Barbour LJ, Nichols PJ, Raston CL and Sandoval CA. *Chem. Eur. J.* 1999; **5**: 990-996. l) Tucci FC, Rudkevich DM and Rebek J. *J. Org. Chem.* 1999; **64**: 4555-4559. m) Schlachter I, Hoeweler U, Iwanek W, Urbaniak M and Mattay J. *Tetrahedron* 1999; **55**: 14931-14940. n) Bhattacharya S, Nayak SK, Chattopadhyay S, Banerjee M and Mukherjee AK. *J. Chem. Soc., Perkin Trans. 2*, 2001; 2292-2297.

25. a) Georghiou PE, Mizyed S and Chowdhury S. *Tetrahedron Lett.* 1999; **40**: 611-614. b) Mizyed S, Georghiou PE and Ashram M. *J. Chem. Soc., Perkin Trans. 2*, 2000; 277-280. c) Mizyed S, Tremaine PR and Georghiou PE. *J. Chem. Soc., Perkin Trans. 2*, 2001; 3-6. d) Mizyed S, Ashram M, Miller DO and Georghiou PE. *J. Chem. Soc., Perkin Trans. 2*, 2001; 1916-1919.

26. a) Steed JW, Junk PC, Atwood JL, Barnes MJ, Raston CL and Burkhalter RS. *J. Am. Chem. Soc.* 1994; **116**: 10346-10347. b) Hardie MJ, Godfrey PD and Raston CL. *Chem. Eur. J.* 1999; **5**: 1828-1833.

27. Paci B, Amoretti G, Arduini G, Ruani G, Shinkai S, Suzuki T, Ugozzoli F and Caciuffo R. *Phys. Rev. B*, 1997; **55**: 5566-5569.

28. a) Ikeda A, Yoshimura M and Shinkai S. *Tetrahedron Lett.* 1997; **38**: 2107-2110. b) Ikeda A, Suzuki Y, Yoshimura M and Shinkai S. *Tetrahedron* 1998; **54**: 2497-2508.

29. Olsen SA, Bond AM, Compton RG, Lazarev G, Mahon PJ, Marken F, Raston CL, Tedesco V and Webster RD. *J. Phys. Chem. A* 1998; **102**: 2641-2649.

30. Islam SD-M, Fujitsuka M, Ito O, Ikeda A, Hatano T and Shinkai S. *Chem. Lett.* 2000; 78-79

31. Bourdelande JL, Font J, Gonzalez-Moreno R and Nonell S. *J. Photochem. Photobiol. A*, 1998; **115**: 69-71

32. Sun D and Reed CA. *Chem. Commun.* 2000; 2391-2392.

33. a) Zhou P, Dong ZH, Rao AM and Eklund PC. *Chem. Phys. Lett.* 1993; **211**: 337-340. b) Wang Y, Holden JM, Dong ZH, Bi XX, and Eklund PC. *Chem. Phys. Lett.* 1993; **211**: 341-345. c) Rao AM, Zhou P, Wang KA, Hager GT, Holden JM, Wang Y, Lee W-T, Bi X-X, Eklund PC, Cornett DS, Duncan MA and Amster IJ. *Science* 1993; **259**: 955. d) Wang Y, Holden JM, Bi X-X and Eklund PC. *Chem. Phys. Lett.* 1994; **217**: 413-417. e) Iwasa Y, Arima T, Fleming RM, Siegrist T, Zhou O, Haddon RC, Rothberg LJ, Lyons KB, Carter HL, Hebard AF, Tycko R, Dabbagh G, Krajewski JJ, Thomas GA and Yagi T. *Science* 1994; **264**: 1570. f) Stephens PW, Bortel G, Faigel G, Tegze M, Janossy A, Pekker S, Oszlanyi G and Forro L. *Nature* 1994; **370**: 636. g) Eklund PC, Rao AM, Zhou P, Wang Y and Holden JM. *Thin Solid Films* 1995; **257**: 185-203.

34. a) Andrews PC, Atwood JL, Barbour LJ, Nichols PJ and Raston CL. *Chem. Eur. J.* 1998; **4**: 1384-1387. b) Croucher PD, Nichols PJ and Raston CL. *J. Chem. Soc., Dalton Trans.* 1999; 279-284.

35. a) Fox OD, Drew MGB, Wilkinson EJS and Beer PD. *Chem. Commun.* 2000; 391-392. b) Fox OD, Drew MGB and Beer PD. *Angew. Chem. Int. Ed.* 2000; **39**: 136-140.

36. Schuster DI, Rosenthal J, MacMahon S, Jarowski PD, Alabi CA and Guldi DM. *Chem Commun.* 2002; 2538-2359.

37. a) Hungerbühler H, Guldi DM and Asmus K-D. *J. Am. Chem. Soc.* 1993; **115**: 3386-3387. b) Guldi DM, Hungerbühler H and Asmus K-D. *J. Phys. Chem.* 1995; **99**: 13487-13493. c) Guldi DM. *J. Phys. Chem. A* 1997; **101**: 3895-3900. d) Guldi DM. *J. Phys. Chem. B* 1997; **101**: 9600-9605.

38. Raston CL, Atwood JL, Nichols PJ and Sudria IBN. *Chem. Commun.* 1996; 2615-2616.

39. Atwood JL, Barnes MJ, Gardiner MG and Raston CL. *Chem. Commun.* 1996; 1449-1450.

40. a) Drovetskaya T, Reed CA and Boyd PDW. *Tetrahedron Lett.* 1995; **36**: 7971-7974. b) Sun Y, Drovetskaya T, Bolskar RD, Bau R, Boyd PDW and Reed CA. *J. Org. Chem.* 1997; **62**: 3642-3649.

41. a) Evans DR, Fackler NLP, Xie Z, Rickard CEF, Boyd PDW and Reed CA. *J. Am. Chem. Soc.* 1999; **121**: 8466-8474. b) Boyd PDW, Hodgson MC, Rickard CEF, Oliver AG, Chaker L, Brothers PJ, Bolskar RD, Tham FS and

Reed CA. *J. Am. Chem. Soc.* 1999; **121**: 10487-10495. c) Olmstead MM, Costa DA, Maitra K, Noll BC, Phillips SL, van Calcar PM and Balch AL. *J. Am. Chem. Soc.* 1999; **121**: 709-7097. d) Ishii T, Aizawa N, Yamashita M, Matsuzaka H, Kodama T, Kikuchi K, Ikemoto I and Iwasa Y. *J. Chem. Soc., Dalton Trans.* 2000; 4407-4412. e) Konarev DV, Neretin IS, Slovokhotov YL, Yudanova EI, Drichko NV, Shul'ga YM, Tarasov BP, Gumanov LL, Batsanov AS, Howard JAK and Lyubovskaya RN. *Chem. Eur. J.* 2001; **7**: 2605-2616.

42. Hochmuth DH, Michel M. SLJ, White AJP, Williams DL, Barrett AGM and Hoffman BM. *Eur. J. Org. Chem.* 2000; 593-596.

43. Kadish KM, Smith KM, and Guilard R (Ed): *The Porphyrin Handbook.* New York: Academic Press; 1999.

44. Guldi DM, Neta P and Asmus K-D, *J. Phys. Chem.* 1994; **98**: 4617-4621.

45. a) Baran PS, Monaco RR, Khan AU, Schuster DI and Wilson SR. *J. Am. Chem. Soc.* 1997; **119**: 8363-8364. b) Kawaguchi M, Ikeda A, Hamachi I and Shinkai S. *Tetrahedron Lett.* 1999; **40**: 8245-8249. c) Schuster DI. *Carbon* 2000; **38**: 1607-1614. d) Guldi DM, Luo C, Prato M, Troisi A, Zerbetto F, Scheloske M, Dietel, Bauer W and Hirsch A. *Chem. Eur. J.* 2003; *in press.*

46. a) Dietel E, Hirsch A, Eichhorn E, Rieker A, Hackbarth S and Röder B. *Chem. Commun.* 1998; 1981-1982. b) Guldi DM, Luo C, Prato M, Dietel E and Hirsch A. *Chem. Commun.* 2000; 373-374. c) Armaroli N, Marconi G, Echegoyen L, Bourgeois J-P and Diederich F. *Chem. Eur. J.* 2000; **6**: 1629-1645. d) Guldi DM, Luo C, Prato M, Troisi A, Zerbetto F, Scheloske M, Dietel E, Bauer W and Hirsch A. *J. Am. Chem. Soc.* 2001; **123**: 9166-9167.

47. a) Tashiro K, Aida T, Zheng J-Y, Kinbara K, Saigo K, Sakamoto S and Yamaguchi K. *J. Am. Chem. Soc.* 1999; **121**: 9477-9478. b) Zheng J-Y, Tashiro K, Hirabayashi Y, Kinbara K, Saigo K, Aida T, Sakamoto S and Yamaguchi K. *Angew. Chem. Int. Ed.* 2001; **40**: 1858-1861.

48. Sun D, Tham FS, Reed CA, Chaker L, Burgess M and Boyd PDW. *J. Am. Chem. Soc.* 2001; **123**: 10704-10705.

49. Tashiro K, Hirabayashi Y, Aida T, Saigo K, Fujiwara K, Komatsu K, Sakamoto S and Yamaguchi K. *J. Am. Chem. Soc.* 2002; **124**: 12086-12087.

50. Guldi DM, Da Ros T, Braiuca P, Prato M and Alessio E. *J. Mater. Chem.* 2002; **12**: 2001-2008.

51. a) Nierengarten JF, Oswald L, Eckert J-F, Nicoud J-F and Armaroli N. *Tetrahedron Lett.* 1999; **40**: 5681-5684. b) Eckert J-F, Byrne D, Nicoud J-F,

Oswald L, Nierengarten JF, Numata M, Ikeda A, Shinkai S and Armaroli N. *New. J. Chem.* 2000; **24**: 749-758.

52. Cyclotriveratrylene-based dendritic structures reveal a similar trend: the association constant increases with the generation number of the surrounding dendritic substituents.

53. Da Ros T and Prato M. *Chem. Commun.* 1999; 663-669.

54. a) Sun Y-P and Bunker CE. *Nature* 1993; **365**: 398. b) Sun Y-P and Bunker CE. *Chem. Mater.* 1994; **6**: 578-580.

55. a) Andriesvsky GV, Kosevich MV, Vovk OM, Shelkovsky VS and Vashchenko LA. *J. Chem. Soc., Chem. Commun.* 1995; **12**: 1281-1282. b) Andrievsky GV, Klochkov VK, Karyakina EL and Mchedlov-Petrossyan NO. *Chem. Phys. Lett.* 1999; **300**: 392-396. c) Deguchi S, Alargova RG and Tsujii K. *Langmuir* 2001; **17**: 6013-6017.

56. Scrivens WA, Tour JM, Creek KE and Pirisi L. *J. Am. Chem. Soc.* 1994; **116**: 4517-4518.

57. A spherical $(C_{60})_{13}$-cluster with a diameter of 2.8 nm was shown theoretically (molecular dynamics approach) to be the smallest stable form among all possible aggregates. Bulavin L, Adamenko I, Prylutskyy Y, Durov S, Graja A, Bogucki A and Scharff P. *Phys. Chem. Chem. Phys.* 2000; **2**: 1627-1629.

58. a) Fujitsuka M, Kasai H, Masuhara A, Okada S, Oikawa H, Nakanishi H, Watanabe A and Ito O. *Chem. Lett.* 1997; 1211-1212. b) Fujitsuka M, Kasai H, Masuhara A, Okada S, Oikawa H, Nakanishi H, Ito O and Yase K. *J. Photochem. Photobiol. A* 2000; **133**: 45-50.

59. In aqueous media the cluster formation is irreversible; addition of salts (*i.e.*, osmotic effects) or surfactants did not lead to a re-transformation into the monomeric components

60. Biju V, Barazzouk S, George Thomas K, George MV and Kamat PV, *Langmuir* 2001; **17**: 2930-2936.

61. Thomas KG, Biju V, Guldi DM, Kamat PV and George MV, *J. Phys. Chem. B*, 1999; **103**: 8864-8869.

62. a) Kamat PV, Barazzouk S, Hotchandani S and George Thomas K. *Chem. Eur. J.* 2000; **6**: 3914-3921. b) Kamat PV, Barazzouk S, George Thomas K and Hotchandani S. *J. Phys. Chem. B*, 2000; **104**: 4014-4017.

63. Niu S and Mauzerall D. *J. Am. Chem. Soc.* 1996; **118**: 5791-5795.

64. a) Murakami H, Shirakusa M, Sagara T and Nakashima N. *Chem. Lett.* 1999; 815-816. b) Sano M, Oishi K, Ishi-i T and Shinkai S. *Langmuir* 2000;

16: 3773-3776. c) Nakashima N, Ishii T, Shirakusa M, Nakanishi T, Murakami H and Sagara T. *Chem. Eur. J.* 2001; **7**: 1766-1772.

65. a) Sawamura M, Nagahama N, Toganoh M, Hackler UE, Isobe H, Nakamura E, Zhou S-Q and Chu B. *Chem. Lett.* 2000; 1098-1099. b) Zhou S, Burger C, Chu B, Sawamura M, Nagahama N, Toganoh M, Hackler UE, Isobe H and Nakamura E. *Science* 2001; **291**: 1944-1947.

66. Cassell AM, Asplund CL and Tour JM. *Angew. Chem. Int. Ed.* 1999; **38**: 2403-2405.

67. Georgakilas V, Pellarini F, Prato M, Guldi DM, Melle-Franco M and Zerbetto F. *Proc. Natl. Acad. Sci.* 2002; **99**: 5075.

68. Prassides K, Keshavarz M, Beer E, Bellavia C, Gonzalez R, Murata Y, Wudl F, Cheetham AK and Zhang JP. *Chem. Mater.* 1996; **8**: 2405-2408.

69. Shi Z, Jin J, Li Y, Guo Z, Wang S, Jiang L and Zhu D. *New. J. Chem.* 2001 **25**: 670-672.

70. a) Wilson SR and Wu Y. *J. Chem. Soc., Chem. Commun.* 1993; 784-786. b) Osterodt J, Zett A and Voegtle F. *Tetrahedron* 1996; **52**: 4949-4962.

71. a) Bourgeois JP, Echegoyen L, Fibbioli M, Pretsch E and Diederich F. *Angew. Chem. Int. Ed.* 1998; **37**: 2118-2121. b) Liu SG and Echegoyen L. *Eur. J. Org. Chem.* 2000; 1157-1163.

72. Guldi DM, Maggini M, Mondini S, Guerin F and Fendler JH. *Langmuir* 2000; **16**: 1311-1318.

73. Sartori E, Garlaschelli L, Toffoletti A, Corvaja C, Maggini M and Scorrano G. *Chem. Commun.* 2001; 311-312.

74. a) Martínez-Díaz MV, Fender NS, Rodríguez-Morgade MS, Gómez-López M, Diederich F, Echegoyen L, Stoddart JF and Torres T. *J. Mater. Chem.* 2002; **12**: 2095-2099. b) Guldi DM, Ramey J, Martinez-Diaz MV, de la Escosura A and Torres T. *Chem. Commun.* 2002: 2774-2775.

75. a) Schubert US, Weidl CH, Rapta P, Harth E and Muellen K. *Chem. Lett.* 1999; 949-950. b) Schubert US, Eschhaumer C, Hien O and Andres PR. *Tetrahedron Lett.* 2001; **42**: 4705-4707.

76. a) Diederich F, Dietrich-Buchecker CO, Nierengarten J-F and Sauvage J-P. *Chem. Commun.* 1995; 781-782. b) Armaroli N, Diederich F, Dietrich-Buchecker CO, Flamigni L, Marconi G, Nierengarten J-F and Sauvage J-P. *Chem. Eur. J.* 1998; **4**: 406-416.

77. a) Armaroli N, Boudon C, Felder D, Gisselbrecht J-P, Gross M, Marconi G, Nicoud J-F, Nierengarten J-F and Vicinelli V. *Angew. Chem. Int. Ed.* 1999;

38: 3730-3733. b) Nierengarten J-F, Felder D and Nicoud J-F. *Tetrahedron Lett.* 1999; **40**: 273-276.

78. Maggini M and Guldi DM; *unpublished results.*

79. a) Maggini M, Dono' A, Scorrano G and Prato M. *J. Chem. Soc., Chem. Commun.* 1995; 845-846. b) Armspach D, Constable EC, Diederich F, Housecraft CE and Nierengarten J-F. *J. Chem. Soc. Chem. Commun.* 1996; 2009-2010. c) Maggini M, Guldi DM, Mondini S, Scorrano G, Paolucci F, Ceroni P and Roffia S. *Chem. Eur. J.* 1998; **4**: 1992-2000. d) Armspach D, Constable EC, Diederich F, Housecroft CE and Nierengarten J-F. *Chem. Eur. J.* 1998; **4**: 723-733. e) Polese A, Mondini S, Bianco A, Toniolo C, Scorrano G, Guldi DM and Maggini M. *J. Am. Chem. Soc.* 1999; **121**: 3456-3452. d) Guldi DM, Maggini M, Martín N and Prato M. *Carbon* 2000; **38**: 1615-1623. e) Guldi DM, Maggini M, Menna E, Scorrano G, Ceroni P, Marcaccio M, Paolucci F and Roffia S. *Chem. Eur. J.* 2001; **7**: 1597-1605.

80. A gold surface and gold nanoparticles were also probed as a template for a C_{60}-bipyridyl precursor. Du C, Xu B, Li Y, Wang C, Shi Z, Fang H, Xiao S and Zhu D. *New J. Chem.* 2001; **25**: 1191-1194.

81. Haino T, Araki H, Yamanaka Y and Fukazawa Y. *Tetrahedron Lett.* 2001; **42**: 3203-3206.

82. El-ghayoury A, Schenning APHJ, van Hal PA, Weidl CH, van Dongen JLJ, Janssen RAJ, Schubert US and Meijer EW. *Thin Solid Films* 2002; **403-404**: 97-101.

83. a) Armaroli N, Diederich F, Echegoyen L, Habicher T, Flamigni L, Marconi G and Nierengarten JF. *New J. Chem.* 1999; 77-83. b) D'Souza F, Deviprasad GR, Rahman MS and Choi J-P. *Inorg. Chem.* 1999; **38**: 2157-2160. c) Da Ros T, Prato M, Guldi DM, Alessio E, Ruzzi M and Pasimeni L. *Chem. Commun.* 1999; 635-636. d) Da Ros T, Prato M, Guldi DM, Ruzzi M and Pasimeni L. *Chem. Eur. J.* 2001; **7**: 816-827. e) D'Souza F, Rath NP, Deviprasad GR and Zandler ME. *Chem. Commun.* 2001; 267-268.

84. Hauke F, Swartz A, Guldi DM. and Hirsch A. *J. Mater. Chem.* 2002; **12**: 2088-2094.

85. a) D'Souza F, Deviprasad GR, Zandler ME, Hoang VT, Klykov A, VanStipdonk M, Perera A, El-khouly ME, Fujitsuka M and Ito O. *J. Phys. Chem. A* 2002; **106**: 3243-3252; b) Wilson SR, MacMahon S, Tat FT, Jarowski PD and Schuster DI. *Chem. Commun.* 2003, 226-227.

86. D'Souza F, Deviprasad GR, El-Khouly ME, Fujitsuka M and Ito O. *J. Am. Chem. Soc.* 2001; **123**: 5277-5284

87. Guldi DM, Luo C, Da Ros T, Prato M, Dietel E and Hirsch A. *Chem. Commun.* 2000; 375-376.

88. Guldi DM, Luo C, Swartz A, Scheloske M and Hirsch A. *Chem. Commun.* 2001; 1066-1067.

89. a) Cassell AM, Scrivens WA and Tour JM. *Angew. Chem. Int. Ed.* 1998; **37**: 1528-1531. b) Takenaka S, Yamashita K, Takagi M, Hatta T, Tanaka A and Tsuge O. *Chem. Lett.* 1999; 319-320. c) Takenaka S, Yamashita K, Takagi M, Hatta T and Tsuge O. *Chem. Lett.* 1999; 321-322.

90. Nakamura E, Isobe H, Tomita N, Sawamura M, Jinno S and Okayama H. *Angew. Chem. Int. Ed.* 2000; **39**: 4254-4257.